Biomineralization

Edited by Edmund Baeuerlein

Other Titles of Interest

G. Krauss
Biochemistry of Signal Transduction and Regulation
1999. XIX, 474 pages with 162 figures and 183 tables.
Hardcover. ISBN 3-527-29771-5

A. X. Trautwein (Ed.)
Bioinorganic Chemistry
1997. XII, 767 pages with 345 figures and 65 tables.
Softcover. ISBN 3-527-27140-6

J. A. Cowan
Inorganic Biochemistry
2. Edition
1997. XIV, 440 pages.
Hardcover. ISBN 0-471-18895-6

A. M. Kossevich
The Crystal Lattice
1999. 326 pages with 85 figures.
Hardcover. ISBN 3-527-40220-9

Biomineralization

From Biology to Biotechnology
and Medical Application

Edited by
Edmund Baeuerlein

WILEY-VCH

Weinheim · New York · Chichester
Brisbane · Singapore · Toronto

Prof. Dr. E. Bäuerlein
Max-Planck-Institute of Biochemistry
Dept. of Membrane Biochemistry
D-82152 Martinsried
Germany

Library of Congress Card No. applied for.

British Library Cataloguing-in-Publication Data: A catalogue record for this book is available from the British Library.

Die Deutsche Bibliothek – CIP Cataloguing-in-Publication-Data
A catalogue record for this publication is available from Die Deutsche Bibliothek

ISBN 3-527-29987-4

Composition: Asco Typesetters, Hong Kong. Printing: Strauss Offsetdruck GmbH, 69509 Mörlenbach. Bookbinding: J. Schäffer GmbH & Co. KG, 67269 Grünstadt.
Printed in the Federal Republic of Germany.

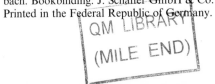

Dedication to

my wife Cornelia
for her permanent encouragement
and her editorial support

and to

Dieter Oesterhelt
the advocate of biotechnology
on the occasion of his 60th birthday

Preface

Biomineralization refers to the processes by which organisms precipitate inorganic minerals. This phenomenon is widespread in the biological world, and occurs in bacteria, single-celled protists, plants, invertebrates and vertebrates. Over 60 biominerals are known, the most abundant of which are calcium carbonates, silica and iron oxides. Most biominerals are organized hierarchically and ordered at many length scales, and often have remarkable physical characteristics. The minerals can be deposited intra- or extracellularly and are intimately connected to cellular metabolic processes. Thus biomineralization as a field of scientific study falls within several scientific disciplines, including biochemistry, biology, condensed matter physics, geology, inorganic chemistry, and molecular biology.

This is not a new scientific field; since the 19th century several thousand papers have been published, due largely to potential applications in areas as diverse as medical and dental science, paleontology and paleogeochemistry, materials science and engineering, evolutionary biology and astrobiology. However, the last two decades have seen the development of a new understanding, based partly on new experimental techniques and partly on conceptual advances. This new understanding has been documented in a number of books and symposium volumes covering the period between 1983 and 1991 and including: *Biomineralization and Biological Metal Accumulation*, edited by P. Westbroek and E. W. de Jong (D. Reidel, Dortdrecht, 1983); *On Biomineralization*, by H. A. Lowenstam and S. Weiner (Oxford University Press, Oxford, New York, 1989); *Biomineralization: Chemical and Biochemical Perspectives*, edited by S. Mann, J. Webb and R. J. P. Williams (VCH, Weinheim, 1989); *Biomineralization*, by K. Simkiss and K. M. Wilbur (Academic Press, New York, 1989); *Iron Biominerals*, edited by R. B. Frankel and R. P. Blakemore (Plenum, New York, 1991) and *Mechanisms and Phylogeny of Mineralization in Biological Systems*, edited by S. Suga and H. Nakahara (Springer Verlag, Tokyo, 1991).

While many researchers have made important contributions to the field, the modern era arguably began with the publication by Heinz Lowenstam of *Minerals Formed by Organisms* (Science **1981**, *211*, 1126–1131). In this paper, Lowenstam emphasized the importance of organic macromolecules in biomineralization, and distinguished between biological-controlled and biological-induced biomineralization processes. The theme of organic–inorganic interactions, and concepts such as directed nucleation, molecular recognition, and molecular tectonics were further developed by R. J. P. Williams, Stephen Mann, Stephen Weiner and others. In fact, the identification of the organic phase and its role in biomineralization in various organisms has been the major theme in biomineralization research over the last two decades. Another important theme, which remains less well developed, is the relationship between biomineralization and metabolism.

The present volume was inspired by a "Workshop on Biomineralization and Nanofabrication", supported by the US Office of Navel Research and organized by one of us (R. B. F). It took place in San Luis Obispo, California, in May, 1996, covered biomineralization phenomena in a number of organisms and looked toward future developments. The other of us (E. B.) was a participant at the workshop and decided to organize the publication of a multi-author volume based on the stimulating presentations. During the planning stage of about three years, progress in the study of proteins involved in biomineralization phenomena by molecular biological methods led to the addition of contributions on silica mineralization in sponges to those on magnetite mineralization in prokaryotes and silica and calcium carbonate mineralization in unicellular eukaryotes. Because biomineralization in unicellular organisms takes place in vesicles, a new membrane biology is developing that may ultimately connect to vesicle-based materials science applications.

The volume begins with a short introduction to biominerals, of which three – magnetite, silicic acid and calcium carbonate – are mineralized by organisms described here. The first part, "Prokaryotes", covers biomineralization phenomena on the surfaces of bacteria, as well as the formation of magnetite (Fe_3O_4) and greigite (Fe_3S_4) nanocrystals in the intracytoplasmatic vesicles (magnetosomes) of magnetotactic bacteria, their role in magnetotaxis, and technical and medical applications of isolated magnetosomes. It also includes *in situ* identification of magnetotactic bacteria, their phylogenetic relationships, and enzymes and related genes involved in their biomineralization processes. The second part, "Eukaryotes", opens with a unified theory of biomineralization in prokaryotes and eukaryotes from evolutionary and paleontological analysis of the Cambrian explosion 525 Myr ago. This retrospection on the evolution of biomineralization is followed by three complementary chapters on the formation of silica nanostructures in unicellular eukaryotes, the diatoms, and a related chapter on polysiloxane synthesis in a marine sponge. These are followed by a chapter on recent research into the protein components of shell nacre. The volume is completed by two chapters on coccolithophores, unicellular eukaryotes that are covered by mineralized scales of calcium carbonate known as coccoliths. It has been possible to study coccolith mineralization by mutation experiments as well as by isolation of the coccolith vesicles.

June, 2000

Richard B. Frankel   Edmund Baeuerlein
San Luis Obispo, CA (USA)   Munich (Germany)

Contents

List of Contributors

Rudolf I. Amann
Max-Planck-Institute for Marine Microbiology
Celsiusstr. 1
28359 Bremen
Germany
Fax: +49-421-2028790
e-mail:
ramann@postgate.mpi-bremen.de
Chapter 4

Edmund Baeuerlein
Max-Planck-Institute for Biochemistry
Dept. of Membrane Biochemistry
82152 Martinsried
Germany
Fax: +49-89-8578-3777
e-mail: baeuerle@biochem.mpg.de
Chapters 1 and 5

Dennis A. Bazylinski
Dept. of Microbiology, 207 Science I
Iowa State University
Ames, IA 50014
USA
Fax: +1-515-2946019
e-mail: dbazylin@iastate.edu
Chapter 3

Angela M. Belcher
Dept. of Chemistry and Biochemistry
Campus Mail Code A5300
University of Texas at Austin
Austin, Texas 78712
USA
Fax: +1-512-471-8696
e-mail: belcher@mail.utexas.edu
Chapter 15

Terrence J. Beveridge
College of Biological Science
Dept. of Microbiology
University of Guelph
Guelph/Ontario, Canada N1G 2W1
Canada
Fax: +1-519-8371802
e-mail: TJB@micro.uoguelph.ca
Chapter 2

Simon Crawford
School of Botany
The University of Melbourne
Parkville, Victoria 3052
Australia
Fax: +61-3-93475460
e-mail:
r.wetherbee@botany.unimelb.edu.au
Chapter 13

Danielle Fortin
Dept. of Earth Sciences
University of Ottawa
140 Louis Pasteur
Ottawa, ON, Canada K1N 6N5
Canada
e-mail: dfortin@science.uottawa.ca
Chapter 2

Richard B. Frankel
Dept. of Physics
California Polytechnic State University

San Luis Obispo, CA 93407
USA
Fax: +1-805-7562435
e-mail: rfrankel@calpoly.edu
Chapter 3

Yoshihiro Fukumori
Dept. of Biology
Faculty of Science
Kanazawa-Univ.
Kakuma-machi
Kanazawa 920-1192
Japan
Fax: +81-76-2645978
e-mail:
fukumor@kenroku.kanazawa-u.ac.jp
Chapter 7

Elma L. González
Dept. of Organismic Biology
Ecology and Evolution
UCLA University of California,
Los Angeles
Los Angeles, CA 90095-1606
USA
Fax: +1-310-2063987
e-mail:
gonzalez@biology.lifesci.ucla.edu
Chapter 17

Erin E. Gooch
Dept. of Chemistry and Biochemistry
University of Texas at Austin
Austin, Texas 78712
USA
e-mail: gooch@mail.utexas.edu
Chapter 15

James W. Hagadorn
Division of Earth & Planetary Sciences
California Institute of Technology
Pasadena, CA91125
USA
Fax: +1-626-5680935
e-mail: hagadorn@caltech.edu
Chapter 10

Mark Hildebrand
UCSD
Marine Biology Research Division,
0202
Scripps Institution of Oceanography
UCSD University of California,
San Diego
9500 Gilman Drive
La Jolla, CA 92093
USA
Fax: +1-619-5347313
e-mail: mhildebrand@ucsd.edu
Chapter 12

Joseph L. Kirschvink
Division of Earth & Planetary Sciences
California Institute of Technology
Pasadena, CA 91125
USA
Fax: +1-626-5680935
e-mail: kirschvink@caltech.edu
Chapter 10

Nils Kröger
Institute for Biochemistry I
University of Regensburg
Universitätsstr. 31
93053 Regensburg
Germany
Fax: +49-941-9432936
e-mail:
nils.kroeger@vkl.uni-regensburg.de
Chapter 11

Mary E. Marsh
Dept. of Basic Sciences
University of Texas Dental Branch
Health Science Center
6516 John Freeman Ave.
Houston, TX 77030
USA
Fax: +1-713-500-4500
e-mail: mmarsh@mail.db.tmc.edu
Chapter 16

Tadashi Matsunaga
Dept. of Biotechnology

Tokyo University of Agriculture and
Technology
2-24-16 Nakacho, Koganei
Tokyo 184-8588
Japan
Fax: +81-42-385-7713
e-mail: tmatsuna@cc.tuat.ac.jp
Chapter 9

Daniel E. Morse
Institute of Molecular Genetics and
Biochemistry
Chairman of the Marine
Biotechnology Center
UCSB University of California,
Santa Barbara
Santa Barbara, CA 93106-9610
USA
Fax: +1-805-8937998
e-mail: d_morse@lifesci.lscf.ucsb.edu
Chapter 14

Paul Mulvaney
School of Chemistry
The University of Melbourne
Parkville, Victoria 3052
Australia
Fax: +61-3-93475460
e-mail:
p.mulvaney@chemistry.unimelb.edu.au
Chapter 13

Regina Reszka
Max-Delbrück-Centrum für
Molekulare Medizin
AG Drug Targeting
Albert-Rössle-Straße 10
13122 Berlin-Buch
Germany
Fax: +49-30-94063213
e-mail: reszka@mdc-berlin.de
Chapter 6

Ramon Rossello-Mora
Max-Planck-Institute for Marine
Microbiology
Celsiusstr. 1
D-28359 Bremen
Germany
Fax: +49-421-2028790
e-mail: rrossell@mpi-bremen.de
Chapter 4

Toshifumi Sakaguschi
Dept. of Biotechnology
Tokyo University of Agriculture and
Technology
2-24-16 Nakacho, Koganei
Tokyo 184-8588
Japan
Fax: +81-42-385-7713
e-mail: sakaguch@cc.tuat.ac.jp
Chapter 9

Dirk Schüler
Max-Planck-Institute for Marine
Microbiology
Celsiusstr. 1
D-28359 Bremen
Germany
Fax: +49-421-2028-580
e-mail:
dschuele@postgate.mpi-bremen.de
Chapters 8 and 4

Katsuhiko Shimizu
Marine Biotechnology Center and the
Dept. of Molecular,
Cellular and Developmental Biology
University of California Santa Barbara
Santa Barbara, CA93106-9610
USA
Fax: +1-805-893-7998
e-mail: shimizu@lifesci.ucsb.edu
Chapter 14

Manfred Sumper
Lehrstuhl Biochemie I
University of Regensburg
Universitätsstr. 31
93040 Regensburg
Germany
Fax: +49-941-9432936
e-mail:

manfred.sumper@vkl.uni-regensburg.de
Chapter 11

Richard Wetherbee
School of Botany
The University of Melbourne
Parkville, Victoria 3052
Australia
Fax: +61-3-93475460
e-mail:
r.wetherbee@botany.unimelb.edu.au
Chapter 13

Abbreviations

AAS	atomic adsorption spectroscopy
ADP	adenosine diphosphate
ATCC	American Type Culture Collection
ATP	adenosine triphosphate
ATPase	adenosine triphosphatase
AFM	atomic force microscope
Bfr	bacterioferritin
BMP	bacterial magnetic particle
CA	carbonic anhydrase activity
CCM	carbon concentrating mechanism
CDF	cation diffusion facilitator
CEA	carcino-embryonal-antigen
CM	cytoplasmic membrane
CN	central nodule
CP	chloroplast
CV	coccolith vesicle
DEAE	diethylaminoethanol
DIC	dissolved inorganic carbon
DSi	dissolved silicon
DSM	Dt. Sammlung für Mikroorganismen
EDTA	ethylenediaminetetraacetic acid
ESI	energy spectroscopic imaging
EL	extracellular loops
ER	endoplasmatic reticulum
FAD	flavin adenine dinucleotide
FESEM	field emission scanning electron microscopy
FISH	fluorescence *in situ* hybridization
FMN	flavin mononucleotide
GA	N-acetylglucosamine
GFP	green fluorescent protein
GUT	grand unified theory
HRTEM	high resolution transmission electron microscopy
HPLC	high pressure liquid chromatography
ICS	intracellular carboxy segment
IgG	immunoglobulin G

IL	intracellular loops
INS	intracellular amino segment
kDa	kiloDalton
LPS	lipopolysaccharide
MA	N-acetyl muramic acid
MM	magnetosome membrane
MMP	many-celled magnetotactic procaryote
MRI	magnetic resonance imaging
MTB	magnetotactic bacteria
Myr	million years
NAD	nicotinamide adenine dinucleotide
NADH	nicotinamide adenine dinucleotide, reduced
NADPH	nicotinamide adenine dinucleotide phosphate, reduced
NMR	nuclear magnetic resonance
OA	ornithineamidelipid
OATZ	oxic–anoxic transition zone
ORF	open reading frame
OM	outer membrane
P	peptidoglycan layer
PC	phosphatidylcholine
PCR	polymerase chain reaction
PE	phosphatidylethanolamine
PET	positron emission tomography
PG	phosphatidylglycerol
PM	plasma membrane
R 123	Rhodamine 123
rRNA	ribosomal ribonucleic acid
SATA	succinimidyl-S-acethylthioacetat
SAED	selected area electron diffraction
SCID	severe combined immunodeficiency
SD	single-magnetic-domain
SDS	sodium dodecyl sulfate
SDS-PAGE	sodium dodecyl sulfate polyacrylamide gel electrophoresis
SDV	silica deposition vesicle
SEM	scanning electron microscopy
SER/THR	serine/threonine
SIT	silicic acid transporters
STEM	scanning transmission electron microscope
STV	silicon transport vesicle
TEM	transmission electron microscope
TEOS	tetraethyleneoxysilane
TMPD	tetramethyl-p-phenylenediamine
TPR	tetratricopeptide repeat
UTP	uridine triphosphate

1 Biominerals – An Introduction

Edmund Baeuerlein

Eukaryotic biomineralization was developed over a period during the Cambrian evolutionary explosion – a period that was short compared to the evolution of life up until today (see Chapter 10). The synthesis of biominerals in eukaryotes probably originated from ancient matrices developed by magnetic bacteria for the production of magnetite (Fe_3O_4) crystals. It has been proposed by Kirschvink and Hagadorn (Chapter 10) that these matrices were transformed by a process called "exaptation" [1] for their new function. The basic inorganic ions incorporated by these early eukaryotic organisms were apparently limited but included at least carbonate, phosphate, silicate, Ca^{2+} and Mg^{2+}.

Although it is difficult to find any fossil evidence, eukaryotic unicells might also have evolved, by the same "exaptation" process, the ability to form mineralized structures. This process includes a series of steps such as ion uptake, transport and formation of inorganic structures, either crystalline or amorphous. Because of this fundamental beginning of eukaryotic biomineralization, which might be based on an "ancient matrix-mediated prokaryotic system", i.e. magnetite (Fe_3O_4) biomineralization, the emphasis of this book is on prokaryotic and eukaryotic unicells, for which methods of protein chemistry and molecular biology are now used. In contrast to previous books on biomineralization, edited about 10 years ago [2–6], where "at last the all-important cells responsible for the whole process are beginning to receive well-deserved attention" [2], in this book they are the focus of most of the contributions.

Unicells, which are described in this book, use a few typical minerals to form structures that are species-specific: (1) magnetic bacteria – magnetite (Fe_3O_4) crystals of different shape; (2) diatoms – their intricate and ornate silicified cell walls; and (3) coccolithophorids – their coccoliths, which are assembled by individual calcite crystals and encase the cells. Two of these biominerals are crystalline and one is amorphous.

In many of the numerous eukaryotic unicells (protoctista), a limited range of minerals has been detected [2, 4]. For example in $CaCO_3$ the anion CO_3^{2-} has been substituted by PO_4^{3-}, $C_2O_4^{2-}$ (oxalate) or SO_4^{2-}. On the other hand, in $CaSO_4$ the cation Ca^{2+} has been replaced by the other earth metals Sr^{2+} or Ba^{2+}. Magnetic bacteria, however, show high specificity in their biominerals, with the exception that sulfur in greigite (Fe_3S_4) substituted for oxygen in magnetite (Fe_3O_4).

An excellent and comprehensive collection of organisms that produce biominerals, either inside or outside, crystalline or amorphous, was presented in 1989

Table 1.1 Biominerals in order of their anions

Anion	Chemical formula	Name
Carbonate	$CaCO_3$	Aragonite
	$CaCO_3$	Calcite
	$Pb_3(CO_3)_2(OH)_2$	Hydrocerussite
	$CaCO_3.H_2O$	Monohydrocalcite
	$CaMg(CO_3)_2$	Protodolomite
	$CaCO_3$	Vaterite
Citrate	$Ca_3(C_6H_6O_7)_2.4H_2O$	Earlandite
Fluoride	CaF_2	Fluorite
Oxalate	$MgC_2O_4.4H_2O$	Glushinskite
	$CaC_2O_4.(2+x)H_2O$ $(x < 0.5)$	Weddelite
	$CaC_2O_4.H_2O$	Whewellite
Oxide	$Na_4Mn_{14}O_{27}.9H_2O$	Birnessite
	$5Fe_2O_3.9H_2O$	Ferrihydrite
	α-$FeO(OH)$	Goethite
	γ-$FeO(OH)$	Lepidocrocite
	$Fe^{2+}Fe_2^{3+}O_4$	Magnetite
	$(Mn^{2+}CaMg)Mn_3^{4+}O_8.H_2O$	Todorokite
Phosphate	$Ca(HPO_4)_2.2H_2O$	Brushite
	$Ca_5(PO_4,CO_3)_3(OH)$	Carbonate–hydroxylapatite
	$Ca_5(PO_4)_3F$	Fluoroapatite (francolite)
	$Ca_5(PO_4)_3(OH)$	Hydroxylapatite
	$Ca_8H_2(PO_4)_6.5H_2O$	Octacalcium phosphate
	$Mg(NH_4)PO_4.6H_2O$	Struvite
	$Fe_3^{2+}(PO_4)_2.8H_2O$	Vivianite
	$Ca_{18}H_2(Mg,Fe)_2^{2+}(PO_4)_{14}$	Whitlockite
Silicate	K_2SiF_6	Hieratite
	$SiO_2.nH_2O$	Opal
Sulfates	$BaSO_4$	Barite
	$SrSO_4$	Celestite
	$CaSO_4.2H_2O$	Gypsum
	$KFe_3^{3+}(SO_4)_2(OH)_6$	Jarosite
Sulfides	PbS	Galena
	$Fe^{2+}Fe_2^{3+}S_4$	Greigite
	$FeS.nH_2O$	Hydrotroilite
	$(Fe,Ni)_8S_8$	Mackinawite
	FeS_2	Pyrite
	ZnS	Sphalerite
	ZnS	Wurtzite

Compositions and names according to Ferraiolo [7] and Lowenstam and Weiner [2].

in the book *On Biomineralization* by Lowenstam and Weiner [2]. This book is still unsurpassed today and some of its topics should be briefly summarized here. The fact that 80% of the known biominerals are crystalline and only 20% are amorphous excites the imagination of material scientists. To promote interest and analogies over and above that, a number of well characterized biominerals are shown in Table 1.1, in order of their anions and with their chemical formulas and common names. As expected from the evolution of eukaryotic biomineralization (see Chapter 10), Ca^{2+} is the cation of about 50% and phosphate the anion of about 25% of all known biominerals. A growing number of biominerals with anions of organic acids such as citrate and oxalate have been described. Comparing the distribution of biominerals in terms of quantities, carbonates and silica (opal) are the most abundant, followed by phosphates and magnetite (Fe_3O_4). The challenge and motivating force for the biomimetic production of new minerals was and is the finding that biominerals are biosynthesized in crystal shapes unknown in inorganic chemistry. Very often membrane vesicles are involved in these processes, which may in the near future lead to vesicle-based production of new materials.

References

[1] S. J. Gould, E. S. Vrba, *Paleobiology* 1982, *8*, 4.
[2] H. A. Lowenstam, S. Weiner (Eds), *On Biomineralization*, Oxford University Press, Oxford, 1989.
[3] S. Mann, J. Webb, R. J. P. Williams (Eds), *Biomineralization: Chemical and Biochemical Perspectives*, VCH, Weinheim, 1989.
[4] J. K. Simkiss, K. M. Wilbur (Eds), *Biomineralization: Cell Biology and Mineral Deposition*, Academic Press, New York, 1989.
[5] R. B. Frankel, R. P. Blakemore (Eds), *Iron Biominerals*, Plenum Press, New York, 1991.
[6] S. Suga, H. Nakahara (Eds), Mechanisms and Phylogeny of Mineralization in Biological Systems, Springer Verlag, Tokyo, 1991.
[7] J. A. Ferraiolo, *Bull. Amer. Mus. Nat. Hist.* 1982, *172*, 1–237.

Prokaryotes

2 Mechanistic Routes to Biomineral Surface Development

Danielle Fortin, Terrance J. Beveridge

2.1 Introduction

Over the years, many studies have shown that bacteria play an important role in the cycling of both major and minor elements in the environment. They are directly involved in the redox reactions of many inorganic species (Fe, Mn, Cr, U, As, etc.) leading to mineral dissolution or precipitation, and they indirectly affect the solubility of mineral species through their metabolic activity (i.e. change in pH, production of ligands, etc.). Bacterial cell surfaces are also highly interactive with ionic species in the environment and are involved in metal binding and subsequent mineral development under various chemical conditions. Bacteria have lived on Earth for at least 3.6 Ga [1, 2] and inhabit a wide range of environments, many of which are inhospitable to other life forms. Through natural selection and evolution, present-day bacteria thrive under a wide range of physicochemical conditions, even in extreme environments such as deserts, salty brines, alkaline lakes, acidic mine tailings, dumps and near hydrothermal vents.

Bacteria are small prokaryotic cells [3, 4] with a volume of around 1.5–2.5 μm^3 [3]. Because of their small size, they have no clearly defined nucleus and their chromosome is free within the cytoplasm [3]. Due to their smallness and simplicity of cell structure, bacteria have rapid growth rates (provided nutrients are not limiting) and adapt easily to harsh conditions. Bacteria have a high surface area-to-volume ratio because they rely on diffusion to obtain their nourishment and dispose of wastes [5]. Bacterial cells are bounded by a tough cell wall which has numerous binding sites for dissolved ionic species. At neutral pH, most anionic sites on bacterial surfaces are ionized, and although bacteria have an overall net negative charge, some positively-charged sites are also available. Bacteria can also have additional layers (such as capsules, sheaths and S-layers) which provide more binding sites on the cell surface. These cells can therefore be thought of as highly reactive bioparticles of small size (but with high surface area) in the aqueous environment. They can bind a variety of dissolved species and provide nucleation surfaces for mineral development.

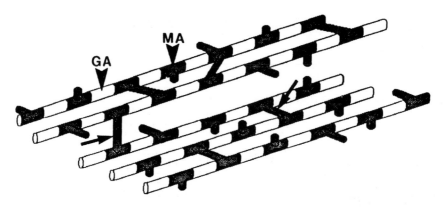

Figure 2.1 Three-dimensional representation of the peptidoglycan structure of Gram-positive cell walls. The backbone of the molecule consists of N-acetylglucosamine (GA) and N-acetylmuramic acid (MA). Covalent bonding occurs between the peptide stems of adjacent MA strands (arrows) so that the entire mesh acts as a giant molecule encircling the cell.

2.2 Bacterial Cell Walls and Other Surface Layers

It is the surface of bacteria that interacts most strongly with the cell's immediate surrounding environment. These cell walls can be divided into two structural formats, Gram-positive and Gram-negative [6], according to a bacterium's reaction to a stain used for light microscopy.

The main components of Gram-positive cell walls are linear polymers of peptidoglycan, rich in carboxylate groups, which are covalently linked together as they assemble around the cell [6]. The backbone of this polymer consists of repeated dimers of N-acetylglucosamine (GA) and N-acetylmuramic acid (MA) (Fig. 2.1). The end result of the covalent linking of all these polymers is a giant macromolecule that encompasses the cell (approximatively 25 nm thick, Fig. 2.2a), giving the bacterial surface great strength and endurance. In fact, it is one of the strongest polymeric substances known and is even stronger than many present-day plastics. Additional components, called secondary polymers (such as teichoic or teichuronic acids), are bound into the peptidoglycan structure. These polymers provide additional phosphoryl groups which add to the net electronegative charge of the cell surface of Gram-positive bacteria. The cell wall is separated from the cell's interior (cytoplasm) by a plasma membrane, a lipid/protein bilayer.

Gram-negative cell walls have a more complex structure (Fig. 2.2b), with a much thinner peptidoglycan layer (3 nm thick), which is devoid of secondary polymers [6]. In these walls, the peptidoglycan is sandwiched between two lipid/protein bilayers, the plasma membrane and the outer membrane, which is the outermost layer facing the external milieu. Its outer face possesses an unusual lipid, lipopolysaccharide

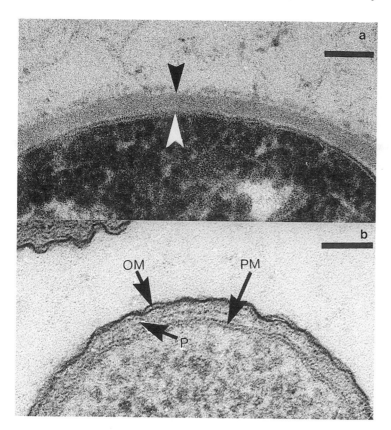

Figure 2.2 (a) Cross-section of the cell wall of *Bacillus subtilis*, a Gram-positive bacterium. The cell wall is composed of multiple layers of peptidoglycan (indicated by arrow heads) to which teichoic acid is attached. (b) Cell envelope structure of *Pseudomonas aeruginosa*, a Gram-negative bacterium, showing the thin peptidoglycan layer (P) sandwiched between the plasma membrane (PM) and the outer membrane (OM). Scale bar in each image is 50 nm.

(LPS), which is highly anionic. The O-sidechain of LPS can extend up to 40 nm away from the core oligosaccharide, which is attached to a lipid A moiety. (The lipid A has several fatty acids chains which are highly hydrophobic and cement the LPS molecule into the bilayer.) The core oligosaccharide and upper regions of the lipid A are rich in phosphates which have a predilection for Mg^{2+} or Ca^{2+} and so form a salt bridge between adjacent LPS molecules [7, 8]. The core also has three keto-deoxyoctonate residues which provide available carboxylate groups for ionization [9]. Many O-sidechains also contain residues rich in carboxylates (e.g. sialic and manuronic acids). Overall, LPS has a much greater charge per surface area than does a molecule of phospholipid.

For Gram-positive cell walls, the multiple layers of peptidoglycan with their numerous carboxylate groups appear to be the main driving force behind mineral

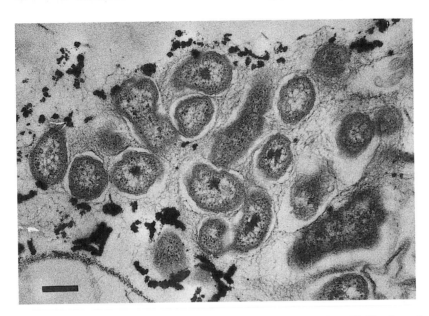

Figure 2.3 Thin section of a bacterial microcolony growing within a biofilm formed on the rock surface of an underground tunnel in the Precambrian shield (Canada). Cells are surrounded by a large matrix of fibrous material. Scale bar is 500 nm. (Courtesy of A. Brown.)

development [10, 11], although if the secondary polymers far outnumber the peptidoglycan (such as in some strains of *B. licheniformis* [12]), they can also have a major influence. For Gram-negative cell walls, the peptidoglycan layer is much thinner and it is shielded from the external environment by the outer membrane. Here, the LPS is the prime functional unit for mineral development because of its rich supply of phosphate and carboxylate groups [9, 13–15].

Gram-positive and Gram-negative bacteria frequently possess additional layers on top of their cell wall, such as capsules, slimes, sheaths and S-layers [3]. Bacteria with large capsules (25–1000 nm thick) are often encountered in the environment (Fig. 2.3). A capsule is an amorphous, hydrated matrix consisting of acidic mucopolysaccharides (or occasionally acidic polypeptides) which are strongly attached to the underlying cell wall [3]. They are loose structures which allow the free flow of aqueous solutions through them. Like cell walls, capsules strongly interact with dissolved ionic species [16, 17]. Slimes are chemically similar to capsules, but they are not attached to the bacterium [3] and so are free to migrate away from the cell, filling much of the immediate surroundings with long polymeric matter that is reactive with metal ions. Slimes are loosely packed and slippery, hence their name. Sheaths are more finite structures in the form of hollow cylinders in which (some) filamentous bacteria reside, e.g. *Beggiatoa*, *Leptothrix* and *Sphaerotilus*, and some cyanobacteria and methanogens [3]. Most eubacterial varieties consist of loosely packed polymeric substances that can interact strongly with metal ions [3, 18].

However, this interaction need not to be passive – *L. discophora* has a protein in its sheath that catalyzes the oxidation of manganese to form Mn oxide precipitates [18–20]. So far, the proteinaceous sheaths of archaeobacteria (e.g. those of *Methanospirillum hungatei* and *Methanosetae concilii* [21]) have not been studied for their metal-binding activity, but their surfaces are anionic and should be reactive [22].

S-layers are proteinaceous paracrystalline surface layers usually composed of a single species of protein or glycoprotein [23, 24]. S-layers can be considered to be ordered networks of proteins with open channels that allow the free flow of aqueous solutions. During their self-assembly, S-proteins are folded and the polar amino acid residues are usually internalized so that, once assembled, the S-layer is more hydrophobic than the underlying cell wall surface. However, some S-proteins require the presence of divalent metal ions (usually Ca^{2+}) for their correct folding during the entropy-driven self-assembly process (e.g. *Aquaspirillum* spp., [25]) and others, such as *Synechococcus* GL24, have enough reactive sites available after the S-layer has been assembled to promote the mineralization of substantial quantities of gypsum or calcite [26].

2.3 Sorption of Ionic Species to Bacterial Surfaces

2.3.1 Metal Sorption

Bacteria act as efficient sorption surfaces for various soluble metal ion species in the aqueous environment. As mentioned previously, their cell walls and additional surface layers consist of various macromolecules that have exposed reactive anionic groups. These anionic groups confer a net negative charge to the cell surface which readily reacts with positively-charged metal species. Metal binding to bacterial surfaces can therefore be seen as an electrostatic interaction between negatively-charged binding sites within the surface structures and positively-charged metal ions in the aqueous environment.

Many qualitative studies have looked at the relative binding capacities of various metal ions with different bacterial cell wall types (i.e. Gram-positive or Gram-negative) or different wall components [10, 11, 13, 15, 27–30]. Several studies on metal–bacteria interactions have shown that metals tend to bind to the carboxyl groups present within the peptidoglycan and to phosphoryl groups present in the secondary polymers (i.e. teichoic and teichuronic acids) of the cell wall of Gram-positive bacteria [11, 12, 27]. Although the thick peptidoglycan appears to be responsible for most metal binding by Gram-positive bacteria, it is unlikely that the same layer plays a significant role in Gram-negative organisms, where it is shielded by an outer membrane. It is in fact the LPS component of the outer membrane that is most implicated in metal binding. For Gram-negative bacteria, such as *Escherichia coli* K-12, the phosphoryl residues of the phospholipids and LPS have been identified as probable binding sites for metals [9, 13]. More recent work by Langley and Beveridge [31] showed that the O-polysaccharide side chain present in the LPS of some Gram-negative bacteria, such as *Pseudomonas aeruginosa*, can contribute to metal binding, but the reactive groups of the core oligosaccharide and lipid A moieties contribute more.

The role of bacterial biofilms on metal sorption has also been investigated, in order to assess the role of extracellular polymers. A study by Nelson *et al.* [32] was indeed able to predict lead sorption to a biofilm, and indicated that the sorption of Pb by extracellular polymers was far greater than that of bacterial cells. In addition, recent work [33] has shown that a bacterium's ability to bind metals is greatly affected by its mode of growth. Biofilms had a greater binding capacity than planktonically grown cells, because of the high production of extracellular exopolymers within the biofilm [33].

However, very little is known about the stability of metal–bacteria surface complexes. The first step in metal ion–bacteria interaction is thought to be passive in nature, because isolated cell walls show similar affinities for metal binding [29]. In fact, Urrutia Mera *et al.* [34] showed that respiring cells possessing an energized membrane (pumping protons into the cell wall) actually competed with metal ions for binding sites; living cells can therefore bind fewer metal ions than dead cells. Even though metabolic processes can influence the surface layers, it is possible to treat bacterial surfaces as chemical interfaces having exposed functional groups characterized by acid/base properties. Recent work by Fein *et al.* [35] and Daughney *et al.* [36] has indeed shown that acid–base titrations of *B. subtilis* and *B. licheniformis* walls fit a three pK model, suggesting the presence of carboxyl, phosphate and hydroxyl sites. The protonation constants of each type of site were, however, slightly different for the two strains, as was the concentration of surface sites. Presumably this is a reflection of the profound differences in cell wall chemistry of the two *Bacillus* species [12, 27]. Using a thermodynamic approach, Fein *et al.* [35] and Daughney *et al.* [36] also showed that carboxyl and phosphate sites were involved in metal (Cd, Pb, Cu and Al) uptake by *B. subtilis* and *B. licheniformis* cell surfaces and calculated stability constants for the different metal–carboxyl surface complexes. Similar results regarding the involvement of carboxyl and phosphate groups in metal sorption were observed for Cu adsorption experiments on spores of the marine *Bacillus* sp. SG-1 [37]. In addition, Daughney and Fein [38] showed that metal adsorption to bacterial surfaces was dependent on the ionic strength of the electrolyte. These results could be a step forward in the prediction of the effects of bacteria on metal adsorption in natural aqueous systems. Yet it is still apparent that metabolism and growth rates subtly alter surface physicochemistry. Indeed, variables such as these, encountered over diverse microbial populations in natural settings, will make accurate modeling extremely difficult.

2.3.2 Anion Sorption

In addition to negatively-charged sites within Gram-positive and Gram-negative cell walls, there are also amine groups (e.g. the diaminopimelic acid of A1 γ peptidoglycan in *B. subtilis* walls and the N-acetylfucosamine of *Pseudomonas aeruginosa* LPS). Recent work has shown that these electropositive groups are involved in the binding of anionic species (such as silicate ions) to *B. subtilus* cell walls at near neutral pH conditons [39, 40]. However, most silicate ion binding is thought to occur through metal ion bridging [41], since the number of available amine groups within the cell wall is relatively low and the silicate binding is high. The proposed mechanism of binding first involves the linkage of a multivalent metal ion (such as Fe) to a negatively-charged site (such as a carboxyl group), followed by the binding of the silicate ion to the same site carrying a net residual positive charge. Silicate anion sorption was shown to be a precursor in the formation of poorly ordered silicate minerals on the cell wall [39]. Interestingly, these silicates do not appear to out-compete dissolved metal species for the available sites remaining in the bacterial wall [41].

2.4 Mechanisms of Mineral Nucleation on Bacterial Surfaces

Various bona fide minerals, such as oxides, sulfides, carbonates and silicates, have been observed associated with bacteria in natural environments, indicating that bacterial surfaces play an important role in metal cycling and immobilization. However, mineral nucleation and precipitation in any aqueous milieu is primarily a function of the degree of saturation of the solution. Because bacteria are small particles possessing a large surface area with multiple reactive sites, they promote mineral development by lowering the total free energy required for precipitation [42–44]. Their metabolic activity can trigger changes in solution chemistry (such as pH and Eh fluctuations or the production of reactive ligands) that lead to over-saturation. Bacteria also play an indirect, more passive role, by acting as chemically reactive solids for metal sorption and subsequent nucleation [29, 35, 36]. In fact, recent work by Warren and Ferris [45] indicated that the use of geochemical and thermodynamic principles allows the prediction and quantification of sorption and nucleation processes onto bacterial surfaces. The study showed that there was a continuum between the adsorption of Fe(III) and the nucleation of Fe oxides on bacterial surfaces and that the onset of Fe oxide precipitation occurred at lower pH values and in greater quantities in the bacterial systems than in abiotic systems. As a result of the direct and indirect role of bacteria, various minerals can precipitate both throughout the cell and in the immediate surroundings.

2.5 Examples of Biogenic Minerals

2.5.1 Iron and Manganese Oxides

Iron and manganese oxides are often observed in association with bacteria in various natural environments (lake and marine waters, soils, sediments, mine tailings, etc.). Their formation is a function of the redox and pH conditions of the aqueous milieu. Under oxic conditions, the oxidation rate of both Fe and Mn is strongly dependent on the pH of the solution. Low pH conditions (i.e. pH < 5) significantly slow down the oxidation rate of both metals. However, some bacteria can catalyze the oxidation of both Mn and Fe, favoring the formation of oxides at lower pH values than those dictated by normal chemical rates. Under neutral pH conditions, bacteria are thought to play a more passive role, by serving as nucleation surfaces for oxide formation [46, 47].

2.5.1.1 Enzymatic Reactions

Enzymatic reactions leading to Fe and Mn oxide precipitation have an overall rate of reaction faster than the non-microbial (chemical) reactions [48]. The well-known iron-oxidizing bacterium, *Thiobacillus ferrooxidans*, can increase the oxidation rate of Fe(II) by as much as six orders of magnitude under very acidic conditions [49]. Thiobacilli populations are widespread in acidic sulfide mine tailings, where they oxidize Fe sulfide minerals [50–52]. *Thiobacillus* fixes atmospheric CO_2 and obtains its energy from the oxidation of reduced iron and sulfur species [53, 54]. The enzymatic oxidation is believed to take place in the vicinity of the cell envelope, where the electrons are transported through the periplasm to the cytochrome oxidase located on the inside surface of the plasma membrane [55]. Other acidophilic bacteria, such as *Leptospirillium ferrooxidans* and strains of *Metallogenium*, have been reported to catalyze the oxidation of Fe(II) at low pH [48]. On the other hand, there is little evidence that neutrophilic bacteria are involved in the enzymatic oxidation of Fe(II) at neutral pH, because chemical oxidation is easily achieved under such conditions. With the exception of *Gallionella ferruginea*, whose growth depends on the presence of Fe(II) (and not on organic carbon), the oxidation of Fe(II) at neutral pH appears to be virtually driven by chemical reactions. Bacteria involved in the enzymatic oxidation of Fe(II) can also serve as interfaces for Fe oxide formation. For example, partially to totally mineralized bacteria have been observed in samples from acidic mine tailing environments [46, 50, 51] (Fig. 2.4). In this case, the precipitates ranged from amorphous to crystalline Fe oxides and occurred as finely granular, fibrous and lath-like minerals on bacterial cell walls.

Mn oxidation in most aqueous systems is catalyzed by bacteria, because the chemical oxidation rate is very slow at pH 7. Unlike Fe(II), enzymatic oxidation of Mn(II) has not been reported under acidic pH conditions [48]. There are several mechanisms by which bacteria can enzymatically oxidize Mn(II). Some bacteria can oxidize dissolved Mn(II) species (with or without energy gain), using oxygen as a terminal electron acceptor, while others can only oxidize Mn(II) when it is bound to the outer cell surface [48]. Oxidation of Mn(II) can also be achieved when catalase hydrolyses H_2O_2 [48]. In addition, bacteria such as *L. discophora* SS-1 excrete a protein which intermingles with the polymeric components of its sheath, giving it the catalytic capacity to oxidize Mn(II) extracellularly [19, 20]. Several Mn-oxidizing bacteria (such as *Gallionella* and *Leptothrix*) are also known to oxidize Fe(II) [48]. As for Fe-oxidizing bacteria, Mn-oxidizing bacteria serve as nucleation surfaces for Mn oxide formation; precipitates usually form on the cell wall (as shown in Fig. 2.5) or on extracellular polymers.

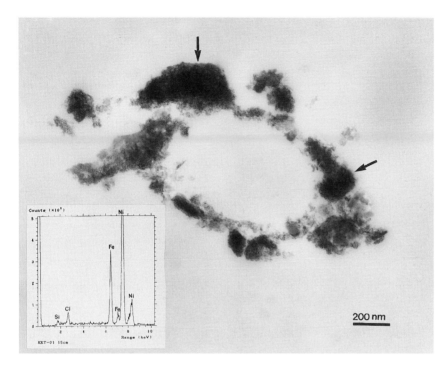

Figure 2.4 Thin section of an acidic (pH 2–3) mine tailings sample (Kidd Creek Mine, Ontario, Canada), showing a partially mineralized bacterium. Precipitates on the cell wall (arrow heads) were identified as poorly ordered Fe oxides by EDS analysis. Scale bar is 200 nm. (Courtesy of B. Davis.)

2.5.1.2 Non-Enzymatic Reactions

As a result of their metabolic activity, bacteria can change the physicochemical conditions of their surroundings. These changes can affect the solubility of both Fe and Mn and are thought of as non-enzymatic reactions. For instance, an increase of Eh or pH in the vicinity of the cell can favor the chemical oxidation of Fe(II) and Mn(II) and the subsequent precipitation of oxides. Increases in pH can be triggered by a wide range of metabolic processes. For example, microorganisms capable of producing ammonia from proteins or utilizing carboxylic acids (such as lactate, malate, citrate, etc.) can raise the pH of the aqueous milieu, allowing the chemical oxidation of Fe(II) and (sometimes) Mn(II) [48]. Cyanobacteria can also promote the oxidation of Fe and Mn by increasing the pH and raising the oxygen level of the waters [48]. Bacterial breakdown of organic ferric complexes has been shown to lead to Fe oxide formation following the release of ionic Fe(III); metabolic end-products can also act as oxidizing agents [48].

Bacterial surfaces and their exopolymers also act as reactive interfaces for oxide nucleation and precipitation. Several studies have shown that Fe and Mn oxides formed in neutral pH marine and freshwater environments are closely associated

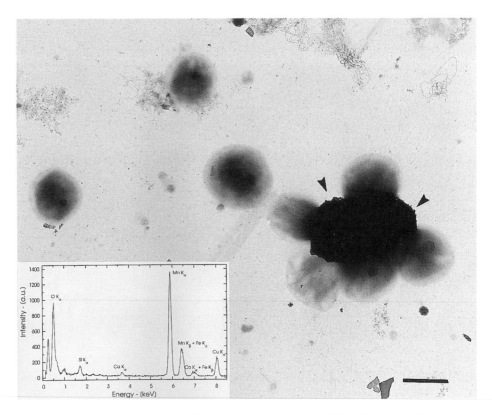

Figure 2.5 Micrograph of particles collected in the water column of Paul Lake, a small meromictic lake located in the upper Michigan peninsula (USA). Bacteria recovered at 5 m depth (where the pH was around 6) were covered with Mn oxides, containing small amounts of Si, Ca and Co [56]. Scale bar is 500 nm. (Courtesy of C.-P. Lienemann.)

with bacteria [57–59]. According to these studies, Fe and Mn oxides precipitate as poorly ordered to crystalline phases on bacterial cell walls and their associated exopolymers.

In natural environments, non-enzymatic reactions can occur simultaneously with enzyme-driven and slower abiotic chemical reactions. Although non-enzymatic and abiotic reactions often form the same mineral; the former will be driven to completion first, because of lower free energy thresholds. This fact, and the large numbers of bacteria present in natural environments, ensures that bacterially-mediated mineral precipitation will play an important part in the synthesis of soils and sediments.

2.5.2 Silicates

Silicate development requires an entirely different set of chemical parameters to those outlined for metal oxide precipitation and yet silica, mixed Fe–Al, and Fe silicates have all been observed in close association with bacteria in a number of

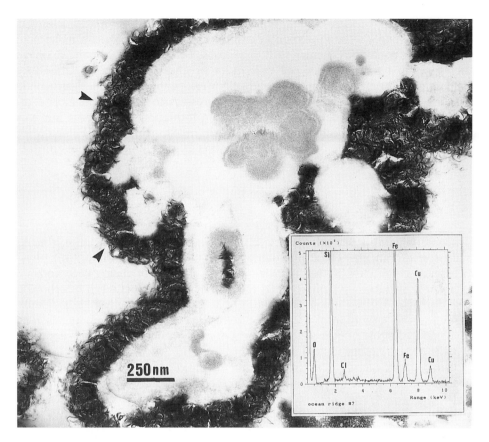

Figure 2.6 Thin section of a sample collected near hydrothermal vents at the Southern Explorer ridge in the north-east Pacific Ocean, showing a microcolony of bacteria surrounded by crystalline Fe silicate precipitates (indicated by arrows). Scale bar is 250 nm. (Courtesy of F. G. Ferris and S. D. Scott.)

different environments. Partially to completely mineralized bacteria have been reported in hot springs, hydrothermal vents, river sediments and mine tailings, and near hydrothermal sea vents [44, 60–67] (Fig. 2.6). These studies showed that amorphous to poorly ordered silicates precipitate on bacterial surfaces. In many cases, the extent of mineralization preserved the shape of cellular structures, a first step in fossilization. The binding of metals (such as Fe) prior to silicification would help preserve the original morphology of the cells [68], because heavy metal ions appear to denature autolysins (a group of surface-associated hydrolases that can rapidly degrade the cell wall and lyse the bacterium [69]).

The binding of metals to bacterial cell walls is also an important process in silicate nucleation on bacterial surfaces [39, 41]. Metals are thought to act as bridging ions between anionic constituents (e.g. carboxyl or phosphate groups) of the cell wall and silicate anions in solution. Laboratory experiments with *B. subtilis* have indicated that poorly ordered fine-grained silicates (Fe–Al) form on the cell wall under neutral pH conditions [39]. Fortin and Beveridge [67] proposed a similar mechanism to explain the formation of amorphous silica on thiobacilli under acid mine leaching conditions. Under such conditions, Fe (abundant in mine drainage waters) was first sorbed to the cell walls prior to silica formation. In laboratory simulations, abiotic conditions and dead cells were unable to produce particulate silica, indicating that the living bacteria were necessary to shift the chemical equilibrium towards silica precipitation [67]. Silicate formation on bacterial surfaces appears to be a passive mechanism (unlike the enzymatic oxidation of Fe and Mn), but bacteria affect the rate of silicate precipitation by providing additional nucleation surfaces in the aqueous milieu.

2.5.3 Carbonates

Carbonate mineral precipitation is closely associated with cyanobacteria in many aqueous environments. These photoautotrophic microorganisms affect the physicochemical conditions of their surroundings as a result of their photosynthetic activity. The metabolic fixation of CO_2 tends to increase the pH when bicarbonate is the dominant carbon species in solution [42, 70–73]. This increase in alkalinity usually leads to saturation of the solution with respect to carbonate minerals. The active metabolic role of cyanobacteria in the formation of carbonates is supported by stable carbon isotope data [43, 74, 75].

Cyanobacteria can serve as templates for nucleation and crystal growth of carbonate minerals, where they usually form on the outer surfaces [42, 70]. In the case of *Synechococcus* GL24, a cyanobacterium found in Fayetteville Green Lake, New York, gypsum and calcite precipitates form on its outer S-layer [26]. This S-layer shows a hexagonal (p6) configuration (Fig. 2.7a) of a 104 kDa protein, so that carboxyl groups are exposed on the low mass axes (or arms) of the folded and assembled S-protein [26]. Under low light, gypsum ($CaSO_4$) first precipitates on the S-layer, but as the cells become photosynthetically active, their immediate environment becomes more alkaline, shifting the mineral stability field from gypsum to calcite ($CaCO_3$) (Fig. 2.7b). During the summer months, this leads to the accumulation of a fine-grained marl sediment in the lake and contributes to the formation of large freshwater bioherms (freshwater "reefs"). Laboratory experiments with the same *Synechococcus* strain also showed that strontianite, celestite, hydromagnesite and mixed calcium – strontium carbonates could also form on the S-layer [42, 70].

Other bacteria can also promote the precipitation of carbonate minerals. Microorganisms involved in nitrate, Fe(III), Mn(IV) and sulfate reduction can trigger the precipitation of carbonate minerals through the local production of alkalinity [76–79].

2.5.4 Sulfides

In low temperature environments (i.e. $< 100\,°C$), microbial sulfate reduction is the main process responsible for the formation of sulfide minerals under anoxic conditions [80, 81]. The activity of sulfate-reducing bacteria in anoxic sediments leads

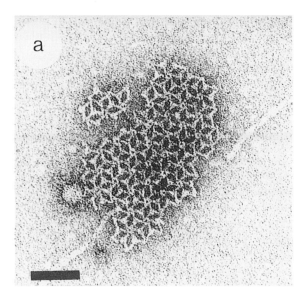

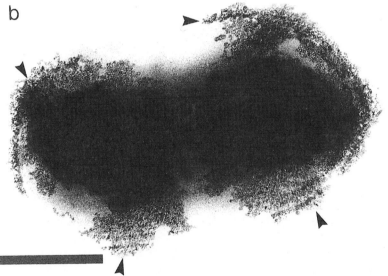

Figure 2.7 (a) Negative stain of the S-layer of *Synechoccocus* GL24, showing the hexagonal arrangement of subunits (scale bar is 50 nm). (b) A cell of *Synechoccocus* showing carbonate mineralization (calcite) of the S-layer (arrow heads). Scale bar is 250 nm. (Courtesy of S. Schultze-Lam.)

to the formation iron monosulfides (FeS), which are eventually transformed into pyrite (FeS_2). Isotopic data signatures from many sulfide mineral deposits support the bacterial origin of reduced sulfur [81].

Bacteria also serve as templates for sulfide formation. Electron microscopy of natural settings and their simulations reveal sulfate-reducing bacteria with iron sulfide precipitates either directly or closely adjacent to their cell walls [44, 50]. In some cases, bacteria become totally mineralized with Fe sulfides, obscuring cell wall structures [67]. In fact, metals bound to bacterial surfaces appear to be more reactive toward sulfide than when they are in solution [82].

2.5.5 Gold

The formation of placer gold has always been explained by geochemical reactions taking place at high temperatures ($\sim 300\,°C$) and over hundreds to thousands of years [83]. However, experiments with *B. subtilis* cells suggest that bacteria could help facilitate the development of gold particles at low temperatures (60 °C) [84, 85]. According to these authors, ionic Au is first immobilized by their representative bacterium, *B. subtilis*, and then precipitated as gold colloids both inside and outside the cells. During these low temperature diagenesis experiments, the release of organics due to autolysis of the bacteria is accompanied by the *in vitro* formation of pseudocrystalline gold particles and crystalline octahedral gold crystals (Fig. 2.8).

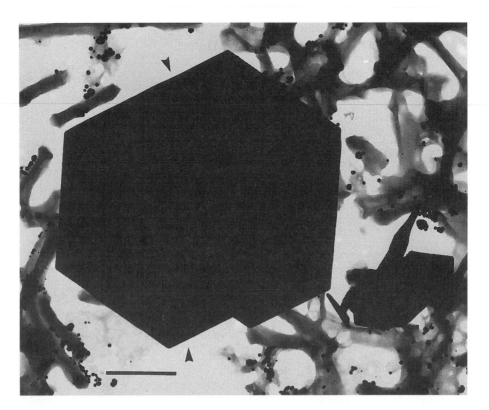

Figure 2.8 Pseudocrystalline Au particles formed at low temperature on *B. subtilis* [79] (arrow head points to a single crystal). Scale bar is 2 μm. (Courtesy of G. Southam.)

Phosphate, a normal constituent of DNA, RNA, phospholipids, teichoic acids and other compounds, was identified as an important initial component of these Au-rich particles [85]. Accordingly, organic phosphate (and possibly sulfur) should play an important role as an Au-complexing agent. As the bacteria lysed, the liberated colloidal particles annealed with other surface-bound particles so that, over time, larger and larger Au crystals grew. After a few weeks, the original ~ 1 nm colloidal gold particles found throughout the cells self-associated to form particles that were visible to the naked eye. Although these were limited laboratory experiments, it is possible that bacteria under suitable natural environments could have an impact in the development of placer gold.

2.6 Surface Reactivity of Biogenic Minerals

Some of the minerals that form on bacterial surfaces, such as Fe and Mn oxides and some hydrous silicates, also play an important role in metal cycling in the aquatic environment. As mentioned previously, biogenic minerals are usually small and poorly ordered and have a large surface area. Such physical and chemical characteristics make them excellent metal sorbents. In fact, Nelson *et al.* [86] showed that biogenic Mn oxides formed by *Leptothrix discophora* SS-1 sorbed five times more Pb (on a dry weight basis) than chemically precipitated Mn oxides and that biogenic oxides possessed a much larger surface area. Natural biogenic Fe oxides formed in the presence of iron-oxidizing bacteria in subterranean environments also displayed similar metal adsorption behavior [87]. As a result, the physical and chemical nature of biogenic minerals probably plays a very important role in metal cycling in many low temperature aquatic environments and it is essential that more research is done on this subject in the near future.

2.7 Conclusions

The evidence is clear that prokaryotes evolved as the first life-forms on Earth and that they were present during the late Archean and Proterozoic periods. This evidence stems from numerous observations of so-called "microfossils" found in ancient, organic-rich cherts and the distinct biochemical fingerprints found in a range of ancient sediments, cherts and shales [1, 2, 30, 88, 89]. Ancient stromatolites, because of their laminated structure, first evoked the idea of a biological origin [88] and are believed to be the mineralized remnants of biofilms composed of filamentous bacteria. Certainly there are many morphological similarities between these ancient microfossils and present-day biofilms; the prokaryotic creatures that inhabit them are approximately the same size and shape. For many years, it was difficult to understand exactly how these ancient remnants of bacteria were so well preserved.

Clearly, they are completely mineralized and their shape is preserved in the rock matrix. However, hardly any organic matter remains (only hard-to-degrade isoprenoid compounds have been detected [89, 90]). The mystery revolves around the unequivocal fact that prokaryotic cells are entirely composed of "soft matter"; bacteria do not possess hard, bony endo- or exoskeletons which would help preserve their cellular structure over vast periods of time.

In this chapter, we have outlined some of the fundamental principles that mediate the collection of metal and silicate ions by bacteria and the subsequent development of fine-grained minerals. Now, a rational explanation for microfossil development is possible – bacteria eventually develop so many minerals that their "soft" substance is turned to rock. The organic matter may eventually disappear over time, but the mineralized inorganic remains endure until drastic geothermal processes rework them into metamorphic rocks. Those few geological horizons possessing microfossils that do not undergo metamorphosis remain unscathed (e.g. the microfossils of the stromatolitic cherts of the 3.5 Ga Western Australia Warrawoona Group).

Now that there is some evidence that microfossils may also be a part of the Martian landscape (i.e. the SNC meteorite (ALH 84001) found in the ice near Allan Hills, Antarctica, in 1984), it is possible that these same fundamental traits of bacteria have been at work in a number of extraterrestrial environments.

References

[1] J. M. Schopf, M. R. Walter, in *Earth's Earliest Biosphere: Its Origin and Evolution* (Ed. J. W. Schopf), Princeton University Press, Princeton, 1983, pp. 214–239.

[2] M. R. Walter, in *Earth's Earliest Biosphere: Its Origin and Evolution* (Ed. J. W. Schopf), Princeton University Press, Princeton, 1983, pp. 187–213.

[3] T. J. Beveridge, in *Bacteria in Nature* (Eds J. S. Poindexter and E. R. Leadbetter), Plenum Press, New York, 1989, pp. 1–65.

[4] T. J. Beveridge, *Annu. Rev. Microbiol.* 1989, *43*, 147–171.

[5] T. J. Beveridge, *Can. J. Microbiol.* 1988, *34*, 363–372.

[6] T. J. Beveridge, *Int. Rev. Cytol.* 1981, *72*, 229–317.

[7] F. G. Ferris, T. J. Beveridge, *Can. J. Microbiol.* 1986, *32*, 594–601.

[8] F. G. Ferris, in *Metal Ions and Bacteria* (Eds T. J. Beveridge and R. J. Doyle), John Wiley & Sons, New York, 1989, pp. 295–323.

[9] F. G. Ferris, T. J. Beveridge, *Can. J. Microbiol.* 1986, *32*, 52–55.

[10] T. J. Beveridge, R. G. E. Murray, *J. Bacteriol.* 1976, *127*, 1502–1518.

[11] R. J. Doyle, T. H. Matthews, N. Streips, *J. Bacteriol.* 1980, *143*, 471–480.

[12] T. J. Beveridge, C. W. Forsberg, R. J. Doyle, *J. Bacteriol.* 1982, *150*, 1438–1448.

[13] T. J. Beveridge, S. F. Koval, *Appl. Environ. Microbiol.* 1981, *42*, 325–335.

[14] B. Hoyle, T. J. Beveridge, *Appl. Environ. Microbiol.* 1983, *46*, 749–752.

[15] F. G. Ferris, T. J. Beveridge, *FEMS Microbiol Lett.* 1984, *24*, 43–46.

[16] G. G. Geesey, L. Jangin. in *Metal Ions and Bacteria* (Eds T. J. Beveridge and R. J. Doyle), John Wiley & Sons, New York, 1989, pp. 325–357.

[17] R. J. C. McLean, D. Beauchemin, T. J. Beveridge, *Appl. Environ. Microbiol.* 1992, *58*, 405–408.

[18] T. J. Beveridge, in *Metal Ions and Bacteria* (Eds T. J. Beveridge and R. J. Doyle), John Wiley & Sons, New York, 1989, pp. 1–29.

[19] L. F. Adams, W. C. Ghiorse, *J. Bacteriol.* 1987, *169*, 1279–1285.

[20] F. C. Boogerd, J. P. M. de Vrind, *J. Bacteriol.* 1987, *169*, 489–494.

[21] M. Sprott, T. J. Beveridge, in *Methanogenesis: Ecology, Physiology, Biochemistry and Genetics* (Ed. J. G. Ferry), Chapman & Hall, New York, 1993, pp. 81–127.

[22] T. J. Beveridge, M. Sara, D. Pum, G. D. Sprott, M. Stewart, U. B. Sleyter, in *Bacterial Cell Surface Layers* (Eds U. B. Sleytr, P. Messner, D. Pum and M. Sara), Springer-Verlag, Berlin, 1988, pp. 26–30.

[23] T. J. Beveridge, *Curr. Opin. Struct. Biol.* 1994, *4*, 204–212.

[24] P. Messner, U. B. Sleytr, *Adv. Microb. Physiol.* 1992, *33*, 213–275.

[25] S. F. Koval, *Can. J. Microbiol.* 1988, *34*, 407–414.

[26] S. Schultze-Lam, G. Harauz, T. J. Beveridge, *J. Bacteriol.* 1992, *174*, 7971–7981.

[27] T. J. Beveridge, R. G. E. Murray, *J. Bacteriol.* 1980, *141*, 876–887.

[28] R. W. Harvey, J. O. Leckie, *Mar. Chem.* 1985, *15*, 33–344.

[29] M. D. Mullen, D. C. Wolf, F. G. Ferris, T. J. Beveridge, C. A. Flemming, G. W. Bailey, *Appl. Environ. Microbiol.* 1989, *55*, 3143–3149.

[30] S. G. Walker, C. A. Flemming, F. G. Ferris, T. J. Beveridge, G. W. Bailey, *Appl. Environ. Microbiol.* 1989, *55*, 2976–2984.

[31] S. Langley, T. J. Beveridge, *Appl. Environ. Microbiol.* 1999, *65*, 489–498.

[32] Y. M. Nelson, L. W. Lion, M. L. Shuler, W. C. Ghiorse, *Environ. Sci. Technol.* 1996, *30*, 2027–2035.

[33] S. Langley, T. J. Beveridge, *Can. J. Microbiol.* 1999, *45*, 1–7.

[34] M. Urrutia Mera, M. Kemper, R. Doyle, T. J. Beveridge, *Appl. Environ. Microbiol.* 1992, *58*, 3837–3844.

[35] J. B. Fein, C. J. Daughney, N. Yee, T. Davis, *Geochim. Cosmochim. Acta* 1997, *61*, 3319–3328.

[36] C. J. Daughney, J. B. Fein, N. Yee, *Chem. Geology* 1998, *144*, 161–176.

[37] L. M. He, B. M. Tebo, *Appl. Environ. Microbiol.* 1998, *64*, 1123–1129.

[38] C. J. Daughney, J. B. Fein, *J. Colloid Interface Sci.* 1998, *198*, 53–77.

[39] M. M. Urrutia, T. J. Beveridge, *J. Bacteriol.* 1993, *175*, 1936–1945.

[40] M. M. Urrutia, T. J. Beveridge, *Chem. Geology* 1994, *116*, 261–280.

[41] M. M. Urrutia, T. J. Beveridge, *Appl. Environ. Microbiol.* 1993, *59*, 4323–4329.

[42] J. B. Thompson, F. G. Ferris, *Geology* 1990, *18*, 995–998.

[43] J. B. Thompson, S. S. Schultze-Lam, T. J. Beveridge, D. J. Des Marais, *Limnol. Oceanogr.* 1997, *42*, 133–141.

[44] D. Fortin, T. J. Beveridge, *Geomicrobiol. J.* 1997, *14*, 1–21.

[45] L. A. Warren, F. G. Ferris, *Environ. Sci. Technol.* 1998, *32*, 2331–2337.

[46] F. G. Ferris, W. S. Fyfe, T. J. Beveridge, *Chem. Geology* 1987, *63*, 225–232.

[47] F. G. Ferris, W. S. Fyfe, T. J. Beveridge, *Geomicrobiol. J.* 1987, *5*, 33–42.

[48] H. L. Ehrlich, *Geomicrobiology*, 3rd edn, Marcel Dekker Inc., New York, 1996.

[49] P. C. Singer, W. Stumm, *Science* 1970, *167*, 1121–1123.

[50] D. Fortin, B. Davis, T. J. Beveridge, *J. Indust. Microbiol.* 1995, *14*, 4391–4404.

[51] D. Fortin, B. Davis, T. J. Beveridge, *FEMS Microbiol. Ecol.* 1996, *21*, 11–24.

[52] G. Southam, T. J. Beveridge, *Appl. Environ. Microbiol.* 1992, *58*, 1904–1912.

[53] J. T. Pronk, D. B. Johnson, *Geomicrobiol. J.* 1992, *10*, 153–171.

[54] A. P. Harrison, *Annu. Rev. Microbiol.* 1984, *38*, 265–292.

[55] W. J. Ingledew, J. C. Cox, P. J. Halling, *FEMS Microbiol. Lett.* 1977, *2*, 193–197.

[56] C-P. Lienemann, M. Taillefert, D. Perret, J-F. Gaillard, *Geochim. Cosmochim. Acta* 1997, *61*, 1437–1446.

[57] J. P. Cowen, K. W. Bruland, *Deep-Sea Res.* 1985, *32*, 253–272.

[58] L. St-Cyr, D. Fortin, P. G. C. Campbell, *Aquatic Botany* 1993, *46*, 155–167.

[59] D. Fortin, G. G. Leppard, A. Tessier, *Geochim. Cosmochim. Acta* 1993, *57*, 4391–4404.

[60] D. Fortin, F. G. Ferris, S. D. Scott, *Am. Mineral.* 1998, *83*, 1309–1408.

[61] F. G. Ferris, T. J. Beveridge, W. S. Fyfe, *Nature* 1986, *320*, 609–610.

[62] S. Schultze-Lam, F. G. Ferris, K. O. Konhauser, R. G. Wiese, *Can. J. Earth Sci.* 1995, *32*, 2021–2026.

[63] N. W. Hinman, R. F. Lindstrom, *Chem. Geology* 1996, *132*, 237–246.

[64] B. Jones, R. W. Renault, *Can. J. Earth Sci.* 1996, *33*, 72–83.
[65] K. O. Konhauser, W. S. Fyfe, F. G. Ferris, T. J. Beveridge, *Geology* 1993, *21*, 1103–1106.
[66] K. O. Konhauser, S. Schultze-Lam, F. G. Ferris, W. S. Fyfe, F. J. Longstaffe, T. J. Beveridge, *Appl. Environ. Microbiol.* 1994, *60*, 549–553.
[67] D. Fortin, T. J. Beveridge, *Chem. Geology* 1997, *141*, 235–250.
[68] F. G. Ferris, W. S. Fyfe, T. J. Beveridge, *Geology* 1988, *16*, 149–152.
[69] R. J. Doyle, A. L. Koch, *Crit. Rev. Microbiol.* 1987, *15*, 169–222.
[70] S. Schultze-Lam, T. J. Beveridge, *Appl. Environ. Microbiol.* 1994, *60*, 447–453.
[71] F. G. Ferris, R. G. Wiese, W. S. Fyfe, *Geomicrobiol. J.* 1994, *12*, 1–13.
[72] F. G. Ferris, C. M. Fratton, J. P. Gerits, S. Schultze-Lam, S. B. Sherwood, *Geomicrobiol. J.* 1995, *13*, 57–67.
[73] W. Stumm, J. J. Morgan, *Aquatic Chemistry*, 3rd edn, John Wiley, New York, 1996.
[74] A. Pentecost, B. Spiro, *Geomicrobiol. J.* 1990, *6*, 129–135.
[75] F. G. Ferris, J. B. Thompson, T. J. Beveridge, *Palaios* 1997, *12*, 213–219.
[76] D. R. Kobluk, R. D. Crawford, *Palaios* 1990, *5*, 134–148.
[77] M. L. Coleman, D. B. Hedrick, D. R. Lovley, D. C. White, K. Pye, *Nature* 1983, *361*, 436–438.
[78] E. E. Roden, D. R. Lovley, *Appl. Environ. Microbiol.* 1993, *59*, 734–742.
[79] C. Vasconcelos, J. A. McKenzie, S. Bernasconi, G. Grujic, A. J. Tien, *Nature* 1995, *377*, 220–222.
[80] R. A. Berner, *Early Diagenesis, A Theoretical Approach*, Princeton University Press, Princeton, 1980.
[81] P. A. Trudinger, L. A. Chambers, J. W. Smith, *Can. J. Earth Sci.* 1985, *22*, 1910–1918.
[82] A. Mohagheghi, D. M. Updegraff, M. B. Goldhaber, *Geomicrobiol. J.* 1985, *4*, 153–173.
[83] H. E. Frimmel, A. P. LeRoex, J. Knight, W. E. Minter, *Econ. Geol.* 1993, *88*, 249–265.
[84] G. Southam, T. J. Beveridge, *Geochim. Cosmochim. Acta* 1994, *58*, 4527–4530.
[85] G. Southam, T. J. Beveridge, *Geochim. Cosmochim. Acta* 1996, *60*, 4369–4376.
[86] Y. M. Nelson, L. W. Lion, W. C. Ghiorse, M. L. Shuler, *Appl. Environ. Microbiol.* 1999 *65*, 175–180.
[87] F. G. Ferris, K. O. Konhauser, B. Lyven, K. Pedersen, *Geomicrobiol. J.* 1999, *16*, 181–192.
[88] E. S. Barghoorn, S. A. Tyler, *Science* 1965, *147*, 563–577.
[89] G. Eglinton, P. M. Scott, T. Belsky, A. L. Burlingame, M. Calvin, *Science* 1964, *145*, 263–264.
[90] J. D. Saxby, *Rev. Pure Appl. Chem.* 1969, *19*, 131–150.

3 Magnetic Iron Oxide and Iron Sulfide Minerals within Microorganisms

Dennis A. Bazylinski, Richard B. Frankel

3.1 Introduction

A number of microorganisms are known to facilitate the deposition of minerals [1]. Mineral deposition by bacteria occurs by two generically different mechanisms [2]. In biologically-induced mineralization, minerals form in the environment in an apparently uncontrolled manner from metabolites produced by the bacteria. The minerals formed – and their properties – depend on environmental conditions, and are generally indistinguishable from minerals formed inorganically under the same chemical conditions. In biologically-controlled mineralization, minerals are deposited on or within organic matrices or vesicles inside the cell, allowing the organism to exert a high degree of control over the composition, size, habit and intracellular location of the minerals [3]. Because the intravesicular pH and Eh can be controlled by the organism, mineral formation is not as affected by external environmental parameters as in the biologically-induced mode.

Both biologically-induced and biologically-controlled deposition of magnetic iron minerals by bacteria are known. Dissimilatory iron-reducing bacteria [4] and dissimilatory sulfate-reducing bacteria [5] export ferrous ions and sulfide ions, respectively, into the environment, inducing the formation of a number of extracellular iron and sulfide minerals, including magnetic iron minerals such as magnetite (Fe_3O_4), greigite (Fe_3S_4) and pyrrhotite (Fe_7S_8) [5, 6]. These mineral particles are characterized by variable composition and crystallinity (depending on environmental conditions), broad size distributions, and lack of a consistent crystal habit [7]. On the other hand, magnetotactic bacteria [8] mineralize magnetosomes [9], which are nanometer-sized magnetite [10] or greigite [11] crystals in intracellular membrane vesicles [12], in a highly controlled manner. The mineral crystals are generally characterized by specific chemical compositions, high crystallinity, narrow size distributions, and well-defined, consistent habits within each bacterial species or strain [13, 14].

In addition to magnetotactic bacteria, a number of eukaryotic microorganisms have been found to contain magnetosomes or magnetosome-like structures [15, 16]. In this chapter we will review the characteristics of magnetic minerals in magnetotactic bacteria and single-celled eukaryotes and the role of the magnetic iron minerals in magnetotaxis. Molecular biological and biochemical aspects of magnetosome formation in bacteria, as well as magnetite in higher organisms, will be discussed in Chapters 5, 7, 8, 9, 10.

3.2 Diversity of Magnetotactic Bacteria

Magnetotactic bacteria are a morphologically and physiologically diverse group of motile Gram-negative prokaryotes, ubiquitous in aquatic environments [8, 13], that orient and migrate along magnetic field lines. They include coccoid, rod-shaped, vibrioid, spirilloid (helical), and even multicellular forms. Physiological forms include denitrifiers that are facultative anaerobes, obligate microaerophiles, and anaerobic sulfate-reducers [14]. Thus, there are probably many species of magneto-tactic bacteria [17], although only a few have been isolated in pure culture [13]. Each magnetotactic bacterium contains magnetosomes and iron can constitute as much as 3% of the dry weight of the cells [18, 19]. The magnetosomes are generally arranged in one or more chains within the cell, aligned parallel to the axis of mo-tility. Although most magnetotactic bacteria exclusively contain either magnetite or greigite magnetosomes, a large, rod-shaped marine bacterium has been found which contains both magnetite and greigite magnetosomes [20, 21]. More detailed discus-sion of the molecular diversity and phylogeny of the magnetotactic bacteria can be found in Chapter 4.

3.3 Ecology of Magnetotactic Bacteria

Magnetotactic bacteria are generally found in aquatic habitats that are close to neutral in pH, and are not thermal, heavily polluted or well oxygenated [8]. They are cosmopolitan in distribution and occur in the highest numbers at the micro-aerobic zone, also known as the oxic–anoxic transition zone (OATZ) [13, 22]. In many freshwater habitats, the OATZ is located at the sediment–water interface or just below it, and this is where the highest concentration of magnetotactic bac-teria occurs [14]. In some brackish-to-marine chemically-stratified systems, the OATZ is found permanently or is seasonally located in the water column as shown in Fig. 3.1. The Pettaquamscutt Estuary (Narragansett Bay, RI, USA) [23] and Salt Pond (Woods Hole, MA, USA) [24] are examples of the latter situation. Hydrogen sulfide, produced by sulfate-reducing bacteria in the anaerobic zone and sediment, diffuses upward, while oxygen diffuses downward from the surface, resulting in a double, vertical, chemical concentration gradient (Fig. 3.1) with a concomitant redox gradient. Typically, a pycnocline and thermocline are also associated with the OATZ in these situations, which help to physically stabilize the chemical stratification.

A variety of magnetotactic bacteria are found at both the Pettaquamscutt Estu-

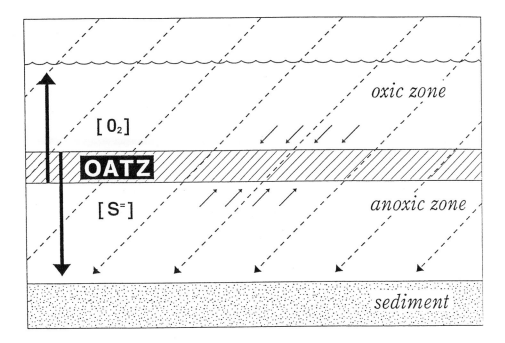

Figure 3.1 Schematic of the oxic–anoxic transition zone (OATZ) in the water column as typified by Salt Pond (Woods Hole, MA, USA). Note the inverse double concentration gradients of oxygen, $[O_2]$ diffusing from the surface and sulfide, $[S^{2-}]$ generated by sulfate-reducing bacteria in the anaerobic zone (vertical arrows). Magnetite-producing magnetotactic bacteria exist in greatest numbers at the OATZ, where microaerobic conditions predominate; greigite producers are found just below the OATZ where S^{2-} becomes detectable. When polar-magneto-aerotactic, magnetite-producing coccoid cells are above the OATZ in vertical concentration gradients of O_2 and S^{2-} (higher $[O_2]$ than optimal), they swim downward (small arrows above OATZ) along the inclined geomagnetic field lines (broken lines). When they are below the OATZ (lower $[O_2]$ than optimal), they reverse direction (by reversing the direction of their flagellar motors) and swim upward (small arrows below the OATZ) along the inclined geomagnetic field lines. The direction of the flagellar rotation is coupled to an aerotactic sensory system that acts as a switch when cells are at a suboptimal position in the gradient, as defined in the text. Magnetotactic spirilla (and other axial magneto-aerotactic microorganisms) align along the geomagnetic field lines and swim up and down, relying on a temporal sensory mechanism of aerotaxis to find and maintain position at their optimal oxygen concentration at the OATZ.

ary and Salt Pond, stratified by depth [21]. Generally, the magnetite-producing bacteria prefer the OATZ proper (Fig. 3.1) and behave as microaerophiles. Two strains of magnetotactic bacteria have been isolated from the Pettaquamscutt Estuary, a vibrio-designated strain MV-2 [25, 26] and a coccus-designated strain MC-1 [25, 27, 28]. Both strains grow as microaerophiles, although strain MV-2 can also grow anaerobically with nitrous oxide (N_2O) as a terminal electron acceptor. Other cultured magnetotactic bacterial strains, including the magnetotactic spirilla [29–33] and rods [34], are microaerophiles or anaerobes or both. The greigite-producing magnetotactic bacteria are probably anaerobes appearing to prefer the more sulfidic waters below the OATZ (Fig. 3.1) where the oxygen concentration is zero [14, 21]. No magnetotactic bacterium with greigite magnetosomes has been grown in pure culture.

3.4 Magnetite Magnetosomes

The mineral composition of magnetite magnetosomes is specific in a given strain, as demonstrated by the fact that cells of several cultured strains continue to synthesize magnetite and not greigite, even when hydrogen sulfide is present in the growth medium [26, 28]. Gorby [35] found that iron was not replaced by other transition metal ions, including titanium, chromium, cobalt, copper, nickel, mercury and lead, in the magnetite crystals of *Magnetospirillum magnetotacticum* when cells were grown in the presence of these ions. Towe and Moench [36] reported very small amounts of titanium in the magnetite particles of an uncultured freshwater magnetotactic coccus, but this has not been confirmed by subsequent studies. Thus, the absence of transition metal ions other than iron is one of the hallmarks of magnetite magnetosomes.

Magnetite magnetosomes from a number of magnetotactic bacteria are shown in the electron micrographs in Fig. 3.2. The habits of the magnetite crystals appear to be consistent within a given species or strain [26, 28], although some variations in shape and size can occur within single magnetosome chains [37]. The presence of smaller and more rounded crystals, common at one or both ends of the chains (Fig. 3.3a) and interpreted as "immature" crystals, has been used as evidence that the magnetosome chain increases in size by the precipitation of new magnetosomes at the ends of the chain following cell division [3]. In addition to the roughly isometric crystal shapes seen in *Magnetospirillum magnetotacticum* and other freshwater magnetotactic spirilla, for example (Fig. 3.2a), several non-isometric shapes have been described in other species or strains [3, 38]. These include prismatic (Fig. 3.2b) and tooth, bullet or arrowhead shapes (Fig. 3.2c) [39–41]. Many crystals of these types have been studied by high-resolution transmission electron microscopy (HRTEM) and/or selected area electron diffraction (SAED), and idealized crystal habits based on simple forms, i.e. combinations of symmetry-related crystal faces, have been proposed (Fig. 3.4a–d) [3, 28]. It should be emphasized that both isometric and non-isometric crystals have the face-centered cubic structure of magnetite, as shown in the interplanar spacings and angles determined by HRTEM or SAED, but have different, species- or strain-specific, crystal habits.

Both magnetite and greigite are in the Fd3m space group. Macroscopic crystals of magnetite display habits of the octahedral {111} form, or more rarely the dodecahedral {110} or cubic {100} forms [42]. The habit of the isometric crystals in *Magnetospirillum magnetotacticum* is cubo-octahedral (Fig. 3.4a), composed of {100} + {111} forms [43], with isometric development of the six symmetry-related faces of the {100} form and of the eight symmetry-related faces of the {111} form.

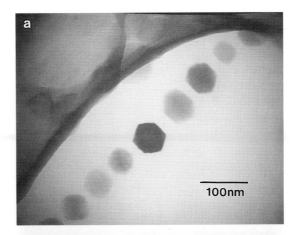

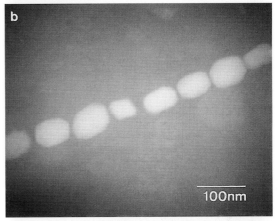

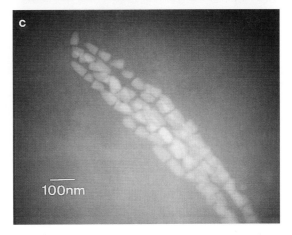

Figure 3.2 Morphologies of intracellular magnetite (Fe_3O_4) particles
produced by magnetotactic bacteria
collected from the OATZ of the Pettaquamscutt Estuary. (a) Brightfield
scanning transmission electron
microscope (STEM) image of a chain
of cubo-octahedra in cells of an
unidentified rod-shaped bacterium,
viewed along a [111] zone axis for
which the particle projections appear
hexagonal. (b) Darkfield STEM image
of a chain of prismatic crystals within
a cell of an unidentified marine vibrio,
with parallelepiped projections.
(c) Darkfield STEM image of tooth-
shaped (anisotropic) magnetosomes
from an unidentified marine rod-
shaped bacterium. Scale bar = 100 nm.

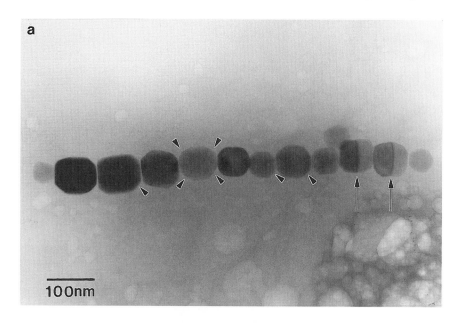

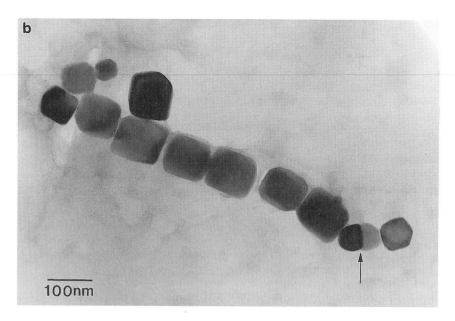

Figure 3.3 Transmission electron microscope (TEM) image of magnetite magnetosome chains within cells of the magnetotactic coccus, strain MC-1, grown autotrophically (a) and heterotrophically (b). Note the presence of smaller, "immature" magnetite crystals at the ends of the chains (most obvious in (a)), the more pronounced truncations at the ends of the crystals in cells grown autotrophically, indicated by the small arrows in (a), and the twinned crystals in (a) and (b), indicated by the long arrows. Scale bar = 100 nm.

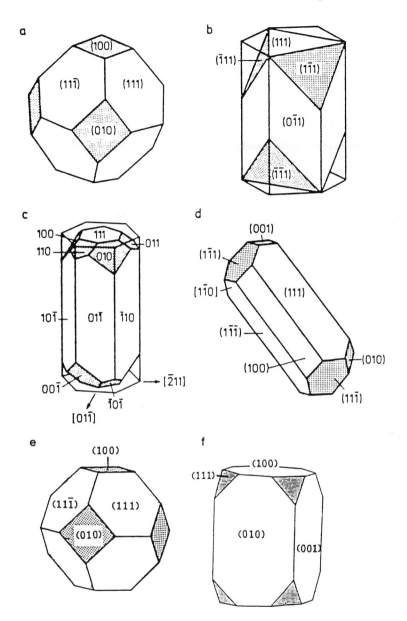

Figure 3.4 Idealized magnetite (a)–(d) and greigite (e)–(f) crystal morphologies derived from HRTEM studies of magnetosomes from magnetotactic bacteria. (a) and (e) cubo-octahedra; (b) and (c) variations of pseudo-hexagonal prisms; (d) elongated cubo-octahedron; (f) elongated truncated cube.

The habits of the non-isometric crystals in strains MV-1 and MC-1, for example, can be described as combinations of {100}, {111} and {110} forms [44]. In these cases, the six, eight and twelve symmetry-related faces of the respective forms that constitute the habits do not develop equally. For example, crystals of both strains MV-1 (Fig. 3.4b) [28] and MC-1 (Fig. 3.4c) [26] have pseudo-hexagonal habits elongated along a $\langle 111 \rangle$ axis, with six well-developed (110) faces parallel to the elongation axis, and capped by (111) planes perpendicular to the elongation axis. In MV-1 crystals, the remaining six (111) faces form truncations of the end caps, and the remaining six (110) faces are missing [28]. In MC-1 crystals, the truncations at each end consist of three (100) faces alternating with three (110) faces. Thus, six (110) faces are larger and six are smaller, and six (111) faces are absent in this habit. Only the six (100) faces are isometric [26]. The motif of six elongated (110) faces capped by (111) faces with differing truncation planes appears to be most common in magnetotactic bacteria with non-isometric magnetosome crystals.

The elongated prismatic habits, corresponding to the anisotropic development of symmetry-related faces, could occur either because of anisotropy in the growth environment (such as for example, concentration and/or temperature gradients) or anisotropy of the growth sites [3]. In the case of the non-isometric magnetosomes, the anisotropy of the environment could derive from an anisotropic flux of ions through the magnetosome membrane surrounding the crystal [12], or from aniso-tropic interactions of the magnetosome membrane with the growing crystal [3]. The most anisotropic habits are those of the tooth, bullet and arrowhead-shaped mag-netite crystals. In one uncultured organism, small and large crystals can have dif-ferent habits, suggesting that growth occurs in two stages. The nascent crystals are cubo-octahedra which subsequently elongate along a [111] axis to form a pseudo-octahedral prism with alternating (110) and (100) faces capped by (111) faces (Fig. 3.4d) [40, 41].

Culturing bacteria under very different growth conditions appears to have little or no effect on the overall crystal habit of magnetite crystals in magnetosomes. How-ever, in one cultured bacterium, strain MC-1, the magnetite particles appear to have more pronounced truncations at the ends of these elongate crystals in cells grown autotrophically versus those grown heterotrophically (Fig. 3.3a, b) [28]. As more cultures of magnetotactic bacteria become available, the effects of growth con-ditions on magnetosome formation can be further tested.

Magnetosome magnetite crystals are typically 35–120 nm long [45]. Statistical analyses of crystal size distributions in cultured strains show narrow asymmetric distributions and consistent width to length ratios within each strain [44]. Whereas the crystal size distributions of inorganic magnetite and magnetite resulting from dissimilatory iron reduction are typically log–normal [46], the shapes of the mag-netosome crystal size distributions are asymmetric, with a sharp high end cut-off, consistent with a transport-controlled Ostwald ripening process [47].

Twinned crystals are common in magnetotactic bacterially-produced magnetites ($\sim 10\%$) with individuals related by rotations of $180°$ around the [111] direction parallel to the chain direction and with a common (111) contact plane (Fig. 3.3a, b). Multiple twins are less common [44].

3.5 Greigite Magnetosomes

Although all freshwater magnetotactic bacteria have been found to mineralize magnetite magnetosomes, many marine, estuarine and salt marsh species produce magnetosomes with iron sulfide mineral crystals [11, 48, 49]. Bacteria with iron sulfide magnetosomes include a many-celled magnetotactic prokaryote (MMP) [25, 50, 51] and a variety of relatively large rod-shaped bacteria [50, 52]. The iron sulfide minerals are either greigite [11, 53, 54] or a mixture of greigite and transient non-magnetic iron sulfide phases that probably represent mineral precursors to greigite. These include non-magnetic mackinawite and probably a sphalerite-type cubic FeS [54, 55]. Based on transmission electron microscopic observations, electron diffraction, and known iron sulfide chemistry [56, 57], the reaction scheme for greigite formation in the magnetotactic bacteria appears to be [54, 55]:

$$\text{cubic FeS} \rightarrow \text{mackinawite (tetragonal FeS)} \rightarrow \text{greigite } (Fe_3S_4)$$

Reports of iron pyrite [11] and pyrrhotite [49] in magnetotactic bacteria have not been confirmed. Interestingly, under the strongly reducing, sulfidic conditions at neutral pH in which the iron sulfide magnetotactic bacteria are found [21], greigite would be expected to transform into pyrite [57]. It is not known how the cells prevent this transformation, but control of the sulfide concentration in the cell and/or in the magnetosome vesicle is probably involved. The size range of the greigite crystals in magnetotactic bacteria is similar to that observed for magnetite.

As with magnetite, several greigite crystal morphologies, composed primarily of {111} and {100} forms, have been observed (Fig. 3.4e, f) [53]. These include cubo-octahedral and pseudo-rectangular prismatic, as shown in Fig. 3.4e, f, and tooth-shaped [55]. Like that of their magnetite counterparts, the morphologies of the greigite particles in the rod-shaped bacteria are also apparently species- and/or strain-specific. In contrast to the rod-shaped bacteria, the MMP contains greigite crystals with a mixture of morphologies, including pleomorphic, pseudo-rectangular prismatic, tooth-shaped and cubooctahedral [55]. Some of these particle morphologies are shown in Fig. 3.5.

The mixed mineral rod-shaped bacterium was found to contain magnetite magnetosomes with arrowhead-shaped crystals, and rectangular prismatic crystals of greigite, co-organized within the same chains of magnetosomes (this organism usually contains two parallel chains of magnetosomes) as shown in Fig. 3.6 [20, 21]. In addition to different, mineral-specific morphologies and sizes, the greigite and magnetite crystals have [100] and [111] crystal axes parallel to the chain direction, respectively. Both crystal morphologies and chain orientations have been found in organisms with single mineral component chains [3, 53, 54]. This suggests different biomineralization processes for the two minerals, possibly due to differences in the respective magnetosome membranes. Thus, it is likely that two separate sets of genes control the biomineralization of magnetite and greigite in this organism [21].

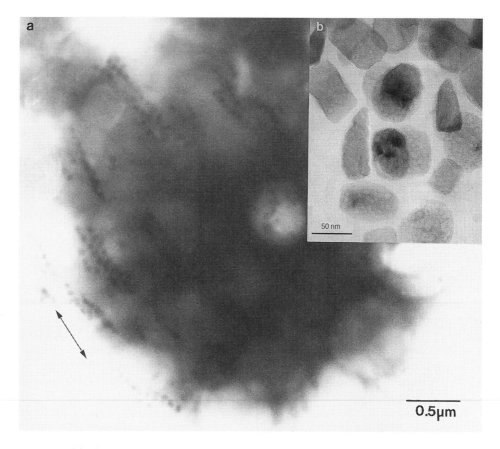

Figure 3.5 (a) TEM image of the MMP, weakly stained with uranyl acetate, showing that most of the magnetosome chains are oriented parallel to the axis designated by the double-ended arrow. Scale bar = 0.5 μm. (b, inset) TEM image of greigite particles within the MMP, including tooth-shaped and rectangular prismatic crystals. Scale bar = 50 nm. ((b) courtesy of M. Pósfai and P. R. Buseck.)

Significant amounts of copper (but not other transition metal ions) have been found in greigite magnetosomes of some magnetotactic bacteria [55, 58]. The amount of copper varied from collection site to collection site. It also varied from magnetosome to magnetosome within an individual cell, ranging in one case from about 0.1 to 10 atomic % relative to iron. This suggests that copper incorporation depends on environmental conditions. The copper appeared in some cases to be mostly concentrated on the surface of the crystals [58]. These observations indicate that mineralization in these organisms is more susceptible to chemical conditions in the external environment, and suggest that the iron sulfide magnetosomes could function in transition metal detoxification.

The conversion of cubic FeS and mackinawite to greigite [52, 55] in iron sulfide magnetosomes involves changes in the stoichiometry of the magnetosome particle.

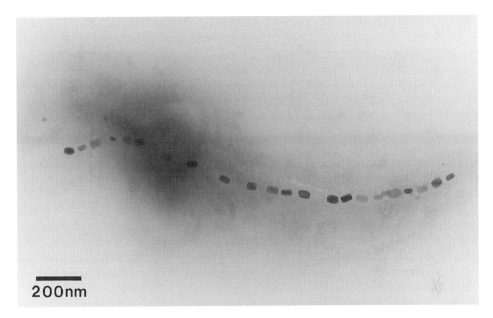

Figure 3.6 TEM image of an unstained cell of the marine magnetotactic vibrio strain MV-1, showing gaps between magnetosomes. Note that even with the relatively large gaps between some magnetosomes, there is a consistent orientation of each individual magnetosome along the direction of the magnetosome chain, suggesting that the separated magnetosomes are co-organized with the others by a chain assembly process. Scale bar = 150 nm.

Based on thermodynamic considerations, cubic FeS and mackinawite transform to greigite under strongly reducing sulfidic conditions at neutral pH [57]. Similar transformations occur inorganically in sediments in the same locales [59]. There is also the possibility that the rate of transformation, including suppression of transformation of greigite to pyrite, might be controlled by the cell.

3.6 Arrangement of Magnetosomes in Cells

In addition to controlled mineralization, magnetotactic bacteria control the placement of magnetosomes in the cell. The arrangement of magnetosomes in the cell, like the crystal habits of the magnetosome mineral particles themselves, also appears to be specific to a particular bacterial species and/or strain.

Typically, magnetosomes are organized into one or more chains in which the particles are arranged along the long axis of the cell (Figs. 3.2, 3.3). In most cases, magnetosomes are directly adjacent to one another but in some bacteria, for example strain MV-1, large gaps can occur between magnetosomes (Fig. 3.6) [21].

The magnetosomes appear to be fixed within the cell, generally adjacent to the cell membrane, by as yet unknown structural elements [21]. There is little evidence, however, that the membrane surrounding the individual magnetosome particles (i.e. the magnetosome membrane) is contiguous with the cell membrane.

Individual magnetite crystals are oriented within the chain with a crystal [111] axis parallel to the chain axis. For bacterial strains in which the magnetite crystals have elongated habits, the [111] axis parallel to the chain axis is also the axis of elongation of the crystals [3].

Several magnetotactic bacteria are known in which the magnetosomes are not arranged in chains. Some magnetite-producing cocci, including *Bilophococcus magnetotacticus*, do not construct magnetosome chains, but instead have a "clump" of magnetosomes at one end of the cell [60].

Most known, uncultured, greigite-producing, magnetotactic rod-shaped bacteria tend to synthesize multiple, usually two to four, parallel chains of magnetosomes adjacent to one another that longitudinally traverse the cell. Unlike magnetite, greigite crystals are oriented with a [100] axis along the chain axis [54]. The iron sulfide magnetosomes in the MMP originally appeared in disrupted organisms to be positioned somewhat randomly in the constituent cells, with some arranged in small chains. However, electron microscopy of intact organisms revealed that most magnetosomes exist in multiple chains which are for the most part aligned parallel to each other within each cell and between cells (Fig. 3.5a). This consensus alignment of magnetosome chains results in a net magnetic dipole moment for the entire organism. Clumps of greigite magnetosomes have also been observed in some rod-shaped bacteria [53].

The *de novo* synthesis of non-magnetic crystalline iron sulfide precursors to greigite, cubic FeS and mackinawite (see section 3.5), aligned along the magnetosome chain, indicates that chain formation and the orientation of the magnetosomes in the chain does not necessarily involve magnetic interactions.

3.7 The Role of Magnetosomes and Magnetosome Chains in Magnetotaxis

The crystal size distributions for both magnetite and greigite magnetosomes peak within the permanent, single-magnetic-domain (SD) size range [45]. Magnetostatic interactions between the SD magnetosomes organized in a chain configuration cause the individual grain moments to orient parallel to each other along the chain direction. Thus the chain has a permanent magnetic dipole approximately equal in magnitude to the sum of the individual magnetosome magnetic moments [61]. This has recently been demonstrated by electron holography of individual magnetotactic bacteria [62]. The permanent magnetic dipole moment of a cell is generally large enough that its interaction with the geomagnetic field overcomes the thermal forces tending to randomize the orientation of the magnetic dipole in its aqueous surroundings [61]. Because the dipole is fixed in the cell, orientation of the dipole results in orientation of the cell. Magnetotaxis results as the oriented cell swims along magnetic field lines (B). The direction of migration along B is determined by the direction of flagellar rotation, which in turn is determined by the aerotactic response of the cell [27].

Some magnetotactic spirilla, such as *Magnetospirillum magnetotacticum*, swim parallel or antiparallel to B and form aerotactic bands [63] at a preferred oxygen concentration [O_2] in oxygen gradient cultures or in suspensions of living cells. In a homogeneous medium, roughly equal numbers of cells swim in either direction along B [18, 63]. Most microaerophilic bacteria form aerotactic bands at a preferred or optimal [O_2] where the proton motive force is maximal [64], using a temporal sensory mechanism [65] that samples the local environment as they swim and compares the present [O_2] with that in the recent past. The change in [O_2] with time determines the sense of flagellar rotation [66]. The behavior of individual cells of *M. magnetotacticum* in aerotactic bands, termed axial magneto-aerotaxis, is consistent with the temporal sensory mechanism [27].

In contrast, the ubiquitous freshwater and marine bilophotrichously-flagellated (having two flagellar bundles on one hemisphere of the cell) magnetotactic cocci, and some other magnetotactic strains, swim persistently in a preferred direction relative to B when examined microscopically in wet mounts [8, 39, 67]. However, magnetotactic cocci in oxygen gradients can swim in both directions along B without turning around [27]. The cocci form microaerophilic, aerotactic bands and seek a preferred [O_2] along the concentration gradient. However, while axial magneto-aerotaxis is consistent with the temporal sensory mechanism, the aerotactic behavior of the cocci is not. Instead their behavior is consistent with a two-state aerotactic sensory model in which the [O_2] determines the sense of the flagellar rotation and hence the swimming direction relative to B [27]. This model, called polar magneto-aerotaxis, accounts for the ability of the magnetotactic cocci to migrate to and maintain position at the preferred [O_2] at the OATZ in chemically-stratified, semi-anaerobic basins (Fig. 3.1). An assay using chemical gradients in thin capillaries has been developed that distinguishes between axial and polar magneto-aerotaxis [68].

For both aerotactic mechanisms, migration along magnetic field lines reduces a three-dimensional search problem to a one-dimensional search problem. Thus, magnetotaxis is presumably advantageous to motile microorganisms in chemically-stratified locales because it increases the efficiency of finding and maintaining an optimal position in vertical concentration gradients – in this case, vertical oxygen gradients [69]. It is possible that there are other forms of magnetically-assisted chemotaxis to molecules or ions other than oxygen, such as sulfide, or magnetically-assisted redox- or phototaxis in bacteria that inhabit the anaerobic zone below the OATZ.

Nevertheless, questions remain concerning the function of magnetotaxis. Many obligately microaerophilic bacteria find and maintain an optimal position at the OATZ without the help of magnetosomes, as do cultured non-magnetotactic mutants that do not make magnetosomes. Moreover, cultured magnetotactic bacteria form microaerophilic bands of cells in the absence of a magnetic field. Thus, it appears that the enormous iron uptake and magnetosome formation is somehow linked to cellular physiology and to other, as yet unknown, cellular functions.

3.8 Chemistry of Magnetosome Formation

Magnetite and greigite synthesis in magnetotactic bacteria begins with the uptake of iron. Because reduced Fe(II) is very soluble (up to 100 mM at neutral pH [70]), it is easily taken up by bacteria, usually by non-specific means. However, Fe(III) is extremely insoluble and many microbes synthesize and rely on low molecular weight (<1 kDa) iron chelators, called siderophores, which bind and solubilize Fe(III) for uptake [71]. These compounds are generally produced under iron-limited conditions and their synthesis is repressed under high iron conditions. Although in many growth media used for cultivation of magnetotactic bactera, iron is supplied as chelated Fe(III) (e.g. Fe(III) quinate [18]), most of these media also contain chemical reducing agents (e.g. thioglycolate, ascorbic acid, etc.) potent enough to reduce Fe(III) to Fe(II) [18]. Thus both oxidized and reduced forms of iron are available to cells. Several studies have focussed on iron uptake in the magnetite-producing, freshwater, magnetotactic spirilla and only one reports siderophore synthesis by a magnetotactic bacterium. A hydroxamate siderophore was found to be produced by cells of *Magnetospirillum magnetotacticum* grown under high – but not under low – iron conditions [72]. The siderophore production pattern here is the reverse of that normally observed. However, subsequent studies have not confirmed this finding. Frankel *et al.* [73] assumed that iron uptake by this organism probably occurred via a non-specific transport system. Nakamura *et al.* [74] did not detect siderophore production by *Magnetospirillum* AMB-1 and concluded that iron was taken up as Fe(III) by a periplasmic binding protein-dependent iron transport system.

Schüler and Baeuerlein [19] found that the major portion of iron for magnetite synthesis in *Magnetospirillum gryphiswaldense* was taken up as Fe(III) and that Fe(III) uptake is an energy-dependent process displaying Michaelis–Menten kinetics. This suggested that Fe(III) uptake by cells of *M. gryphiswaldense* involves a low affinity but high velocity transport system [19]. Fe(II) was also taken up by cells in a slow, diffusion-like process. Schüler and Baeuerlein also showed that magnetite formation was induced in non-magnetotactic cells by a low threshold oxygen concentration of about 2–7 µM (at 30 °C) and was tightly linked to Fe(III) uptake [19]. This finding is consistent with the results of Blakemore *et al.* [75], who found that microaerobic conditions and molecular O_2 were required for magnetite synthesis in *M. magnetotacticum.*

Frankel *et al.* [73] examined the nature and distribution of major iron compounds in *Magnetospirillum magnetotacticum* using [57]Fe Mössbauer spectroscopy, and proposed a model in which magnetite formation is preceded by formation of amorphous hydrous Fe(III) oxide in the magnetosome vesicles. Reduction of one-third of the Fe(III) ions by electron transport into the vesicle, or addition of Fe(II), results in the formation of magnetite. An amorphous electron-dense material was observed in magnetosomes by electron microscopy [3]. However, Schüler and Baeuerlein [76] showed that in cells of *M. gryphiswaldense*, Fe(III) is taken up and rapidly converted to magnetite without any apparent delay, suggesting that there is no significant accumulation of a precursor to magnetite inside the cell, at least under the conditions of the experiment, which appeared optimal for magnetite production by that organism.

It is not known how the crystal habits and crystallographic orientation in the magnetosome chain is controlled, but it is thought to involve specific proteins in the magnetosome membrane which both nucleate and constrain crystal growth [3].

3.9 Other Intracellular Iron Oxides and Sulfides in Bacteria

Iron-rich inclusions, other than magnetosomes, have also been observed in prokaryotic cells. Many bacteria have bacterioferritin to store intracellular iron [77]. The iron is sequestered in an 8 nm cavity in the protein as an amorphous hydrous iron phosphate (P/Fe $\sim$ 1) [78]. Some sulfate-reducing bacteria, including *Desulfovibrio* and *Desulfotomaculum* species, have been shown to produce intracellular electron-dense "particles" of an iron sulfide when grown in a medium containing high concentrations of iron ($\sim$1 mM) [79]. The "particles" were arranged randomly in the cell and appeared to represent poorly ordered (amorphous) deposits of iron. In addition, the "particles" could not be separated from lysed cells using density gradient centrifugation. The function of these inclusions is not known. Although magnetic iron oxide deposits other than magnetite magnetosomes have not been identified or described in prokaryotic cells, there is no reason why they cannot exist. There is one report of a magnetic, iron-rich, bacterial inclusion that may be an iron oxide. Vainshtein *et al.* [80] reported the presence of an intracellular magnetsensitive inclusion in cells of the purple photosynthetic species, *Rhodospeudomonas palustris*, *R. rutilis* and *Ectothiorhodospira shaposhnikovii*, when grown with relatively high concentrations of iron. The inclusions were spherical particles containing an electron-transparent core surrounded by an electron-dense matrix. These particles could be separated from lysed cells and X-ray microanalysis showed that the inclusions are iron-rich but do not contain sulfur. Interestingly, the inclusions, like magnetosomes, appear to form a chain or a "thread of beads" arranged along the long axis of the cell. The cells do show a magnetic response but are not necessarily magnetotactic. The authors speculated that these inclusions function similarly to the magnetosomes.

3.10 Magnetic Iron Oxides and Sulfides in Microorganisms Other Than Bacteria

Magnetosome or magnetosome-like magnetite particles have been observed in a number of Protistan species. They were first described in a Euglenoid alga in which

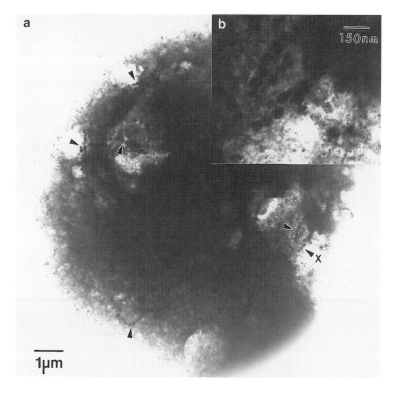

Figure 3.7 (a) Low magnification TEM image of a magnetically-responsive dinoflagellate showing numerous chains of magnetite crystals throughout the cell (at arrows). Because of the difficulty of finding the particles in such a thick cell, a very high beam voltage was used which destroyed much of the cellular structure of the organism. Scale bar = 1 μm. (b, inset) High magnification TEM image of magnetite crystals located at arrow labeled "X". Scale bar = 150 nm.

several adjacent parallel chains of tooth-shaped magnetite particles were arranged in rows running down the length of the cell [15].

Recently, we reported [81] the presence of diverse populations of magnetically-responsive protozoans in the chemically-stratified coastal pond, Salt Pond (Woods Hole, MA, USA; discussed earlier). These microorganisms included dinoflagellates, biflagellated Cryptomonads, and a ciliate of the genus *Cyclidium*. Magnetite was identified as the mineral component in the dinoflagellate and a large Cryptomonad. Magnetite particles were arranged in chains arranged somewhat randomly within the cell and were about 55–75 nm in diameter (Fig. 3.7). The particles were indistinguishable from those produced by some magnetotactic bacteria found at the site. Determination of the precise crystal habit of the particles was not possible due to the thickness of the protozoal cell. The mineral particles in other magnetically-responsive protists were not identified so the possibility of greigite crystals in these microorganisms cannot be eliminated.

It is not known whether these protists biomineralize their magnetic particles or whether they ingest magnetotactic bacteria or their magnetosomes as they are released from lysing cells. It is known, however, that some species of protozoa ingest iron particles, thereby becoming magnetic [82]. Iron is well recognized as a limiting factor in primary production in some oceanic areas and is often present in seawater in particulate and colloidal forms [83]. Barbeau *et al.* [83] showed that digestion of colloidal iron in the food vacuoles of protozoans during grazing of particulate and colloidal matter might generate more bioavailable iron for other species such as phytoplankton. In the case of Salt Pond, protists that ingest magnetotactic bacteria may play an important role in iron cycling by solubilizing iron in magnetosomes which might then contribute to the high ferrous iron concentration at the OATZ and the high microbial cell concentrations there. Much more work is required to determine the biogeochemical roles these microorganisms play in iron cycling in chemically stratified and other aquatic habitats.

3.11 Biogenic Iron Oxides and Sulfides in Modern and Ancient Environments, Their Use as Biomarkers, and Their Presence in Higher Organisms

Nanophase magnetite grains have been recovered from a number of soils, freshwater sediments, and modern and ancient deep sea sediments [84], and have been referred to as "magnetofossils" [85] (Fig. 3.8). In some cases, they were identified as biogenic by their shape and size similarity to crystals in magnetotactic bacteria. Live bacteria have sometimes been recovered with the magnetite grains. McKay *et al.* [86] included ultra-fine-grained magnetite, pyrrhotite and possibly greigite in the rims of carbonate inclusions in the Martian meteorite ALH84001 as evidence for ancient life on Mars. The magnetite crystals ranged from about 10 to 100 nm, and were cuboid, tear-drop and irregular in shape. Some of them were comparable to those in magnetotactic bacteria. The iron sulfide particles varied in size and shape, with the pyrrhotite particles ranging up to 100 nm. Because pyrrhotite has never been confirmed as a biologically-controlled mineralization product, and the presence of greigite in ALH84001 has not been confirmed, attention has been focussed on the magnetite crystals [87] as potential biomarkers for life. Based on studies of magnetite in magnetotactic bacteria [44], biologically-controlled magnetite has the following idealized characteristics: (a) chemical and structural purity; (b) high structural perfection; (c) consistent habits within a given species or strain, most commonly isometric cubo-octahedra or non-isometric, pseudo-hexagonal prisms with (110) side faces and truncated (111) end caps, elongated along the [111] axis perpendicular to the end caps; (d) a certain fraction ($\sim 10\%$) of twinned crystals characterized by rotations of 180° around the [111] axis with a common (111) contact plane; (e) a consistent width to length ratio; and (f) an asymmetric crystal size distribution with a sharp cut-off for larger sizes within the single-magnetic-domain size range [44].

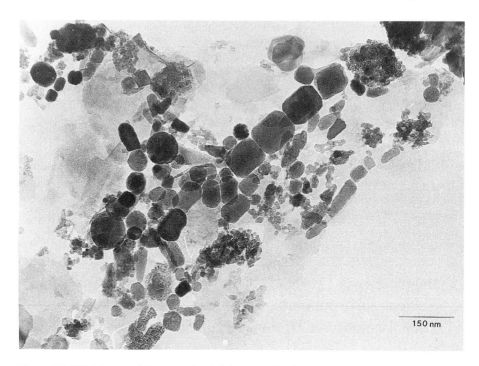

150 nm

Figure 3.8 TEM image of "magnetofossils" in magnetically separated material from surface sediments collected from the Irish Sea. Cubo-octahedral, parallelepipedal and tooth-shaped magnetic crystals, presumably derived from magnetotactic bacteria, are clearly present. Scale bar = 200 nm. (Micrograph courtesy of Z. Gibbs-Eggar.)

Although statistical analyses of the sizes and shapes of fine-grained magnetite crystals [44], as well as the characteristics listed above, might prove to be robust criteria for distinguishing between biogenic and non-biogenic magnetite, there remains a need for additional criteria by which to distinguish between biogenic and non-biogenic nanophase magnetic iron minerals. In a recent study, Mandernack *et al.* [88] found a temperature-dependent fractionation of oxygen isotopes in magnetite produced by *Magnetospirillum magnetotacticum* and strain MV-1 which closely matched that for extracellular magnetite produced by a bacterial consortium containing thermophilic iron-reducing bacteria [89]. However, no detectable fractionation of iron isotopes was observed in the bacterial magnetite. In contrast, Beard *et al.* [90] found enrichment of Fe^{54} compared to Fe^{56} in the soluble ferrous iron produced by a dissimilatory iron-reducing bacterium, *Shewanella alga*, growing with the iron oxide mineral ferrihydrite. How this isotopic fractionation is reflected in magnetite formed by this organism and other iron-reducing bacteria remains to be seen, but it could provide a means of distinguishing biologically-induced magnetite from abiotic magnetite.

Magnetic effects have been reported in a number of higher organisms [91, 92]. For example, magnetite crystals with morphologies similar to those produced by the magnetotactic bacteria have been found in the ethmoid tissues of salmon [16] and in the human brain [93]. The fact that many higher creatures biomineralize single-magnetic-domain magnetite crystals suggests the intriguing idea that all these organisms share the same or a similar set of genes responsible for magnetite bio-mineralization that would probably have originated in the magnetotactic bacteria. Thus, studying how magnetotactic bacteria biomineralize crystals of iron oxides and sulfides might have a scientific impact far beyond the studies of microbiology and geology.

Acknowledgements

We thank M. Pósfai and P. R. Buseck for Fig. 3.5b, Z. Gibbs-Eggar for Fig. 3.8 and P. R. Buseck, B. Devouard, K. W. Mandernack, M. Pósfai and D. Schüler for valuable discussions. We deeply appreciate the interest, insights and generosity of the late H. W. Jannasch. We acknowledge financial support from the United States National Science Foundation (NSF), grant CHE-9714101, and United States National Aeronautics and Space Administration (NASA) Johnson Space Center, grant NAG 9-1115.

References

[1] J. F. Banfield, K. H. Nealson (Eds) *Geomicrobiology: Interactions Between Microbes and Minerals*, Mineralogical Society of America, Washington DC, 1997.
[2] (a) H. A. Lowenstam, *Science* 1981, *211*, 1126–1131.
    (b) H. A. Lowenstam, S. Weiner (Eds), *On Biomineralization*, Oxford University Press, Oxford, 1989.
[3] S. Mann, R. B. Frankel, in *Biomineralization: Chemical and Biochemical Perspectives* (Eds S. Mann, J. Webb and R. J. P. Williams), VCH, Weinheim, 1989, pp. 389–426.
[4] D. R. Lovley, *Microbiol. Rev.* 1991, *55*, 259–287.
[5] D. T. Rickard, *Stockholm Contrib. Geol.* 1969, *20*, 50–66.
[6] D. R. Lovley, in *Iron Biominerals* (Eds R. B. Frankel and R. P. Blakemore) Plenum Press, New York, 1990, pp. 151–166.
[7] N. H. C. Sparks, S. Mann, D. A. Bazylinski, D. R. Lovley, H. W. Jannasch, R. B. Frankel, *Earth Planet. Sci. Lett.* 1990, *98*, 14–22.
[8] R. P. Blakemore, *Annu. Rev. Microbiol.* 1982, *36*, 217–238.
[9] D. L. Balkwill, D. Maratea, R. P. Blakemore, *J. Bacteriol.* 1980, *141*, 1399–1408.
[10] R. B. Frankel, R. P. Blakemore, R. S. Wolfe, *Science* 1979, *203*, 1355–1356.
[11] S. Mann, N. H. C. Sparks, R. B. Frankel, D. A. Bazylinski, H. W. Jannasch, *Nature*, 1990, *343*, 258–260.
[12] Y. A. Gorby, T. J. Beveridge, R. P. Blakemore, *J. Bacteriol.* 1988, *170*, 834–841.
[13] D. A. Bazylinski, *ASM News*, 1995, *61*, 337–343.
[14] D. A. Bazylinski, B. M. Moskowitz, in *Geomicrobiology: Interactions Between Microbes and Minerals* (Eds J. F. Banfield and K. H. Nealson), Mineralogical Society of America, Washington DC, 1997, pp. 181–223.

[15] F. F. Torres de Araujo, M. A. Pires, R. B. Frankel, C. E. M. Bicudo, *Biophys. J.* 1986, *50*, 375–378.
[16] S. Mann, N. H. C. Sparks, M. M. Walker, J. L. Kirschvink, *J. Exp. Biol.* 1988, *140*, 35–49.
[17] S. Spring, R. Amann, W. Ludwig, K.-H. Schleifer, N. Petersen, *Syst. Appl. Microbiol.* 1992, *15*, 116–122.
[18] R. P. Blakemore, D. Maratea, R. S. Wolfe, *J. Bacteriol.* 1979, *140*, 720–729.
[19] D. Schüler, E. Baeuerlein, *Arch. Microbiol.* 1996, *166*, 301–307.
[20] D. A. Bazylinski, B. R. Heywood, S. Mann, R. B. Frankel, *Nature* 1993, *366*, 218.
[21] D. A. Bazylinski, R. B. Frankel, B. R. Heywood, S. Mann, J. W. King, P. L. Donaghay, A. K. Hanson, *Appl. Environ. Microbiol.* 1995, *61*, 3232–3239.
[22] S. Spring, R. Amann, W. Ludwig, K.-H. Schleifer, H. van Gemerden, N. Petersen, *Appl. Environ. Microbiol.* 1993, *59*, 2397–2403.
[23] P. L. Donaghay, H. M. Rines, J. M. Sieburth, *Arch. Hydrobiol. Beih. Ergebn. Limnol.* 1992, *36*, 97–108.
[24] S. G. Wakeham, B. L. Howes, J. W. H. Dacey, R. P. Schwarzenbach, J. Zeyer, *Geochim. Cosmochim. Acta* 1987, *51*, 1675–1684.
[25] E. F. DeLong, R. B. Frankel, D. A. Bazylinski, *Science* 1993, *259*, 803–806.
[26] F. C. Meldrum, B. R. Heywood, S. Mann, R. B. Frankel, D. A. Bazylinski, *Proc. R. Soc. Lond. B* 1993, *251*, 237–242.
[27] R. B. Frankel, D. A. Bazylinski, M. Johnson, B. L. Taylor, *Biophys. J.* 1997, *73*, 994–1000.
[28] F. C. Meldrum, B. R. Heywood, S. Mann, R. B. Frankel, D. A. Bazylinski, *Proc. R. Soc. Lond. B* 1993, *251*, 231–236.
[29] D. Maratea, R. P. Blakemore, *Int. J. Syst. Bacteriol.* 1981, *31*, 452–455.
[30] K.-H. Schleifer, D. Schüler, S. Spring, M. Weizenegger, R. Amann, W. Ludwig, M. Kohler, *Syst. Appl. Microbiol.* 1991, *14*, 379–385.
[31] T. Matsunaga, F. Tadokoro, N. Nakamura, *IEEE Trans. Magnet.* 1990, *26*, 1557–1559.
[32] T. Matsunaga, T. Sakaguchi, F. Tadokoro, *Appl. Microbiol. Biotechnol.* 1991, *35*, 651–655.
[33] D. Schüler, S. Spring, D. A. Bazylinski, *System. Appl. Microbiol.* 1999, *22*, 466–471.
[34] T. Sakaguchi, J. G. Burgess, T. Matsunaga, *Nature* 1993, *365*, 47–49.
[35] Y. A. Gorby, PhD thesis, University of New Hampshire, New Hampshire, USA, 1989.
[36] K. M. Towe, T. T. Moench, *Earth Planet. Sci. Lett.* 1981, *52*, 213–220.
[37] D. A. Bazylinski, A. J. Garratt-Reed, R. B. Frankel, *Microsc. Res. Tech.* 1994, *27*, 389–401.
[38] T. Matsuda, J. Endo, N. Osakabe, A. Tonomura, T. Arii, *Nature* 1983, *302*, 411–412.
[39] R. P. Blakemore, R. B. Frankel, A. J. Kalmijn, *Nature* 1980, *236*, 384–385.
[40] S. Mann, N. H. C. Sparks, R. P. Blakemore, *Proc. R. Soc. Lond. B* 1987, *231*, 469–476.
[41] S. Mann, N. H. C. Sparks, R. P. Blakemore, *Proc. R. Soc. Lond. B* 1987, *231*, 477–487.
[42] C. Palache, H. Berman, C. Frondel, *Dana's System of Mineralogy*, Wiley, New York, 1944.
[43] S. Mann, R. B. Frankel, R. P. Blakemore, *Nature* 1984, *310*, 405–407.
[44] B. Devouard, M. Pósfai, X. Hua, D. A. Bazylinski, R. B. Frankel, P. R. Buseck, *Am. Mineral.* 1998, *83*, 1387–1398.
[45] (a) B. M. Moskowitz, *Rev. Geophys. Supp*, 1995, 123–128.
        (b) M. Farina, B. Kachar, U. Lins, R. Broderick, H. Lins de Barros, *J. Microsc.* 1994, *173*, 1–8.
[46] B. M. Moskowitz, R. B. Frankel, D. A. Bazylinski, H. W. Jannasch, D. R. Lovley, *Geophys. Res. Lett.* 1989, *16*, 665–668.
[47] D. D. Eberl, V. A. Drits, J. Srodon, *Am. J. Sci.* 1998, *298*, 499–533.
[48] D. A. Bazylinski, R. B. Frankel, A. J. Garratt-Reed, S. Mann, in *Iron Biominerals* (Eds R. B. Frankel and R. P. Blakemore), Plenum Press, New York, 1990, pp 239–255.
[49] M. Farina, D. M. S. Esquivel, H. G. P. Lins de Barros, *Nature* 1990, *343*, 256–258.
[50] M. Farina, H. Lins de Barros, D. M. S. Esquivel, J. Danon, *Biol. Cell* 1983, *48*, 85–88.
[51] F. G. Rogers, R. P. Blakemore, N. A. Blakemore, R. B. Frankel, D. A. Bazylinski, D. Maratea, C. Rogers, *Arch. Microbiol.* 1990, *154*, 18–22.
[52] M. Pósfai, P. R. Buseck, D. A. Bazylinski, R. B. Frankel, *Science* 1998, *280*, 880–883.
[53] B. R. Heywood, D. A. Bazylinski, A. J. Garratt-Reed, S. Mann, R. B. Frankel, *Naturwiss.* 1990, *77*, 536–538.

[54] B. R. Heywood, S. Mann, R. B. Frankel, in *Materials Synthesis Based on Biological Processes* (Eds M. Alpert, P. Calvert, R. B. Frankel, P. Rieke and D. Tirrell), Materials Research Society, Pittsburgh, 1991, pp. 93–108.

[55] M. Pósfai, P. R. Buseck, D. A. Bazylinski, R. B. Frankel, *Am. Mineral.* 1998, *83*, 1469–1481.

[56] R. A. Berner, *J. Geol.* 1964, *72*, 293–306.

[57] R. A. Berner, *Am. J. Sci.* 1967, *265*, 773–785.

[58] D. A. Bazylinski, A. J. Garratt-Reed, A. Abedi, R. B. Frankel, *Arch. Microbiol.* 1993, *160*, 35–42.

[59] A. R. Lennie, S. A. T. Redfern, P. E. Champness, C. P. Stoddart, F. F. Schofield, D. J. Vaughn, *Am. Mineral.* 1997, *82*, 302–309.

[60] (a) T. T. Moench, W. A. Konetzka, *Arch. Microbiol.* 1978, *119*, 203–212.
(b) T. T. Moench, *Antonie van Leeuwenhoek* 1988, *54*, 483–496.

[61] R. B. Frankel, *Annu. Rev. Biophys. Bioeng.* 1984, *13*, 85–103.

[62] R. E. Dunin-Borkowski, M. R. McCartney, R. B. Frankel, D. A. Bazylinski, M. Pósfai, P. R. Buseck, *Science* 1998, *282*, 1868–1870.

[63] A. M. Spormann, R. S. Wolfe, *FEMS Lett.* 1984, *22*, 171–177.

[64] I. B. Zhulin, V. A. Bespelov, M. S. Johnson, B. L. Taylor, *J. Bacteriol.* 1996, *178*, 5199–5204.

[65] J. E. Segall, S. M. Block, H. C. Berg, *Proc. Natl. Acad. Sci. USA* 1986, *83*, 8987–8991.

[66] B. L. Taylor, *Trends Biochem. Sci.* 1983, *8*, 438–441.

[67] R. P. Blakemore, *Science* 1975, *190*, 377–379.

[68] R. B. Frankel, D. A. Bazylinski, D. Schüler, *Supramol. Sci.* 1998, *5*, 383–390.

[69] R. B. Frankel, D. A. Bazylinski, *Hyperfine Interactions* 1994, *90*, 135–142.

[70] J. B. Neilands, *Biol. Metals* 1984, *4*, 1–6.

[71] M. L. Guerinot, *Annu. Rev. Microbiol.* 1994, *48*, 743–772.

[72] L. C. Paoletti, R. P. Blakemore, *J. Bacteriol.* 1986, *167*, 73–76.

[73] R. B. Frankel, G. C. Papaefthymiou, R. P. Blakemore, W. O'Brien, *Biochim. Biophys. Acta* 1983, *763*, 147–159.

[74] C. Nakamura, T. Sakaguchi, S. Kudo, J. G. Burgess, K. Sode, T. Matsunaga, *Appl. Biochem. Biotechnol.* 1993, *39/40*, 169–176.

[75] R. P. Blakemore, K. A. Short, D. A. Bazylinski, C. Rosenblatt, R. B. Frankel, *Geomicrobiol. J.* 1985, *4*, 53–71.

[76] D. Schüler, E. Baeuerlein, *J. Bacteriol.* 1998, *180*, 159–162.

[77] (a) E. I. Steifel, G. D. Watt, *Nature* 1979, *279*, 81–83.
(b) S. C. Andrews, J. B. C. Findlay, J. R. Guest, P. M. Harrison, J. N. Keen, J. M. A. Smith, *Biochim. Biophys. Acta* 1991, *1078*, 111–116.

[78] G. D. Watt, R. B. Frankel, D. Jacobs, H. Heqing, G. C. Papaefthymiou, *Biochemistry* 1992, *31*, 5672–5679.

[79] H. E. Jones, P. A. Trudinger, L. A. Chambers, N. A. Pyliotis, *Z. Allg. Mikrobiol.*, 1976, *16*, 425–435.

[80] M. Vainshtein, N. Suzina, V. Sorokin, *Syst. Appl. Microbiol.* 1997, *20*, 182–186.

[81] D. A. Bazylinski, D. R. Schlezinger, B. H. Howes, R. B. Frankel, S. S. Epstein, 2000, Chem. Geol., in press.

[82] J. L. Rifkin, R. Ballentine, *Trans. Am. Microsc. Soc.* 1976, *95*, 189–197.

[83] K. Barbeau, J. W. Moffett, D. A. Caron, P. L. Croot, D. L. Erdner, *Nature* 1996, *380*, 61–64.

[84] (a) N. Petersen, T. von Dobeneck, H. Vali, *Nature* 1986, *320*, 611–615.
(b) J. F. Stolz, S.-B. R. Chang, J. L. Kirschvink, *Nature* 1986, *321*, 849–850.
(c) S.-B. R. Chang, J. L. Kirschvink, *Annu. Rev. Earth Planet. Sci.* 1989, *17*, 169–195.
(d) J. Akai, S. Takaharu, S. Okusa, *J. Electron Microsc.*, 1991, *40*, 110–117.

[85] (a) J. L. Kirschvink, S.-B. R. Chang, *Geology* 1984, *12*, 559–562.
(b) S.-B. R. Chang, J. F. Stolz, J. L. Kirschvink, *Precambrian Res.* 1989, *43*, 305–315.

[86] D. S. McKay, E. K. Gibson Jr, K. L. Thomas-Keprta, H. Vali, C. S. Romanek, S. J. Clemett, X. D. F. Chillier, C. R. Maechling, R. N. Zare, *Science* 1996, *273*, 924–930.

[87] K. L. Thomas-Keprta, S. J. Wentworth, D. S. McKay, D. A. Bazylinski, M. S. Bell, C. S. Romanek, D. C. Golden, E. K. Gibson Jr, *Lunar Planet. Sci. Conf.* 1999, XXX, Abstract #1856.

[88] K. W. Mandernack, D. A. Bazylinski, W. C. Shanks, T. D. Bullen, *Science* 1999, *285*, 1892–1896.
[89] C. Zhang, S. Liu, T. J. Phelps, D. R. Cole, J. Horita, S. M. Fortier, *Geochim. Cosmochim. Acta* 1997, *61*, 4621–4632.
[90] B. L. Beard, C. M. Johnson, L. Cox, H. Sun, K. H. Nealson, C. Aguilar, *Science* 1999, *285*, 1889–1892.
[91] A. J. Kobayashi, J. L. Kirschvink, in *Electromagnetic Fields: Biological Interactions and Mechanisms. ACS Advances in Chemistry Series No. 250*, (Ed. M. Blank), American Chemical Society, Washington DC, 1995, pp. 367–394.
[92] M. M. Walker, C. E. Diebel, C. V. Haugh, P. M. Pankhurst, J. C. Montgomery, *Nature* 1997, *390*, 371–373.
[93] J. L. Kirschvink, A. Kobayashi-Kirschvink, B. J. Woodford, *Proc. Natl. Acad. Sci. USA* 1992, *89*, 7683–7687.

4 Phylogeny and *in Situ* Identification of Magnetotactic Bacteria

Rudolf Amann, Ramon Rossello-Mora, Dirk Schüler

4.1 Microbial Diversity and the Problem of Culturability

In the last decade, molecular biological data have reinforced what is common knowledge to microbiologists: that it is difficult to grow bacteria in pure culture! Some of the bacteria that are most conspicuous under the microscope have until now resisted all attempts at enrichment and cultivation, including not only symbiotic prokaryotes such as those chemolithoautotrophic bacteria found in marine invertebrates, the many bacteria and archaea dwelling in protozoa, slow-growing bacteria adapted to life in oligotrophic environments, but also magnetotactic bacteria [1]. The comparative analysis of bacterial 16S rRNA sequences directly retrieved from various environments by techniques pioneered by Woese [2] has proved that the ∼ 4500 validly described bacterial species represent only a small part, probably less than 1 %, of the extant bacterial diversity [1, 3]. The combination of cultivation-independent rRNA gene retrieval, comparative sequence analysis and fluorescence *in situ* hybridization has been shown to enable phylogenetic affiliation and *in situ* identification of previously uncultured bacteria [4]. In this chapter we will review the application of this methodology to magnetotactic bacteria.

4.2 The rRNA Approach to Microbial Ecology and Evolution

The full potential of the rRNA approach to microbial ecology and evolution was first described in 1986 by a group of scientists including Norman Pace and David Stahl [5]. The method is based on the comparative sequence analysis of ribosomal ribonucleic acid (rRNA) [2]. The different rRNA molecules – in bacteria the 5S, 16S and 23S rRNAs with approximate lengths of 120, 1500 and 3000 nucleotides – are essential components of all ribosomes. They are the cellular protein factories present in every cell in high copy numbers. Their sequences are evolutionary quite conserved but also contain regions in which mutations accumulate more rapidly. Due to their ubiquity, conserved function and lack of lateral gene transfer, the longer 16S and 23S rRNA molecules in particular are ideal chronometers for the reconstruction of bacterial evolution [2]. Furthermore, these two molecules contain highly conserved sites that allow their amplification from the rRNA genes present in environmental DNA by the polymerase chain reaction (PCR) [1, 3]. In the currently most widespread format, almost full-length 16S rRNA genes are amplified from conserved sites existing at the $5'$ and $3'$ ends of this molecule. The resulting mixed amplificate should reflect the natural bacterial community. It is subsequently ligated into a plasmid vector and cloned into *Escherichia coli* using standard molecular biology techniques. The cloning step allows segregation of the different fragments. This is necessary for sequencing and is comparable in its effect to the segregation of individual strains by growth on agar plates. The 16S rRNA sequences thereby retrieved from environmental samples without cultivation of the original bacteria are then compared to large databases that contain more than 90% of the 16S rRNA sequences of the previously cultured and validly described bacteria.

The closest known sequence can be identified by comparative analyses and a 16S rRNA-based evolutionary tree can be constructed which places the new sequence either in a known phylogenetic group or on a new branch of the universal tree. The comparative analysis also allows identification of sequence idiosyncrasies that, like a fingerprint, may serve to identify the new sequence. This sequence can then be the target for an oligonucleotide probe, which is a short, single-stranded piece of nucleic acid labeled with a marker molecule. A short oligonucleotide of 15–25 nucleotides is often sufficient to discriminate the 16S rRNA sequence retrieved from the environment from all other known sequences by hybridization. Hybridization is the binding of a probe to a fully or partially complementary target. Under optimized conditions hybridization is specific, meaning that the probe only binds to the target nucleic acid and not to other (non-target) nucleic acids. The probes can be used to quantify the target nucleic acids in a mixture of environmental nucleic acids. In one particular technique, fluorescence *in situ* hybridization (FISH), the nucleic acid probe is labeled with a fluorescent dye molecule and incubated with fixed, permeabilized environmental samples. During an incubation of one to several hours at defined conditions, the probes diffuse into the cells and bind specifically to their complementary target sites. Because 16S rRNA is fairly abundant in bacterial cells (for example, a rapidly growing *E. coli* cell contains about 70 000 copies of the molecule), even fluorescein-monolabeled oligonucleotides are sensitive enough to visualize individual bacterial cells in epifluorescence microscopes [1]. Using FISH, the 16S rRNA sequence retrieved from the environment is linked to defined cells, with a certain abundance, shape and spatial distribution. By this so-called full cycle rRNA approach, bacteria can be phylogenetically affiliated and identified without prior cultivation.

4.3 Application of the rRNA Approach to Magnetotactic Bacteria

The potential of the rRNA approach for analysis of previously uncultured bacteria has also been demonstrated on various conspicuous bacteria, including the magnetotactic bacteria. Although these bacteria have typical shapes and sizes, they only become conspicuous when live mounts of marine or freshwater surface sediments are exposed to changing magnetic fields. A fraction of cells decisively swims along the lines of the magnetic field and immediately follows any change in its orientation. Following the discovery of magnetotactic bacteria in 1975 by Blakemore [6], methods of visualization and enrichment have been developed. The unique advantage of working with these bacteria is that they can be readily separated from sediment particles and other bacteria due to their magnetotaxis. However, of the many morphotypes detected, including spirilla, cocci, vibrios, ovoid, rod-shaped and even multicellular bacteria, only a few bacteria have so far been grown as a pure culture (for review see [7]).

Some members of this interesting group of bacteria, with their ferromagnetic crystalline inclusions, have been investigated by the cultivation-independent rRNA approach. The main questions of interest were: (1) Is the morphological diversity reflected in a corresponding diversity at the 16S rRNA level? There are two possibilities – it may be that in fact there are only a few species of magnetotactic bacteria that have variable morphology (pleomorphism), or several species may be hidden behind one common morphotype. (2) Do the magnetotactic bacteria form a monophyletic group or are ferromagnetic crystalline inclusions found in different phylogenetic groups? (3) Has this specific trait developed only once or several times independently during bacterial evolution?

4.4 The Genus *Magnetospirillum*, Culturable Magnetotactic Bacteria

The obvious starting point was to determine the 16S rRNA sequences of the pure cultures available. Schleifer *et al.* [8] studied the two pure cultures available in 1991 and created the genus *Magnetospirillum*, including the two species *Magnetospirillum magnetotacticum* (formerly *Aquaspirillum magnetotacticum* [9]) and *Magnetospirillum gryphiswaldense* [8]. In parallel the sequence of *A. magnetotacticum* was determined by Eden and coworkers [10]. The 16S rRNA sequences of the two species affiliated them with the alpha-subclass of the class Proteobacteria, whereas the type species of the genus *Aquaspirillum* falls into the beta-subclass of Proteobacteria. *M. magnetotacticum* and *M. gryphiswaldense* strain MSR-1 share a similarity of 94.1%, while the corresponding similarity values to the other proteobacterial sequences available at that time were between 84 and 89%. The two culturable magnetospirilla have a very similar cell size (0.2–0.7 μm by 1–3 μm) and ultrastructure with respect to the arrangement (single chain of up to 60 magnetosomes), size (diameter ~ 40–45 nm) and cubo-octahedral crystal structure of their magnetosomes, as well as similar modes of flagellation (single flagella at each pole). However, there are also differences, such as oxidase and catalase activities, that are found only in strain MSR-1 which has an increased oxygen tolerance. The mol % G + C content of MSR-1 was originally reported to be 71%, considerably higher than that of *M.*

magnetotacticum (64.5%) [8], but was recently re-examined by an HPLC technique and found to be 62.7%, close to a new value for *M. magnetotacticum* of 63% [11].

In 1993 a group of workers including Matsunaga [12] published the evolutionary relationships between the two facultatively anaerobic strains of magnetic spirilla (AMB-1 and MGT-1) and the genus *Magnetospirillum*. The 16S rRNAs of AMB-1 and MGT-1 share 98–99% similarity with that of *M. magnetotacticum*, but only 95–96% similarity to *M. gryphiswaldense*. AMB-1 and MGT-1 clearly fall in the genus *Magnetospirillum* and their proximity to *M. magnetotacticum* on the 16S rRNA level does not exclude the placement of these strains in this species. The authors also note that there are clearly two groups of magnetospirilla: the one around *M. magnetotacticum*, including MGT-1 and AMB-1, and the one with *M. gryphiswaldense*. These are about as far apart as they are from some non-magnetotactic photo-organotrophic spirilla, e.g. *Phaeospirillum* (formerly *Rhodospirillum*) *fulvum* and *P. molischianum*.

Further diversity of magnetospirilla was recently revealed in a study by Schüler *et al.* [13] in which a new two-layer isolation medium with opposing oxygen and sulfide gradients was used for cultivation. Using this technique, seven strains of microaerophilic magnetotactic spirilla were isolated from one freshwater pond in Iowa, USA. While the 16S rRNA sequences of five of the isolates (MSM-1, MSM-6, MSM-7, MSM-8 and MSM-9) were very similar to both *M. gryphiswaldense* and *M. magnetotacticum* (>99.7%), two (MSM-3 and MSM-4) probably represent a third phylogenetic cluster and at least one additional species. There appears to be considerable diversity within this genus of culturable magnetic bacteria.

4.5 Phylogenetic Diversity and *in Situ* Identification of Uncultured Magnetotactic Cocci from Lake Chiemsee

The sequences of the two cultivated *Magnetospirillum* strains were subsequently compared to sequences originating from the upper sediment layers of Lake Chiemsee, a large mesotrophic freshwater lake in Upper Bavaria, Germany [14]. The sediment was stored in a 30 litre aquarium for several weeks on a laboratory shelf protected from direct light. At that time, large numbers of magnetotactic bacteria could be detected in wet mounts of subsamples taken right beneath the water–sediment interface. An enrichment was obtained based on magnetotactic bacteria swimming into sterile water or diluted agarose. This contained four distinct morphotypes: cocci, two large rods of distinct morphology (one slightly bent and therefore originally referred to as "big vibrio" [14]) and small vibrios. 5' end 16S rRNA gene fragments of about 800 nucleotides were PCR-amplified directly from the enriched cells without further DNA isolation and segregated by cloning. Within the 54 clones analyzed, 21 different sequence types could be discriminated. Most of them grouped with 16S rRNA sequences of alpha-subclass proteobacteria, several with other subclasses of this class and one sequence, later shown not to originate from a magnetotactic bacterium, was found to be identical to the 16S rRNA of *Mycobacterium chitae*.

Three probes, constructed complementary to signature regions of the most frequent alpha-subclass sequences, all bound to discrete subpopulations of the cocci which accounted for about 50% of all cells in the magnetotactic enrichment investigated. Simultaneous application of two differentially labeled (red, green) probes for these magnetotactic cocci indicated differences in abundance and tactic behavior of the different populations. Genotype CS308 accounted for approximately 80% of all magnetotactic cocci and was therefore more frequent than the genotypes CS103 and CS310. Under the influence of a magnetic field, cells of genotype CS103 were predominantly entrapped closest to the agarose solution/air interface.

By comparative analysis the partial 16S rRNA sequences of the three types of magnetotactic cocci were shown not to be closely related to any known sequence. The similarities were greatest between the three types, but even these were only moderate (89–93%). The three newly retrieved sequences form a separate lineage of descent within the alpha-subclass of Proteobacteria. Surprisingly, even though the genus *Magnetospirillum* also falls into this subclass, the magnetococci have more sequence similarity with other non-magnetic representatives of this subclass than with the culturable magnetospirilla.

The study by Spring *et al.* [14] is interesting for several reasons. From a methodological point of view it was one of the first studies in which problems with the rRNA approach became apparent. Although three additional morphotypes were present in the enrichment, together accounting for about 50% of all magnetotactic bacteria, their sequences were obviously not among those retrieved. This might have been caused by preferential PCR amplification of the partial 16S rRNA gene fragment of the magnetotactic cocci. Alternatively, because several non-magnetotactic bacteria were also readily amplified in the experiment, the other magnetotactic bacteria might have been discriminated in any one of the following steps – cell lysis, DNA release, amplification or cloning. With regard to the diversity of magnetotactic bacteria, the discrimination of three genotypes within the magnetotactic cocci and the lack of binding of oligonucleotide probes for the cultivated magnetospirilla and the magnetotactic cocci to the other morphotypes indicated that the genotypic diversity of this bacterial group is higher than the morphological diversity. Furthermore, the first hints of a polyphyletic origin of the magnetotactic bacteria were obtained because the next known relatives of both the cultivated magnetospirilla and the Chiemsee magnetococci show no magnetotaxis. Interestingly, even though the magnetotactic cocci are quite abundant in Lake Chiemsee and can be readily enriched from its sediment, they have until now resisted all attempts to bring them into pure culture (S. Spring, personal communication). This underlines the importance of the cultivation-independent rRNA approach in the study of magnetotactic bacteria.

4.6 Magnetotactic Bacteria are Polyphyletic with Respect to Their 16S rRNA

The magnetosomes of most magnetotactic bacteria contain iron oxide particles, however some magnetotactic bacteria collected from sulfidic, brackish-to-marine

aquatic habitats instead contain iron sulfide. When DeLong and coworkers analyzed three magnetotactic bacteria of both types by the rRNA approach [15], they
found the two isolates with iron oxide magnetosomes, a magnetotactic coccus and a
magnetotactic vibrio, to be affiliated with the alpha-subclass of Proteobacteria. The
coccus actually fell in the group of Chiemsee magnetococci, whereas the vibrio was
closer to the magnetospirilla, even though based on different tree reconstructions
it could not be shown whether it was closer to *Rhodospirillum rubrum* or to the
genus *Magnetospirillum*. These findings were in line with those of Spring *et al.* [14].
However, the 16S rRNA sequence retrieved from an uncultured, many-celled,
magnetotactic prokaryote (MMP) with iron sulfide magnetosomes, collected at
various coastal sites in New England, was specifically related to the dissimilatory
sulfate-reducing bacteria within the delta-subclass of Proteobacteria. The closest
relative is *Desulfosarcina variabilis* with a 16S rRNA similarity of 91 % [15]. This
clearly indicates a polyphyletic origin for the magnetotactic bacteria. The authors
also argue that their findings suggest that magnetotaxis based on iron oxide and
iron sulfide magnetosomes evolved independently. They state that the biochemical
basis for biomineralization and magnetosome formation for iron oxide type and
iron sulfide type bacteria are probably fundamentally different and speculate that in
two independent phylogenetic groups of bacteria analogous solutions for the problem of effective cell positioning along physicochemical gradients were found based
on intracellular particles with permanent magnetic dipole moments [15].

4.7 "Magnetobacterium bavaricum"

The polyphyletic distribution of magnetotaxis in bacteria was further corroborated
by the phylogenetic affiliation and *in situ* identification of the large rod-shaped
magnetic bacterium from Lake Chiemsee sediment, which was found to belong to a
third independent lineage [16]. This bacterium was conspicuous because of its large
size (5–10 μm long; ~ 1.5 μm in diameter) and high number of magnetosomes. Up
to 1000 hook-shaped magnetosomes with a length of 110–150 nm are found in
several chains. The large cells are Gram-negative and often contain sulfur globules.
The cells are mobile due to one polar tuft of flagella. This morphotype, tentatively
named "Magnetobacterium bavaricum", has so far only been enriched from the
calcareous sediments of a few freshwater lakes in Upper Bavaria [17]. As is the case
for many other magnetotactic bacteria, microbiologists have been unable to grow
this bacterium in pure culture until now. This morphotype was abundant in the
magnetotactic enrichment previously investigated by Spring and coworkers [14], but
its 16S rRNA sequence could not be retrieved in the presence of the magnetotactic
cocci. "M. bavaricum" cells were therefore sorted from this enrichment by flow
cytometry, based on the high forward and sideward light scattering caused by the
large cell size and the high numbers of magnetosomes. From the sorted cells an
almost full-length 16S rRNA sequence could be retrieved that was proven by FISH
to originate from "M. bavaricum" (Fig. 4.1). Unlike the magnetotactic cocci, this

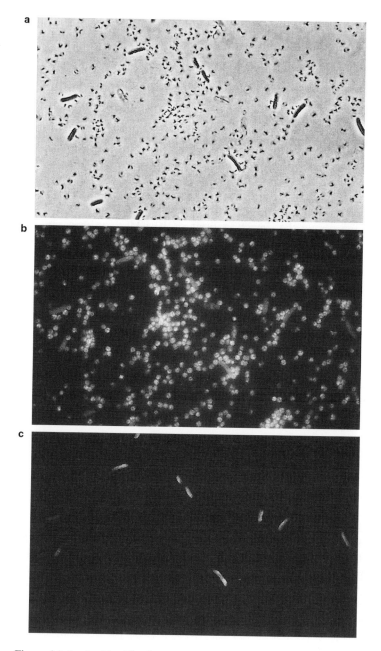

Figure 4.1 *In situ* identification of the previously uncultured "Magnetobacterium bavaricum" by FISH with a specific, 16S rRNA-targeted oligonucleotide probe [16]. (a) Phase contrast micrograph. (b) Visualization of hybridization of bacterial probe EUB338-fluorescein. (c) Selective visualization of "M. bavaricum" by a specific tetramethylrhodamine-labeled oligonucleotide probe. Identical microscopic fields are shown in (a)–(c).

conspicuous morphotype consisted of only one genotype, which was not affiliated with either the alpha- or the delta-subclass of Proteobacteria but with a different line of descent, tentatively referred to as the Nitrospira phylum because it encompasses the cultured *Nitrospira moscovensis*. The 16S rRNA of "M. bavaricum" has similarities of less than 80% with any other known sequence of magnetotactic bacteria. Recently, the magnetosomes were shown to consist of the iron oxide magnetite (N. Petersen, personal communication), suggesting that there are also multiple phylogenetic origins for iron oxide/magnetite-based magnetotaxis.

"M. bavaricum" could best be enriched from a reddish-brown layer at a depth of 5–8 mm below the sediment surface. Its abundance in the sediment was quantified by FISH and correlated with physicochemical gradients determined by needle electrodes. Up to 7×10^5 mobile cells per cm^3 were present in the reddish-brown zone. This layer coincided with the microaerobic zone. Using sulfide electrodes no free sulfide could be detected above the detection limit of 10 μM. However, sulfate-reducing bacteria were present in the microaerobic zone and the authors argue that low levels of sulfide might be continuously produced. They suggested that "M. bavaricum" has an iron-dependent method of energy conservation that depends on balanced gradients of oxygen and sulfide [16]. Based on its relative abundance of $0.64 \pm 0.17\%$ and a large average cell volume of 25.8 ± 4.1 μm^3 it was estimated that "M. bavaricum" made up approximately 30% of the bacterial biovolume in the reddish-brown zone. This demonstrates how hypotheses on the physiology and ecology of previously uncultured bacteria can be constructed, based on the combined application of microscopic techniques, the rRNA approach and the *in situ* characterization of the microhabitat of the bacterium of interest.

4.8 Evidence for Further Diversity of Magnetotactic Bacteria

In the 1990s it became standard practice to infer evolutionary relationships of bacteria by phylogenetic analysis. In the following we will briefly review further publications reporting 16S rRNA sequences from both cultured strains of magnetotactic bacteria and magnetic enrichments.

In 1994 Spring and coworkers used the cultivation-independent approach to retrieve another three partial and seven almost full-length 16S rRNA gene sequences from freshwater sediments at various sites in Germany [18]. Using FISH all sequences were assigned to magnetotactic bacteria, nine to magnetotactic cocci and one to the second rod-shaped magnetotactic morphotype ("large vibrio") originally described in Lake Chiemsee [14]. The magnetotactic rod shared a 16S rRNA similarity of 90–92% with the magnetotactic cocci, which among themselves mostly had similarity values below 97%. All sequences could be grouped with those previously retrieved from the uncultured Chiemsee magnetotactic cocci [14]. The authors point out that the finding that most magnetotactic cocci have 16S rRNA similarities below 97% has important taxonomic implications. In several studies on culturable bacteria it has been shown that a significant DNA–DNA similarity, which would justify assignment to one species, only exists above 97%. Therefore, upon isolation the different magnetococci could be placed in different species. This work by Spring *et al.* [16] not only confirmed the fact that the diversity of magnetotactic cocci is fairly high, but it also showed that the "Lake Chiemsee magnetococci" branch does not exclusively consist of cocci. This once again demonstrates the limited value of cell morphology in bacterial systematics.

In 1995 Matsunaga and coworkers published two reports related to the diversity and distribution of magnetotactic bacteria. In their first study, a PCR primer set specific for the 16S rRNA gene of the Lake Chiemsee magnetotactic cocci [14] was used to amplify DNA from magnetically isolated cocci. Comparative sequence analysis of the amplified 16S rDNA fragments proved their affiliation to the Lake Chiemsee magnetotactic cocci [19]. This demonstrated that this group of magnetotactic bacteria not only occurs in Germany but also in Japan. The authors used the primer set to investigate the distribution of magnetotactic cocci in laboratory enrichments. 16S rRNA gene fragments of magnetotactic cocci were readily amplified from a water column above the sediment kept in an anoxic environment, but little was amplified from a water column kept in an oxic environment. These results suggest that the magnetotactic cocci found in the anoxic water column had migrated there from the sediment in response to the micro-oxic or anoxic conditions, rather than having been present previously in a non-magnetic form and having become magnetic due to the change in conditions.

In their second report, the group described the phylogenetic analysis of a novel sulfate-reducing magnetic bacterium, RS-1 [20]. The almost complete 16S rRNA gene of the pure culture was amplified and partially sequenced. The comparative sequence analysis placed it with the sulfate-reducing bacteria of the delta-subclass of the Proteobacteria, within the genus *Desulfovibrio*. Interestingly, *Desulfovibrio* sp. RS-1 was the first isolate containing magnetite inclusions to be reported outside the alpha-subclass of Proteobacteria [21]. It therefore disrupts the correlation between the alpha- and delta-proteobacterial magnetotactic bacteria and iron oxide (magnetite) and iron sulfide (greigite) magnetosomes, respectively, as suggested by DeLong and coworkers [15].

This list of applications of the rRNA approach to the phylogeny and *in situ* identification of magnetotactic bacteria ends with a 1998 publication by Spring *et al.* [22]. In this study, natural enrichments of magnetic bacteria from the Itaipu lagoon near Rio de Janeiro in Brazil were analyzed. These were dominated by coccoid-to-ovoid morphotypes. Some of the cells produced unusually large magnetosomes that, with a length of 200 nm and a width of 160 nm, were almost twice as large as those found in other magnetotactic bacteria [23]. Partial sequencing of 16S rRNA genes revealed two clusters (Itaipu I and II) of closely related sequences within the lineage of magnetotactic cocci [14, 18, 19]. For a detailed phylogenetic analysis several almost full-length 16S rRNA gene sequences were determined. A new methodology was used in order to link at high resolution the ultrastructure of the enriched cells with their 16S rRNA sequence. Instead of light microscopic FISH with fluorescent oligonucleotide probes, *in situ* hybridizations with polynucleotide probes on ultra-thin sections of embedded magnetotactic bacteria were examined by electron microscopy. For this, one representative clone of each of the two closely

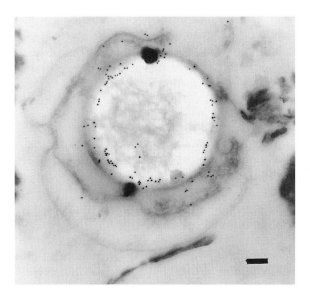

Figure 4.2 Electron micrograph of a hybridized thin section of magnetically enriched bacteria from the Itaipu lagoon (Rio de Janeiro, Brazil). Magnetosomes are visible as big black inclusions. The digoxigenin-labeled poly-nucleotide probe mabrj58 specific for the morphotype Itaipu I was detected with anti-DIG antibodies conjugated with 15 nm gold particles (small black dots). Bar = 0.5 μm.

related 16S rRNA clusters was used as a template for *in vitro* transcription of a variable region of 230 nucleotides at the 5' end of the 16S rRNA. The resulting RNA probe was labeled with digoxigenin- and fluorescein-labeled UTP during *in vitro* transcription. The bound polynucleotide probe was detected by incubation of the sections with gold-labeled antibodies specific for fluorescein or digoxigenin. The gold labels could then be detected under the electron microscope (Fig. 4.2). This enabled for the first time a detailed description of the morphological variety and ultrastructure of *in situ* identified, uncultured magnetic bacteria. Using this technique it was possible to link the presence of unusually large magnetosomes in ovoid magnetotactic bacteria to the Itaipu I 16S RNA type.

4.9 Current View of the Phylogeny of Magnetotactic Bacteria

Our current view of the phylogeny of magnetotactic bacteria is shown in a 16S rRNA-based tree reconstruction in Fig. 4.3. Magnetotactic bacteria can be found in the Nitrospira phylum and within the alpha- and delta-subclasses of the class Proteobacteria. Whereas the diversity in the Nitrospira branch of the magnetotactic bacteria consists only of the still uncultured "Magnetobacterium bavaricum" [16], the two proteobacterial branches are based on rRNA sequences of pure cultures and of still uncultured magnetotactic bacteria. Within the delta-subclass is the sequence of the pure culture *Desulfovibrio* sp. RS-1 [19] and the sequence MMP [15], which

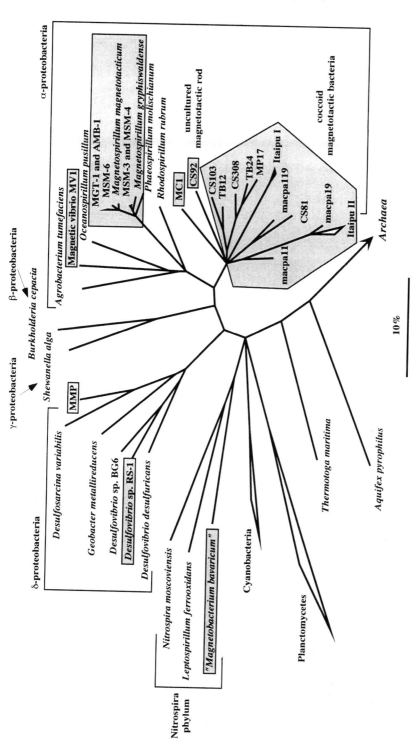

Figure 4.3 16S rRNA-based tree reconstruction showing the phylogeny of magnetotactic bacteria (shaded). The tree is based on parsimony analyses and was corrected according to the results of maximum likelihood and neighbor joining methods, using the ARB program package [25]. Multifurcations were drawn whenever branching orders were not stable [26]. The tree topology is based on almost full-length 16S rRNA sequences. The partial sequences of strains CS103, MC-1, MV-1, MMP and RS-1 have been added with the parsimony tool without allowing changes to the overall tree topology [26].

by the rRNA approach was assigned to the previously uncultured multicellular magnetotactic prokaryote. The MMP sequence is also the only one that is available of a magnetotactic bacterium with iron sulfide/greigite magnetosomes. The vast majority of currently known sequences of magnetotactic bacteria (> 90% of those we found in the publicly available databases) fall within the alpha-subclass of Proteobacteria. Of this large diversity only a few members of the genus *Magnetospirillum*, the magnetic coccus MC-1 and the magnetic vibrio MV-1, have been cultured. In the last decade cultivation has stagnated and only few additional strains of magnetospirilla have been described [12, 13]. What was found in terms of new diversity over this period was mostly from uncultured magnetotactic bacteria. However, new information obtained by the cultivation-independent approach since the phylogeny of magnetotactic bacteria was last reviewed by Spring and Schleifer in 1995 [24] is also limited. There were only a few new sequences affiliated with the already described branch of Lake Chiemsee magnetotactic cocci for which the reports from Germany [14, 18], Japan [19] and Brazil [22] now suggest global distribution and considerable intra-group diversity.

What is the reason for this stagnation? One possibility is that the extant diversity of magnetotactic bacteria is now fully described. The other possibility is that our methods are selective. We know that this is the case for the cultivation methods, but it must be realized that it is also true for the rRNA approach, especially if it starts from standard laboratory enrichments which themselves are selective. The methodology currently applied is biased towards motile, aero-tolerant bacteria. New diversity might be detected if magnetotactic bacteria from various habitats were directly retrieved without prior storage of the sediments in the laboratory. Are there strictly anaerobic, non-motile bacteria that form intracellular magnetosomes? Also, primer sets other than the standard "bacterial" ones should be tested for 16S rRNA retrieval from magnetic enrichments. It is known that every primer set has preferences and the example of the discrimination of the "Magnetobacterium bavaricum" sequence against the magnetotactic cocci has been described before. In this case, it was only the large size and the extraordinarily high magnetosome content that allowed further purification of the initial magnetotactic enrichment by flow cytometric sorting. This is not possible for less conspicuous magnetotactic bacteria. Magnetotactic bacteria might occur in other bacterial lineages. It would also be interesting to check whether archaebacterial sequences can be retrieved from magnetotactic enrichments. Further attempts should be made to identify and characterize new magnetotactic bacteria. These should make use of the potential synergistic effects of cultivation-independent *in situ* and traditional cultivation approaches. If a cultured close relative can be identified in the 16S rRNA tree, then the affiliation of a "new magnetic sequence" may provide important clues to its enrichment and cultivation. Also, data on the *in situ* microhabitat of magnetotactic bacteria should be obtained and used for the formulation of suitable media.

In the last decade two of the three questions raised above have been answered. There exists a large diversity of magnetotactic bacteria that goes beyond that indicated by the many morphotypes detected in the 1970s and 1980s, and the magnetotactic bacteria are polyphyletic. The third question, however, over whether the biomineralization of magnetosomes, or at least the intracellular formation of magnetite for example, is monophyletic, remains open. It would be highly interesting to investigate by comparative analysis of genes involved in magnetosome formation whether lateral gene transfer, e.g. from the alpha-proteobacterial magnetotactic bacteria to "M. bavaricum", contributed to the spread of magnetite-based magnetotaxis or whether the mechanisms of magnetosome formation have developed independently in the different phylogenetic groups. Studies of this type will not necessarily rely on cultured strains because there is a rapidly increasing potential to directly retrieve large DNA fragments from the environment and analyze them. If these fragments contain 16S rRNA genes or can be linked by overlaps to such fragments, environmental genomics allows for the comparative genome analysis of identified, uncultured bacteria [27].

Acknowledgements

The authors would like to thank Stefan Spring for critically reading earlier versions of this manuscript and for helpful discussions. The original research reviewed in this manuscript and in which the authors were involved was supported by the Deutsche Forschungsgemeinschaft and by the Fonds der chemischen Industrie.

References

[1] R. I. Amann, W. Ludwig, K. H. Schleifer, *Microbiol. Rev.* 1995, *59*, 143–169, and references therein.

[2] C. R. Woese, *Microbiol. Rev.* 1987, *51*, 221–271.

[3] N. R. Pace, *Science* 1997, *276*, 734–740, and references therein.

[4] R. Amann, N. Springer, W. Ludwig, H. D. Görtz, K. H. Schleifer, *Nature (London)* 1991, *351*, 161–164.

[5] G. J. Olsen, D. J. Lane, S. J. Giovannoni, N. R. Pace, D. A. Stahl, *Annu. Rev. Microbiol.* 1986, *40*, 337–365.

[6] R. P. Blakemore, *Science* 1975, *190*, 377–379.

[7] (a) R. P. Blakemore, *Annu. Rev. Microbiol.* 1982, *36*, 217–238.
    (b) S. Spring, K. H. Schleifer, *Syst. Appl. Microbiol.* 1995, *18*, 147–153.
    (c) R. S. Wolfe, R. K. Thauer, N. Pfennig, *FEMS Microbiol. Ecol.* 1987, *45*, 31–35.

[8] K. H. Schleifer, D. Schüler, S. Spring, M. Weizenegger, R. Amann, W. Ludwig, M. Köhler, *Syst. Appl. Microbiol.* 1991, *14*, 379–385.

[9] R. P. Blakemore, D. Maratea, R. Wolfe, *J. Bacteriol.* 1979, *140*, 720–729.

[10] P. A. Eden, T. M. Schmidt, R. P. Blakemore, N. R. Pace, *Int. J. Syst. Bacteriol.* 1991, *41*, 324–325.

[11] T. Sakane, A. Yokota. *Syst. Appl. Microbiol.* 1994, *17*, 128–134.

[12] J. G. Burgess, R. Kawaguchi, T. Sakaguchi, R. H. Thornhill, T. Matsunaga, *J. Bacteriol.* 1993, *175*, 6689–6694.

[13] D. Schüler, S. Spring, D. Bazylinski, *Syst. Appl. Microbiol.* 1999, *22*, 466–471.

[14] S. Spring, R. Amann, W. Ludwig, K.-H. Schleifer, N. Petersen, *Syst. Appl. Microbiol.* 1992, *15*, 116–122.

[15] E. F. DeLong, R. F. Frankel, D. A. Bazylinski, *Science* 1993, *259*, 803–806.
[16] S. Spring, R. Amann, W. Ludwig, K. H. Schleifer, H. van Gemerden, N. Petersen, *Appl. Environ. Microbiol.* 1993, *59*, 2397–2403.
[17] H. Vali, O. Förster, G. Amarantidis, N. Petersen, *Earth Planet. Sci. Lett.* 1987, *86*, 389–426.
[18] S. Spring, R. Amann, W. Ludwig, K.-H. Schleifer, D. Schüler, K. Poralla, N. Petersen, *Syst. Appl. Microbiol.* 1994, *17*, 501–508.
[19] R. H. Thornhill, J. G. Burgess, T. Matsunaga, *Appl. Environ. Microbiol.* 1995, *61*, 495–500.
[20] R. Kawaguchi, J. G. Burgess, T. Sakaguchi, H. Takeyama, R. H. Thornhill, T. Matsunaga, *FEMS Microbiol. Lett.* 1995, *126*, 277–282.
[21] T. Sakaguchi, J. G. Burgess, T. Matsunaga, *Nature (London)* 1993, *365*, 47–49.
[22] S. Spring, U. Lins, R. Amann, K.-H. Schleifer, L. C. S. Ferreira, D. M. S. Esquivel, M. Farina, *Arch. Microbiol.* 1998, *169*, 136–147.
[23] M. Farina, B. Kachar, U. Lins, R. Broderick, H. Lins de Barros, *J. Microsc.* 1994, *173*, 1–8.
[24] S. Spring, K. H. Schleifer, *Syst. Appl. Microbiol.* 1995, *18*, 147–153.
[25] O. Strunk, O. Gross, B. Reichel, M. May, S. Hermann, N. Stuckmann, B. Nonhoff, M. Lenke, A. Ginhart, A. Vilbig, T. Ludwig, A. Bode, K. H. Schleifer, W. Ludwig, http://www.mikro.biologie.tu-muenchen.de/pub/ARB, Department of Microbiology, Technische Universität München, Munich, Germany, 1999.
[26] W. Ludwig, O. Strunk, S. Klugbauer, N. Klugbauer, M. Weizenegger, J. Neumaier, M. Bachleitner, K. H. Schleifer, *Electrophoresis* 1998, *19*, 554–568.
[27] J. L. Stein, T. L. Marsh, K. Y. Wu, H. Shizuya, E. F. DeLong, *J. Bacteriol.* 1996, *178*, 591–599.

5 Single Magnetic Crystals of Magnetite (Fe$_3$O$_4$) Synthesized in Intracytoplasmic Vesicles of *Magnetospirillum gryphiswaldense*

Edmund Baeuerlein

5.1 A Challenge to Membrane Biochemistry

Bacteria are not only able to bind metal ions on their surfaces and to thereby partic- ipate in mineral development (see Chapter 2), they can also transport huge amounts of iron ions into special phospholipid vesicles in their cytoplasm. This iron transport is probably very different from that required for growth, as it leads to the formation of magnetite crystals in these vesicles. In addition these iron ions have to pass through three membranes, the outer, the cytoplasmic and the vesicle membrane, where various morphologies of magnetite crystals have been found, specific to each bacterial species (Chapter 3). This was probably the first example of biominerali- zation in Earth's history (Chapter 10) and its secret is apparently hidden in the proteins and phospholipids of the vesicle membrane, and consequently in the related genes. To solve this mystery, the physiology of one magnetic bacterium, designated *Magnetospirillum gryphiswaldense*, was first studied in detail [1] (Fig. 5.1).

5.2 The Difficulties of Cultivating Magnetic Bacteria

It is now 25 years ago since Blakemore [2–4] discovered the first magnetic bacteria in marine marsh muds, which were described as *Aquaspirillum magnetotacticum* (now *Magnetospirillum magnetotacticum* [5]). It was not difficult to enrich various magnetic bacteria by magnetic methods. This was possible because the magnetic bacteria found were all motile by means of their flagella motors and therefore mi- grated at different speeds along a magnetic field in a chamber containing aquatic agar [6]. Until 1996, however, only eight pure cultures of magnetic bacteria had been described.

These were four spirilla: (i) *Magnetospirillum magnetotacticum*, strain MS-1 [2], (ii) *Magnetospirillum gryphiswaldense* [5–7], (iii) *Magnetospirillum*, strain AMB-1 [8], (iv) *Magnetospirillum* MGT-1 [9]; (v) one coccoid strain: MC-1 [10], (vi) and

a

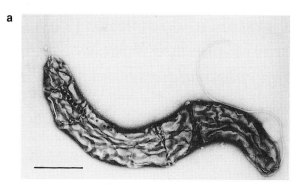

b

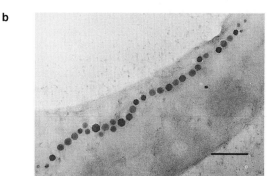

Figure 5.1 Transmission electron micrograph *of Magnetospirillum gryphiswaldense*, (a) with a chain of cubo-octahedral magnetite crystals and a single flagellum at each pole (bar: 500 nm); (b) magnification of a cell with more than 30 cubo-octahedral crystals, maximal diameter 42–45 nm (bar: 200 nm).

(vii) two vibriod strains: MV-1 [11, 12] and MV-2 [12], and (viii) one rod-like to helical strain: RS-1 [13].

Until now, essentially three magnetic bacteria, all spirilla, have been used in the various studies on biomineralization, because these were the only ones that could be cultivated in sufficient yields. Only two are available as commercial cell collections, *M. magnetotacticum* MS-1 (ATCC 31632) and *M. gryphiswaldense* (DSM 6361). Recently two new *Magnetospirilla*, strains MSM-3 and MSM-4, have been isolated and cultivated [14].

Magnetic bacteria are found mainly in freshwater sites near the sediment–water interface [1, 15] and in marine and brackish muds [2] under microaerobic to anaerobic conditions.

Given these origins of magnetic bacteria it is obvious that not only might it be difficult to find a growth medium for each magnetic species, but also to recreate the precise conditions necessary for bacterial growth and magnetite biomineralization. The third challenge is to produce magnetite crystals, 35–120 nm in length and, therefore, of a single-magnetic-domain [16], in high yields by cultivating them in 100–1000 liter fermenters for medical [17] (Chapter 6) and technical applications [18, 19].

5.3 A Simple Spectroscopic Method of Following Magnetization of the Magnetite-Forming Bacteria

An important step in the study of magnetite crystal formation was the development of a simple time-resolved analytical method of following magnetization of these bacteria [20]. The basic observation was that cells of *M. gryphiswaldense* in a test tube changed their pattern of dark and bright areas when the angle of a permanent magnet on its surface was varied. To monitor this magnetically-induced differential light scattering, a light beam was passed through a cell suspension that was exposed to a homogenous magnetic field (Fig. 5.2). The magnetic field was applied to the sample both parallel and perpendicular to the light beam. The ratio of the two scattering intensities, $C_{mag} = E_{max}/E_{min}$, was used to characterize the average orientation of the cells. C_{mag} corresponded well to the average number of magnetic particles in various cell populations of *M. gryphiswaldense* [20]. The high sensitivity of this rapid method allowed magnetization to be studied in parallel with iron uptake or oxygen concentration.

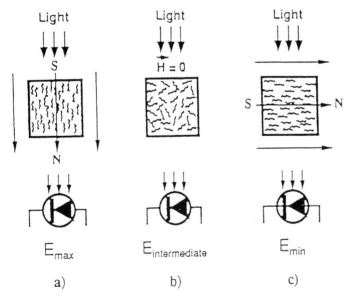

Figure 5.2 Effects of different magnetic orientation to the incident light beam on the light scattering of a suspension of magnetic spirilla (detected by measurement of extinctions E). Maximum scattering occurs if the applied magnetic field is directed parallel to the light beam (a), whereas perpendicular orientation with respect to the light results in minimal scattering (c). Without a magnetic field ($H = 0$), magnetic cells are randomly oriented resulting in an intermediate scattering intensity (b) [20].

5.4 The Exceptional Iron Uptake of Magnetic Bacteria

The growth and magnetism of *M. gryphiswaldense* were close to a maximum at 10–20 μM iron, which was added to the medium as $Fe(II)SO_4$ [21]. Iron did not precipitate as ferric hydroxide even by the end of the incubation time, although it was oxidized to ferric iron. The iron concentration range for unlimited growth of *M. gryphiswaldense* was up to 10-fold that necessary for *E. coli* [22] and for non-magnetic cells of *M. gryphiswaldense*. Surprisingly a weak magnetism was measured under limited growth, even when the iron concentration was below 1 μM. Some crystals of less than 30 nm, which were superparamagnetic, and therefore without a permanent magnetic moment [16], were formed in competition for metabolic iron, obligatory for cell growth [21]. These results demonstrate that iron supply to *M. gryphiswaldense* is crucial to growth and magnetite formation, and that iron may be accumulated in magnetic crystals from a very poor environment.

Magnetic bacteria exist in the largest numbers at the oxic–anoxic transition zone (Chapter 3), generally located at sediment–water interfaces, and so the very soluble Fe(II) would be expected to be transported preferentially. *M. gryphiswaldense*, however, took up Fe(II) only at low rates, apparently by a diffusion-like process [21]. In place of it free oxidized Fe(III), which is normally very insoluble in water at biological pH with a maximum concentration of the order of 10^{-18} M, was transported very efficiently in the presence of spent culture fluids into cells of *M. gryphiswaldense*. Maximal iron uptake was attained if cells were grown at moderate iron depletion. Fresh medium and transport buffer were unable to promote iron uptake (Fig. 5.3). This Fe(III) uptake followed Michaelis–Menten kinetics, the kinetic constants of which were determined [21] to be $K_m = 3$ μM Fe and $V_{max} = 0.86$ nmol Fe min^{-1} (mg dry weight)$^{-1}$. These data fit to a low affinity, but high velocity

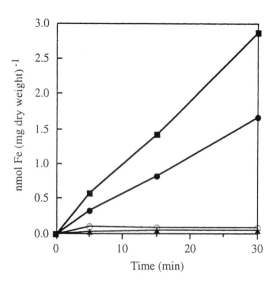

Figure 5.3 Uptake of ^{55}Fe(III) by cells of *M. gryphiswaldense* from transport buffer (▲), fresh growth medium (O), spent culture fluids (■), and spent culture fluids plus 0.1 mM sodium nitrilotriactate (●). The extracellular concentration of iron used in the experiment was 2.5 μM. (After [21] with permission of Springer-Verlag GmbH & Co. KG.)

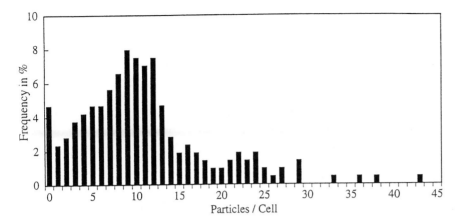

Figure 5.4 Distribution of particle numbers in a typical example of *M. gryphiswaldense*. The average number in the given example was 11.6 magnetosomes per cell. Values were obtained by counting electron-dense particles in micrographs from a total of about 220 cells [20].

transport system [23]. Its saturating iron concentration of 10–20 µM corresponds to the average free iron concentration in the natural habitats of magnetic bacteria such as freshwater sediments [3, 24]. Although the spent medium obviously enhanced the iron uptake of *M. gryphiswaldense*, no siderophores (ferric-specific iron complexing compounds of molecular weights close to 1000 Da and formation constants of $10^{30}–10^{50}$ [23, 25]) have so far been detected and isolated [21]. The conditions for siderophore biosynthesis in non-magnetic bacteria, implying growth under low iron stress, are obviously opposite to those for high ferric iron uptake in *M. gryphiswaldense*, requiring saturating ferric iron concentrations.

The enormous iron uptake resulted in the formation of up to 60 magnetite crystals within the cytoplasm of *M. gryphiswaldense*. The number of magnetosomes – the membrane-enveloped crystals – varied between the individual cells within a culture at a given time (Fig. 5.4). This was also true for the average number of magnetosomes in various cultures of *M. gryphiswaldense*. These variable rates of magnetite biosynthesis are the main challenge in cultivating magnetic bacteria, and was addressed in the studies discussed below.

5.5 Specific Microaerobic Conditions for Magnetite Formation in *M. gryphiswaldense*

Magnetic bacteria can exist in conditions ranging from microaerobic to anaerobic. Within these bounds *M. gryphiswaldense* is an obligate microaerophile that is not able to grow anaerobically and requires oxygen for its energy production.

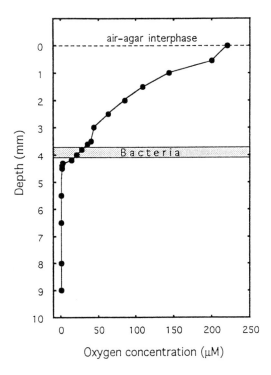

Figure 5.5 Oxygen profile of a semi-solid agar culture of *M. gryphiswaldense*. The oxygen gradient was formed by oxygen diffusion through the air–agar interface and oxygen consumption by the respiring cells. Cells arranged themselves at their preferred oxygen concentration by aerotactic migration, within a sharp band 4 mm below the surface corresponding to a concentration of 15–25 μM O_2 (25 °C).

5.5.1 Aerotactic Orientation in an Aquatic, Spatial Oxygen Gradient

In a spatial oxygen gradient of a semi-solid agar column, the cells accumulated within an aerotactic band, which was found at a depth of 4 mm. By means of a microelectrode [27, 28], an oxygen concentration of about 15–25 μM oxygen could be assigned to this thin layer of cells (Fig. 5.5).

5.5.2 Initial Oxygen Concentration in the Gas Phase and Its Effect on Growth Yield and Magnetite Synthesis

Table 5.1 shows that final cell yield and cell magnetism are dependent on the initial oxygen concentration in the headspace of batch cultures grown in sealed vials. Maximal magnetite production occurred at the lowest initial oxygen level of 0.1 % oxygen, whereas levels higher than 9 % oxygen resulted in non-magnetic cells. Growth was inhibited at more than 15 % oxygen. (D. Schüler, E. Baeuerlein, unpublished results). An important observation was made when a large inoculum of an initial cell concentration of 5×10^6 cells ml^{-1} was introduced instead of the small inoculum of 2×10^5 cells ml^{-1}. Cells of *M. gryphiswaldense* now grew in a liquid medium with air-free gas exchange at moderate agitation [29], which allowed the introduction of fermenters for mass cultivation [17].

Table 5.1 Effect of the initial O_2 concentration on growth yield and Fe_3O_4 synthesis of *M. gryphiswaldense**

% O_2 in the headspace	Final cell density (OD_{400})	Magnetism (C_{mag})
0	no growth	–
0.1	0.16	0.83
1	0.32	0.55
2	0.50	0.37
3	0.70	0.22
4	0.74	0.13
5	0.69	0.14
6	0.66	0.10
7	0.65	0.07
8	0.65	0.04
9	0.65	0.00
10	0.64	0.00
12	0.61	0.00
15	no growth	–

* Cells were inoculated at low initial cell density (2×10^5 cells ml^{-1}) into sealed vials containing various oxygen concentrations (0–21 % O_2) in the headspace. Growth and cellular magnetism (C_{mag}) were determined after 48 h incubation at 30 °C. The additional experimental conditions as in [29].

5.5.3 The Concentration of Dissolved Oxygen and the Induction of Magnetite Formation

Five-liter laboratory fermenter vessels (Bioflow III; New Brunswick Scientific) were introduced to follow oxygen concentration continuously. Instead of the slow free air exchange by the medium surface, air was blown through the stirrer into the medium as small bubbles. Cell growth was determined by optical density at 400 nm parallel to the average magnetic orientation of the cell suspension, designated as magnetism, which was measured by the time-resolved light scattering method, described in Section 5.3. The experiments were started by inoculation with non-magnetic cells, which had been grown in the absence of an added iron source, to an initial cell concentration of 5.0×10^6 cells ml^{-1} in a growth medium containing 30 µM ferric citrate. In the first experiment (Fig. 5.6a) the aeration rate of 0.5 l air min^{-1} created an initial period of aerobic growth because of the high concentration of oxygen. With increasing growth the oxygen concentration of the medium decreased. Immediately after microaerobic conditions were obtained (2–7 µM O_2 at 30 °C, which corresponds to $\sim$1–3 % saturation), cells became magnetic because of the biosynthesis of magnetite crystals. In a second experiment (Fig. 5.6b) a higher cell yield resulted from a more intensive aeration of 2.0 l air min^{-1}. The aerobic growth phase was therefore extended, which led to a shortened microaerobic phase and conse-

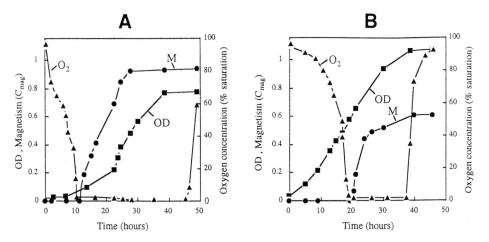

Figure 5.6 Dissolved oxygen concentration (O_2, ▲), cell density (OD, ■) and magnetism (M, ●) in cultures of *M. gryphiswaldense* during growth at different aeration rates. Aeration was kept constant throughout each experiment: (a) 0.5 l air min^{-1}, (b) 2.0 l air min^{-1}. The experiments were started by the inoculation of non-magnetic cells into the growth medium. (After [29] with permission of the American Society of Microbiology.)

quently to lower magnetite production. Growth was inhibited completely when the aeration rate exceeded 3.0 l air min^{-1}.

Although it is difficult to make precise measurements of oxygen at low concentrations using oxygen electrodes, there is a surprising discrepancy between the aerotactically preferred concentrations of 15–25 µM O_2 and the concentration of 2–7 µM O_2 essential for the magnetite biomineralization of *M. gryphiswaldense*. What is the purpose of magnetotaxis, when non-magnetic cells synthesize magnetite deeper in marine marsh muds, at 2–7 µM O_2, whereas magnetic cells form layers at 15–25 µM O_2?

5.6 Dynamics of Iron Uptake and Magnetite Formation in *M. gryphiswaldense*

The formation of magnetite crystals in intracytoplasmic vesicles took several hours under physiological conditions. The essential oxygen concentration in the medium could not be determined precisely, if at all, because of the oxygen gradient in this growing bacterial population, but it was narrowed down by various experiments. One of the most intriguing questions now in both biomineralization biology and inorganic biomimetic chemistry is whether magnetite crystal formation in the vesicles is coupled to an increased iron uptake or whether iron is accumulated continuously up to supersaturation with subsequent precipitation of magnetite in the vesicles.

Table 5.2 Effect of the time of addition of iron on final cell yield, iron content and magnetism in cultures of *M. gryphiswaldense**

Time of addition of iron (hours)	Cell density at time of addition (OD_{400})	Final cellular magnetism (C_{mag})	Final intracellular iron content [nmol Fe (mg dry weight)$^{-1}$]
0	0.06	0.36	24.20
16	0.25	0.37	28.72
17	0.31	0.66	36.68
18	0.40	0.91	42.39
19	0.55	0.98	48.00
20	0.63	0.73	35.79
24	0.69	0.59	34.46

* Identical amounts of radioactive $^{55}FeCl_3$ were added to a final concentration of 30 μM after various time intervals, corresponding to different growth stages. The cells were incubated under otherwise identical conditions. All cultures were inoculated simultaneously with non-magnetic cells. Intracellular iron content and magnetism were determined at the end of growth.

5.6.1 Iron Addition – Point of Time and Its Effect on Magnetism and Iron Content

In a first approach, iron-starved non-magnetic cells were inoculated into seven identical batch cultures containing 30 ml medium without added ferric citrate. The initial cell concentration was 5.0×10^6 cells ml^{-1}. Cultures were incubated in loosely capped flasks at 30 °C with moderate agitation [29]. Radioactive $^{55}FeCl_3$ was added at various time points to a final concentration of 20 μM $^{55}FeCl_3$. Cell density was determined at the time of iron addition, and intracellular iron content and magnetism [29] were measured after 28 h, the end of growth (Table 5.2). The ultimate result was that the final iron content and the cell magnetism were generally higher if iron was added at a late state of growth, with a maximum close to 19 h after inoculation with non-magnetic cells.

5.6.2 Magnetite Formation in *M. gryphiswaldense* is Closely Coupled to an Increased Iron Uptake

In two crucial experiments [29] an effort was made to answer the question of whether supersaturation in the intracytoplasmic vesicles is a prerequisite for crystallization, or if the crystal formation is coupled to increased iron transport into the vesicles. The experimental conditions were similar to those in the preceding time point experiments. In experiment A (Fig. 5.7a), radioactive $^{55}FeCl_3$ was added to the medium immediately after inoculation with non-magnetic bacteria to a final concentration of 30 μM. With the exception of a transient peak in iron uptake during initial growth, the iron content of the non-magnetic bacteria remained essen-

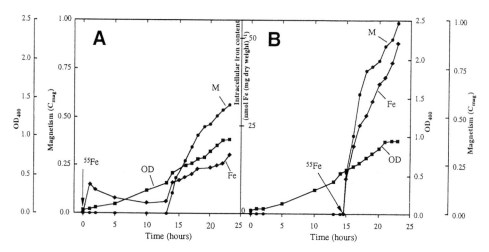

Figure 5.7 Cell density (OD, ■), cellular magnetism (M, ●) and intracellular iron content (Fe, ◆) of *M. gryphiswaldense* during growth. The experiment was started by simultaneous inoculation of non-magnetic cells to (a) and (b). Iron was added to 30 μM as ^{55}FeCl₃ (in 0.1 N HCl) at the time of inoculation (a) and after 14.5 h (b) as indicated by arrows. The figures given do not accurately reflect the iron content of the cells because the iron content of the non-magnetic inoculum was not taken into account. (After [29] with permission of the American Society of Microbiology.)

tially low, between 0.02 and 0.06% (dry weight). This iron concentration represents the same order of magnitude as for the non-magnetic *E. coli* when grown under iron-sufficient conditions [22]. Apparently growth and iron uptake were balanced while cells did not synthesize magnetite. A sudden increase in cellular magnetism was measured after 14 h, when microaerobic conditions were attained (see also Fig. 5.6a, b).

In experiment B (Fig. 5.7b) radioactive iron was added about 15 h after inoculation with non-magnetic bacteria, when microaerobic conditions had been attained, as determined from the preceding experiment (Fig. 5.7a). In the initial 15 h, the original iron concentration in the medium of about 1 μM was sufficient for cell growth, but not for measurable magnetism. However, in this case. not only was iron taken up immediately, but also magnetism increased almost in parallel with it. This result strongly supports the assumption that the majority of iron was transported into the vesicles, the magnetosomes.

A further piece of evidence for this close coupling was obtained, as in experiment B (Fig. 5.7b), when magnetite formation was measured 5–10 min after the addition of iron by a change in light scattering of the culture in a magnetic field, as described in Section 5.3, i.e. by the development of magnetism. Samples of cells were taken after 30 min and killed by formaldehyde, for examination by transmission electron microscopy. Intracellular electron-dense particles were detected on the electron micrographs which apparently began to form a chain (Fig. 5.8b), which is known to be the final stage of magnetite biomineralization in *M. gryphiswaldense* (Fig. 5.8a).

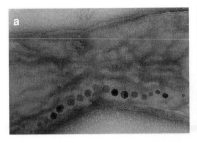

Figure 5.8 Electron micrographs of magnetosomes in cells of *M. gryphiswaldense*. (a) Chain of magnetosomes from a cell grown in the presence of a steady concentration of 30 μM FeCl$_3$. Mature crystals, mostly located in the middle of the chain, are cubo-octahedral and 45 nm in diameter. (b) Crystals present in cells about 30 min after the induction of Fe$_3$O$_4$ biomineralization by the addition of 30 μM FeCl$_3$. The immature particles are 5–20 nm in diameter and fit predominantly into the superparamagnetic size range. (After [29] with permission of the American Society of Microbiology.)

The diameter of these early particles varied between 5 and 20 nm, which is within the superparamagnetic size range [16], whereas the diameters of mature crystals fell in a narrow size distribution of 42–45 nm (Fig. 5.8b).

5.7 One Single-Magnetic-Domain Crystal of Magnetite is Formed in Each Phospholipid Vesicle of a Chain in *M. gryphiswaldense*

Up to 60 cubo-octahedral magnetite crystals were found to be arranged in a chain in the vesicles [7]. A more elaborate method was developed to obtain purified vesicles, designated as magnetosomes, by introducing a special magnetic separation column [17, 29 and Chapter 8]. This method allowed to prepare 300 mg of *M. gryphiswaldense*, the yield from a 100 l fermenter culture. Single magnetic crystals can be seen in electron micrographs, each apparently enveloped by a membrane (Fig. 5.9a). After addition of the detergent Triton X100, the stable suspension of magnetosome chains collapsed to an aggregation of magnetite crystals (Fig. 5.9b), suggesting repulsion by surface charges on the membranes.

Using high-resolution transmission electron microscopy, magnetite crystals of *M. gryphiswaldense* were found to be almost without defects [1a], like most of the different crystals detected in other magnetotactic bacteria [30]. Visible lattice planes were oriented parallel to each other throughout the whole crystal (4.7 Å spacing) [1a]. Elemental maps of magnetite crystals inside a cell of *M. gryphiswaldense* were obtained by electron cryomicroscopy operating under low-dose conditions. The three-window method of energy spectroscopic imaging (ESI) was used in an auto-

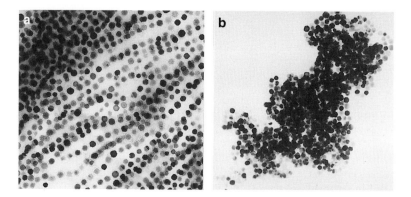

Figure 5.9 Purified magnetosomes from *M. gryphiswaldense*. (a) Isolated particles are forming stable suspensions because they are surrounded by the intact magnetosome membrane which prevents them from agglomeration. (b) The magnetosome membrane is solubilized by detergent treatment, resulting in immediate agglomeration and precipitation of inorganic magnetite cores from suspension.

mated procedure [31]. Both the iron and the oxygen maps fitted well with the unfiltered TEM bright-field image (Fig. 5.10A–C). The elemental composition was analyzed from a subarea of several crystals by an energy loss spectrum which supported their high purity (Fig. 5.10d) [32].

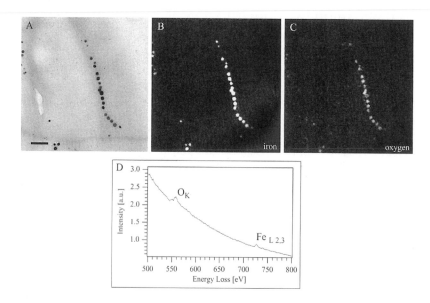

Figure 5.10 Magnetosome crystals inside a bacterial cell. (A) Unfiltered TEM brightfield image. Scale bar = 200 nm. (B) Iron map using the $L_{2,3}$-edge at 708 eV. (C) Oxygen map using the K-edge at 532 eV. The maps were obtained by the three-window method [31]. (D) Energy loss spectrum recorded from a subarea of (a), containing several crystals.

5.7.1 Fe(II)–Fe(III) – Spinels with Substitution?

The most frequent question asked at the 7th International Conference on Ferrites [1c] was whether spinel-type ferrites with substitution may be synthesized in the magnetosomes rather than of magnetite. Cultivation of *M. gryphiswaldense* in the presence of 10 or 20 µM copper, cobalt, chromium, nickel or titanium salts confirmed the analyses of magnetite from various magnetic bacteria, in that none of these metals was found in their magnetosomes. Thus iron uptake of the bacterial cells is highly specific and, therefore, these bivalent metal ions did not substitute for Fe(II) in magnetite (Fe_3O_4) either stoichiometrically to MFe_2O_4 or unstoichiometrically to $M_xFe_{3-x}O_4$ [24].

In a more detailed study [33], copper (II) sulfate was added at various time points to cultures of *M. gryphiswaldense* at various concentrations and copper to iron ratios. First, the copper toxicity threshold was determined by growth curves (not shown here). In the presence or absence of 50 µM iron (III) citrate, growth was inhibited completely by 100 µM copper (II) sulfate, whereas it was reduced only slightly by 50 µM. As shown in Table 5.3, 10 or 50 µM copper ions were added to the medium either before inoculation or during the logarithmic growth phase, the beginning of magnetite synthesis. Copper and iron were determined by atomic absorbtion spectroscopy (AAS). In none of these various experiments could copper be detected in the magnetosomes, even when the copper to iron ratio was 5:1.

The highest uptake of copper into the bacterial cells, 0.029% of dry cell weight, was measured after addition of 50 µM copper sulfate to the medium during the exponential growth phase (Expt D). This amount represents a 20-fold enrichment compared to cells grown under normal conditions (Expt C). When 50 µM copper sulfate was added before inoculation, a six-fold accumulation of copper was found (Expt B), which excluded a predominant binding of copper to the cell wall [35].

These experiments do not definitely exclude the formation of iron spinels with substitution, because the highest uptake of 0.029% copper (dry cell weight) corre-

Table 5.3 Content of copper and iron in cells and magnetosomes after their addition at two different growth phases of *M. gryphiswaldense*

Expt	Addition before inoculation	Addition during logarithmic phase	% Cu in magnetosomes	% Cu in dry cell mass	% Fe in dry cell mass
A	10 µM Cu	50 µM Fe	–	0.004	0.17
B	50 µM Cu	50 µM Fe	–	0.008	0.18
C	–	50 µM Fe	–	0.0014	0.46
D	–	50 µM Fe 50 µM Cu	–	0.029	0.27
E	–	50 µM Cu	–	0.022	0.047
F	50 µM Cu	10 µM Fe	–	0.0072	0.20

From each 1 l culture of experiments A–E 2 ml were taken and dried at 60 °C yielding about 5 mg dry cell weight. The remaining main volume of each experiment was used to isolate about 200 µg magnetosomes. The elements cooper and iron were determined by atomic absorption spectroscopy [34].

sponds roughly to an estimated maximum concentration of 4 µM copper in the cell. These 4 µM copper would probably not only represent free, but also bound copper; the remaining concentration of free copper, therefore, might be too low to be taken up into the magnetosomes. Isolated magnetosomes will allow to analyse the direct uptake of copper at various concentrations because copper now has not to pass two membranes, the outer and the inner.

5.7.2 The Phospholipid Profiles of the Magnetosome and Cytoplasmic Membrane are Different

In a previous report by Gorby *et al.* [36], the magnetosome membrane of *Magnetospirillum magnetotacticum* (the former *Aquaspirillum magnetotacticum* [5]) was characterized not only by freeze etching and thin sections, but also by its lipid profile. This analysis revealed fractions which included (i) neutral lipids and free acids (8%), (ii) glycolipids, sulfolipids and phosphatides (30%) and (iii) phospholipids (making up 62% of the total lipids by weight).

Phosphatidylethanolamine and phosphaditylserine were shown by thin layer chromatography to be part of the phospholipid fraction. Because it was concluded from these results that "the ratio of abundance of the lipid components is that expected for a biological membrane" [36], the phospholipid profiles of all three membranes (the outer, the cytoplasmic and the magnetosome membrane of *M. gryphiswaldense*) were determined quantitatively (M. Gassmann, C. Benning, D. Schüler, E. Baeuerlein, unpublished results). This was possible because D. Schüler had developed a special method for purifying magnetosomes (see Chapter 8). The magnetosomes were first separated from the other membranes by ultracentrifugation and then applied to a magnetizable column filled with a ferromagnetic matrix. Using an external magnetic field, the magnetosomes were bound to the column, could be separated from non-magnetic material and washed intensively. After removal of the magnet from the column, the magnetosomes were eluted and purified in a second step using sucrose gradient ultracentrifugation. The outer and cytoplasmic membranes were obtained using the method of Flip *et al.* [34]. The extraction of lipids from whole cells, outer cytoplasmic and magnetosome membranes was performed essentially according to Bligh and Dyer [37].

Using the method of Benning *et al.* [38], the lipids were obtained quantitatively by a two-dimensional thin layer chromotography. The fatty acids were released by acidic methanolysis and the resulting fatty acid methylesters determined quantitatively by gas chromatography [33].

As shown in Table 5.4, five lipids were found in the total extract of whole cells. Two of them, ornithineamidelipid (OA) and an unknown, apparently new, lipid X-NH₂, were not detectable in the magnetosome membrane. (The absence of both compounds was useful later as a marker for the purity of a magnetosome preparation.) The concentration of phophatidylcholine (PC) was eight-fold (1.1–8.0 mol %) and phosphatidylglycerol (PG) three-fold (12.6–38.3 mol %) higher than those in the cytoplasmic membrane. The main phospholipid in all three membranes is phosphatidylethanolamine (PE) which is present at a level about 20% lower in the

Table 5.4 Lipid composition of membrane fractions in *M. gryphiswaldense*

Lipid	Outer membrane (mol %)	Cytoplasmic membrane (mol %)	Magnetosome membrane (mol %)
Phosphatidylethanolamine	59.9 ± 5.1	70.7 ± 0.5	52.8 ± 5.5
Phosphatidylglycerol	13.9 ± 0.9	12.6 ± 1.9	38.3 ± 5.7
Ornithineamidelipid	18.6 ± 5.8	4.7 ± 0.4	–
Phosphatidylcholine	–	1.1	8.9 ± 5.0
X-NH$_2$	7.6 ± 0.7	7.2 ± 1.9	–

X-NH$_2$: unknown, ninhydrin-positive lipid with probably two branched hydroxy fatty acids (further analysis in process). For the quantitative analysis of lipids a two-dimensional thin layer chromatography was followed by gas chromatography of the fatty acid methylesters of each lipid according to Benning *et al.* [38]. (After [1b] with permission of the Harwood Academic Publishers, the Netherlands.)

magnetosome than in the cytoplasmic membrane (52.8–70.7 mol %). On the basis of these results we will discuss in Section 5.8 whether the magnetosome membrane originates from the cytoplasmic membrane by endocytosis or is created by a separate biosynthesis.

5.8 Mechanism(s) of Magnetite Crystal Formation in *M. gryphiswaldense*

5.8.1 The First Step: Iron Uptake

Unlike *Magnetospirillum magnetotacticum* (Chapter 7) and *Magnetospirillum* AMB-1 (Chapter 9), *M. gryphiswaldense* does not need an iron complexing agent such as as citrate or quinate in the medium for iron uptake under microaerobic conditions. It was however, stimulated considerably by spent medium, obtained from a stationary culture grown at moderate iron depletion (Fig. 5.3). When iron (II) sulfate was added as the iron source, no iron (III) hydroxide was precipitated during the rapid oxidation by air, indicating a complexing compound in the spent medium. None of the various siderophores tested in the transport assay, however, was found to be efficient in promoting iron uptake. These results were supported by Nakamura *et al.* [39], who were also unable to find a siderophore in spent medium of Magnetospirillum sp. AMB-1.

In *M. gryphiswaldense*, three iron-regulated outer membrane proteins (OMPs) have been identified. Two of them (16 and 38 kDa) were expressed more strongly with decreasing concentration of Fe(III) and one of them (70 kDa) was an iron-inducible OMP (M. Gassmann, D. Schüler, E. Baeuerlein, unpublished results).

As described in Section 5.4, iron (III), but not iron (II), is taken up by a low affinity but high velocity transport system, as deduced from the K_M and V_{max} values [21].

5.8.2 The Second Step: Passing to the Cytoplasm

A still unanswered question is whether iron (III) is transported into the cytoplasm of *M. gryphiswaldense* or whether it is first reduced to iron (II) by a ferrireductase. Recently Fukumori *et al.* [40] isolated and characterized from *M. magnetotacticum* a ferric iron reductase (NADH-Fe^{3+} oxidoreductase), a soluble cytoplasmic enzyme which is not periplasmic but is assumed to be bound to the inner side of the cytoplasmic membrane. It nevertheless appears to be a possible candidate for the reduction of ferric iron on the way to magnetite formation, because its inhibition by zinc (II) was accompanied by a reduction in the average magnetosome number in the cells and an increase in non-magnetic cells.

Possible mechanisms for magnetite formation are dominated by the question of which species, iron (II) or iron (III), will be transported into the empty magnetosome vesicles (see Section 5.8.3). Instead of being first reduced, iron (III) might be guided directly into them, if the chain of vesicles is bound to the cytoplasmic membrane. In this case, "magnetic formation is preceded by formation of amorphous hydrous iron (III) oxide in the magnetosome vesicles", which Frankel *et al.* [41] found in *M. magnetotacticum* using ^{57}Fe Mössbauer spectroscopy (the complete model is described in Chapter 3).

The convincing hypothesis that magnetosome membranes may be invaginations of the cytoplasmic membrane, could not be verified, because their different phospholipid profiles in *M. gryphiswaldense* (Table 5.4) are strong indications for two distinct membranes. Magnetosome chains, nevertheless, appear to be fixed within the cell [42], which is a prerequisite for its magnetotaxis. Electrostatic interactions might bind the magnetosome chain in *M. gryphiswaldense* to the cytoplasmic membrane, as the charge difference of the two phospholipid profiles provides the basis for this. Strong interactions, here repulsions, led to stable suspensions of isolated magnetosomes, forces which also support the proposed binding *in vivo* (Fig. 5.9a, b).

If huge amounts of iron are transported into the cytoplasm, intermediate clusters of iron should be detected – this has not been the case until now. Several observations of the formation of greigite (Fe_3S_4) in iron sulfide magnetotactic bacteria [42], indicate a highly reduced state of the cytoplasm [43], the consequence of which is the presence of free Fe(II) ions at neutral pH (Chapter 7; p. 405).

5.8.3 The Final Step: Formation of Single-Magnetic-Domain Magnetite Crystals

In cultures of *M. gryphiswaldense* reaching oxygen concentrations of 2–7 μM, a sudden increase in iron uptake is closely coupled to the generation of magnetism, i.e. the growth of magnetite crystals (Fig. 5.7b). Whereas this biomineralization process takes about 4 h, magnetism could be measured by the light scattering method (see Section 5.3) even 5–10 min after the addition of iron. After 30 min, electro-dense particles of 5–20 nm were detected in electron micrographs of these cells, particles that are in the superparamagnetic size range [16]. Although not yet neighbouring mature crystals of about 45 nm (Fig. 5.8a), the small particles were apparently placed within an invisible chain (Fig. 5.8b), which probably consists of empty vesicles.

The rapid development of magnetism, within 5–10 min, together with the detection of chain-like organized particles after 30 min, provides evidence for a prior biosynthesis of empty vesicles. T. Beveridge was able for the first time to show these empty vesicles in electron micrographs of thin sections from *M. magnetotacticum* [33, 44]. Later D. Schüler succeeded in detecting similar structures in *M. gryphiswaldense* [24]. Moreover the differences in the phospholipid profiles of the cytoplasmic and magnetosome membranes (Table 5.4) indicate a separate biosynthetic pathway for the phospholipids of the magnetosome membrane. Variations in this biosynthesis may be responsible for the different habits of magnetite crystals within the various species or strains of magnetic bacteria.

If iron (II) is transported into the vesicles and then two iron (II) ions are oxidized to iron (III), i.e. because Fe_3O_4 is composed of one iron (II) and two iron (III) ions, one H^+ would result from each iron (II) oxidized:

$$Fe^{2+} + H_2O \qquad [Fe^{3+}OH^-] + H^+ + Red.$$

This oxidation reaction would result in acidification of the intravesicular phase. Without discussing control of pH and the membrane proteins involved (for discussion see Chapters 7, 8 and 9), we can only summarize here some similarities and three central characteristics of prokaryotic and eukaryotic unicells.

The encouragement for this simplified view arose from reading Chapter 10 of this book, the contribution of J. L. Kirschvink and J. W. Hagadorn. In this chapter, magnetite biomineralization is described as "the most ancient matrix-mediated system which may have served as the ancestral template for exaptation. Complete sequencing of the genome of a magnetotactic bacterium, and identifying the magnetite operon, might provide a 'road map' for unraveling the genetics of biomineralization in higher organisms, including humans (see Section 10.1)."

On the lower level of comparing the physiology of prokaryotic and eukaryotic unicells, such as magnetic bacteria, diatoms and coccolithophorids, these organisms reveal that they (i) use vesicles for production of inorganic materials (magnetosomes in magnetic bacteria, silica deposition vesicles in diatoms, coccolith vesicles in coccolithophorids); (ii) accumulate inorganic ions in these vesicles from lower concentrations outside; and (iii) acidify the intravesicular phase (not measured in magnetic bacteria).

The final step in synthesizing a single-magnetic-domain magnetite crystal is unknown, until the "magnetite operon" can be identified and sequenced. Because such a single-magnetic-domain crystal should grow from a single nucleation center, a first indication may be the electron-dense particles found in vesicles of *M. gryphiswaldense* 30 min after the addition of iron (Fig. 5.8b).

Acknowledgements

I would like to thank my colleagues D. Schüler, C. Benning, L.-O. Essen, M. Gassmann, R. Grimm, N. Petersen, S. Spring, A. Stahlberg, R. Uhl and U. Weser for their valuable cooperation and discussions.

I also greatly appreciate the support of my son Felix in providing various hardware and software for revising the many manuscripts of this book.

This work was supported by the Deutsche Forschungsgemeinschaft (Schwerpunktsprogramm: Bioinorganic Chemistry: Transition Metals in Biology and their Coordination Chemistry) and by the Max-Planck-Society, München.

References

[1] (a) D. Schüler, E. Baeuerlein, in *Bioinorganic Chemistry: Transition Metals in Biology and their Coordination Chemistry.* (Deutsche Forschungsgemeinschaft) (Ed. A. X. Trautwein), Wiley-VCH, Weinheim, 1997, pp. 24–36.
(b) D. Schüler, E. Baeuerlein, in *Transition Metals in Microbial Metabolism* (Eds G. Winkelmann and C. J. Carrano), Harwood Academic Publishers, Amsterdam, 1997, pp. 159–185.
(c) D. Schüler, E. Baeuerlein, in *Proceeedings of the 7th International Conference on Ferrites*, Bordeaux 1996, Colloque C1 (Eds V. Cagan and M. Guyot), *J. Phys. IV*, 7, 647–650, Les Editions de Physique, Les Ulis Cedex A 1997.
[2] R. P. Blakemore, *Science* 1975, *190*, 377–379.
[3] R. P. Blakemore, D. Maratea, R. S. Wolfe, *J. Bacteriol.* 1979, *140*, 720–729.
[4] (a) R. P. Blakemore, R. B. Frankel, A. J. Kalmijn, *Nature* 1980, *236*, 384–385.
(b) for review: R. B. Frankel, R. P. Blakemore (Eds) *Iron Biominerals*, Plenum Press, New York, 1991.
[5] K. H. Schleifer, D. Schüler, R. S. Spring, M. Weizenegger, R. Amann, W. Ludwig, M. Köhler, *Syst. Appl. Microbiol.* 1991, *14*, 379–385.
[6] D. Schüler, Diploma thesis, University of Greifswald, Germany, 1990.
[7] D. Schüler, M. Köhler, *Zentralbl. Mikrobiol.* 1992, *142*, 150–151.
[8] T. Matsunaga, T. Sakaguchi, F. Tadokoro, *Appl. Microbiol. Biotechnol.* 1991, *35*, 651–655.
[9] J. G. Burgess, R. Kawaguchi, T. Sakaguchi, R. H. Thornhill, T. Matsunaga, *J. Bacteriol.* 1993, *175*, 6689–6694.
[10] R. B. Frankel, D. A. Bazylinski, M. Johnson, B. L. Taylor, *Biophys. J.* 1997, *73*, 994–1006.
[11] D. A. Bazylinski, R. B. Frankel, H. W. Jannasch, *Nature* 1988, *334*, 518–519.
[12] E. F. DeLong, R. B. Frankel, D. A. Bazylinski, *Science* 1993, *259*, 803–806.
[13] T. Sakaguchi, J. G. Burgess, T. Matsunaga, *Nature* 1993, *365*, 47–49.
[14] D. Schüler, R. S. Spring, D. A. Bazylinski, *Syst. Appl. Microbiol.* 1999, *22*, 466–471.
[15] R. S. Spring, R. Amann, W. Ludwig, K. H. Schleifer, H. van Gemerden, N. Petersen, *Appl. Environ. Microbiol.* 1999, *59*, 2397–2403.
[16] R. F. Butler, S. K. Banerjee, *J. Geophys. Res.* 1975, *80*, 252–259.
[17] E. Baeuerlein, D. Schüler, R. Reszka, S. Päuser, (a) German Patent DE 197 16 732 C 2; (b) International Application PCT/DE 98/00668, 1998.
[18] T. Matsunaga, F. Tadokoro, N. Nakamura, *IEEE Trans. Magnet. Mag.* 1990, *26*, 1557–1559.
[19] D. Schüler, R. B. Frankel, *Appl. Microbiol. Biotechnol.* 1999, *52*, 464–473.
[20] D. Schüler, R. Uhl, E. Baeuerlein, *FEMS Microbiol. Lett.* 1995, *132*, 139–145.
[21] D. Schüler, E. Baeuerlein, *Arch. Microbiol.* 1996, *166*, 301–307.
[22] A. Hartmann, V. Braun, *Arch. Microbiol.* 1981, *130*, 353–356.

[23] G. Winkelmann, in: *CRC Handbook of Microbial Iron Chelates* (Ed. G. Winkelmann), CRC Press, Boca Raton, 1991, pp. 65–106.

[24] D. Schüler, PhD thesis, Technical University, Munich, 1994.

[25] J. B. Neilands, *J. Biol. Chem.* 1995, *270*, 26723–26726.

[26] G. Winkelmann (Ed.) *CRC Handbook of Microbial Iron Chelates*, CRC Press, Boca Raton, 1991.

[27] S. Spring, R. Amann, W. Ludwig, K. M. Schleifer, H. van Gemerden, N. Petersen, *Appl. Environ. Microbiol.* 1993, *59*, 2397–2403.

[28] P. T. Vischer, J. Beukema, H. van Gemerden, *Limnol. Oceanogr.* 1991, *36*, 1465–1480.

[29] D. Schüler, E. Baeuerlein, *J. Bacteriol.* 1998, *180*, 159–162.

[30] S. Mann, N. H. C. Sparks, R. P. Blakemore, *Proc. R. Soc. Lond. B*, 1987, *231*, 477–487.

[31] R. Grimm, A. J. Koster, U. Ziese, D. Typke, W. Baumeister, *J. Microsc.* 1996, *183*, 60–68.

[32] D. Grimm, D. Schüler, E. Baeuerlein, unpublished results.

[33] M. Gassmann, Diploma thesis, University of Tübingen, Germany, 1996.

[34] C. Flip, G. Fletcher, I. L. Wulff, C. F. Earhart, *J. Bacteriol.* 1973, *11*, 717–722.

[35] M. Gassmann, D. Schüler, U. Weser, E. Baeuerlein, unpublished results.

[36] Y. A. Gorby, T. J. Beveridge, R. P. Blakemore, *J. Bacteriol.* 1988, *170*, 834–841.

[37] E. G. Bligh, W. J. Dyer, *Can. J. Biochem. Physiol.* 1951, *37*, 911–917.

[38] C. Benning, Z. Huang, D. Gage, *Arch. Biochem. Biophys.* 1995, *317*, 103–111.

[39] C. Nakamura, T. Sakaguchi, S. Kudo, I. G. Burgess, K. Sude, T. Matsunaga, *Appl. Biochem. Biotechnol.* 1993, *39/40*, 169–177.

[40] Y. Noguchi, T. Fujiwara, K. Yoshimatsu, Y. Fukumori, *J. Bacteriol.* 1999, *181*, 2142–2147.

[41] R. B. Frankel, G. C. Papaefthymion, R. P. Blakemore, W. O'Brien, *Biochem. Biophys. Acta* 1983, *763*, 147–159.

[42] D. A. Bazilinski, R. B. Frankel, B. R. Heywood, S. Mann, J. W. King, P. L. Donaghay, A. V. Hanson, *Appl. Environ. Microbiol.* 1995, *61*, 3232–3239.

[43] S. C. Adam, *Adv. Microbiol. Physiol.* 1998, *40*, 281–341.

[44] T. J. Beveridge, in: *Metal Ions and Bacteria* (Eds T. Beveridge and R. J. Doyle), John Wiley & Sons, New York, 1998, pp. 1–30.

6 Applications for Magnetosomes in Medical Research

Regina C. Reszka

6.1 Introduction

Since the first publication on magnetotactic bacteria by Richard P. Blakemore in 1975 [1], the subject has stimulated the research interests of scientists from various disciplines, leading to numerous co-operations. It has been suggested that, in particular, the magnetosomes detectable in all magnetotactic bacteria (MTB) and characterized as magnetic iron mineral particles surrounded by membrane vesicles [2], have the potential for further biotechnological and medical applications [3]. These magnetosomes, which enable the bacteria to orient themselves along the lines of a magnetic field (magnetotaxis [4]), offer a great opportunity in the diagnostic and therapeutic field, for magnetic drug targeting in a wide range of diseases [5–7].

Compared to the commercially available and scientifically used synthetic magnetic particles [8], the magnetosomes produced in bacteria have several advantages in terms of controlled crystal growth and structural properties [9].

As far as scaling up to a future industrial scale is concerned, biomineralization offers the possibility of producing highly uniform magnetite crystals without the need for extreme conditions such as high pressure, temperature and pH [10].

The small size of the magnetosomes (35–100 nm [11]) provides a large surface-to-volume ratio and makes them attractive as carriers for the immobilization of pharmacologically active substances, with the option of separation by magnetic fields.

Shortly after the discovery of magnetosomes, their biotechnological potential was recognized by Japanese scientists. A group including Matsunaga was the first to describe the use of magnetic particles isolated from magnetotactic bacteria for the immobilization of enzymes [12] and antibodies [13–16], and as sensing devices in biosensors [14].

They also developed a fully automated immunoassay for the determination of human insulin in blood serum using antibody–ProteinA–magnetosome complexes [17].

A report by Sode *et al.* [18] showed that antisense oligonucleotides could be immobilized on magnetosomes for mRNA detection. This paved the way to the new fields of gene transfer [19] and high-throughput genotyping [20]. In a later study,

Takeyama and co-workers [21] described the use of magnetosomes as carriers for the magneto-ballistic transfer of plasmid DNA into the marine cyanobacterium *Synechococcus* using the gene gun.

Although the biotechnological potential of bacterial magnetite particles has been demonstrated, it is difficult to generate valid data for widespread application. Due to the problems with the mass cultivation of magnetotactic bacteria on a production scale, and the lack of a fundamental understanding of the biochemical and genetic principles behind the process of bacterial magnetite biomineralization, the amount of magnetosomes required is to date the limiting step for animal experiments [3].

However, the potential of magnetosomes as drug carrier systems is extremely high, justifying their development as pharmaceutical and biotechnological formulations in a wide range of medical applications. In this chapter, we would like to describe the initial experiments we have carried out to study the efficacy of magnetosomes in combination with cationic lipids and immobilized antibodies, formulated as a new non-viral gene transfer system.

Over the last 10 years more than 300 clinical trials of gene therapy approaches have been carried out. Various monogenetic diseases such as severe combined immunodeficiency (SCID/ADA), haemophilia A and B, familial hypercholesterolaemia, cystic fibrosis, Gaucher's disease and acquired diseases such as cancer, cardiovascular and infectious diseases were included in these studies. The data obtained clearly demonstrate that the gene delivery aspect is one of the weaknesses of all these therapies. Verma and Somia [22] summarized the current problems of viral and non-viral gene delivery methods and characterized the ideal vector as a powerful chimera of both the viral and synthetic systems, including the following properties:

· high concentration, allowing many cells to be infected;
· convenience and reproducibility of production;
· ability to integrate in a site-specific location in the host chromosome, or be successfully maintained as a stable episome;
· a transcriptional unit that can respond to manipulation of its regulatory elements;
· no components that elicit an immune response.

All of these properties exist in different delivery systems and it is necessary to assemble the desired elements.

The magnetosomes offer a unique opportunity to combine a therapeutic and a diagnostic tool [23]. Due to their properties magnetosomes can be used as site-specific targeted gene delivery systems. The genuine anionic lipid bilayer facilitates complexation with cationic liposomes, driven by electrostatic forces. Moreover the system allows *in vivo* follow-up of the gene transfer using magnetic resonance imaging (MRI). The aim of our experiments was to utilize magnetosomes in combination with cationic liposomes to generate a novel delivery system for improved and efficient gene therapy.

6.2 Gene Transfer Using Cationic Lipid–Magnetosome– DNA Complexes

6.2.1 Preparation of Cationic Lipid–Magnetosome–DNA Complexes

Over the last 5 years we have developed various novel cationic lipids for use in gene transfer, including the cholesterol derivatives, e.g. DAC-Chol [24]; lipospermine derivatives, e.g. DOCSPER [25]; and others [26–28]. DAC-Chol/DOPE (3:7 w:w), called DAC-30™, and DAC-40 (DAC-Chol/DOPE; 4:6 w:w) were used in the initial experiments.

The magnetotactic bacterium *M. gryphiswaldense* [29, 30] was cultivated and the magnetosomes isolated [31–33]. The structural integrity was checked by transmission electron microscopy (for further details see Chapters 5 and 8 of this book).

In an aqueous medium isolated magnetosomes form a colloidal and monodisperse suspension. The properties of the suspension are determined by the magnetic and electrostatic interactions between the particles as well as by the presence or absence of the magnetosome phospholipid membrane. This membrane contains mainly phosphatidylethanolamine (PE), which accounts for more than 50 mol % of the total lipid in the membrane, and phosphatidylglycerol (PG, 38 mol %) and phosphatidylcholin (PC, 8 mol %) ([34]; M. Gassmann, C. Benning, D. Schüler, E. Baeuerlein, unpublished results).

To prepare the cationic lipid–magnetosome–DNA complex, the dried lipid film of DAC-40 was dispersed in the magnetosome suspension. Green fluorescent protein (GFP) (pEGFPC1; cytoplasmatic expression; Clontech Heidelberg, Germany) or LacZ (pUT 651; nuclear localized; Cayla, Toulouse, France), cloned under a CMV promoter, were used as marker gene DNA. Supercoiled plasmid DNA was transformed into *E. coli* using a new method [35] and purified by alkaline lysis and column chromatography using Qiagen-tips from Qiagen GmbH (Hilden, Germany).

6.2.2 Immobilization of Anti-Carcino-Embryonal Antigen (CEA) Antibody to the Magnetosome Membrane

According to the Thy 1.1-antibody immobilization onto liposome membranes we used in previous experiments [36, 37], we chemically modified phosphatidylethanolamine in the magnetosome membrane using N-succinimidyl 4-(p-maleimidophenyl)-butyrate (SMPB). Thereafter the monoclonal anti-CEA antibody (D11DG2) [38] was modified using succinimidyl-S-acetylthioacetate (SATA) followed by deacetylation to activate the antibody by creation of free SH groups. The amount of SH groups on the anti-CEA antibody was determined used Ellman's Reagent. Finally the activated SATA-modified anti-CEA antibody was crosslinked to the MPB-modified magnetosomes.

6.2.3 Cell Transfection

Rat glioblastoma cells (F98) and human colon carcinoma cells (LoVo) were seeded onto 24-well or glass cover slips (fluorescence microscopy), and grown for 24 h in Dulbeccos-modified Eagle medium (DMEM) supplemented with 10% fetal calf serum and antibiotics at 37 °C and 5% CO_2 until 60–70% confluence was achieved [35]. Thereafter the medium was removed and the different cationic lipid–magnetosome–DNA complexes were added to the cells for comparison with the controls in DMEM without FCS. After 4 or 24 h incubation the cells were washed twice and supplemented with complete growth medium. The cells were further incubated and after 72 h X-Gal staining for the detection of cells expressing the LacZ gene was performed [35] (Figs. 6.1, 6.2). The cytotoxicity of the cationic lipid–magnetosome–DNA or antibody complexes was determined using the well-established MTT (3-[4,5-dimethylthiazol-2-yl]-2,5-diphenyltetrazolium bromide) test [39].

Using the LacZ reporter gene, gene expression and cytotoxicity can be determined simultaneously using a novel dual test. β-Galactosidase activity serves as a marker for transfection efficiency and acidic phosphatase activity as a marker for *in vitro* toxicity. Both values can be measured in parallel by UV spectroscopy [40].

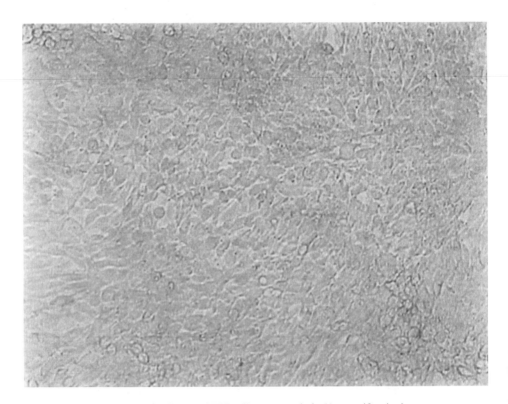

Figure 6.1 Light microscopic picture of F98 cells as controls (×40 magnification).

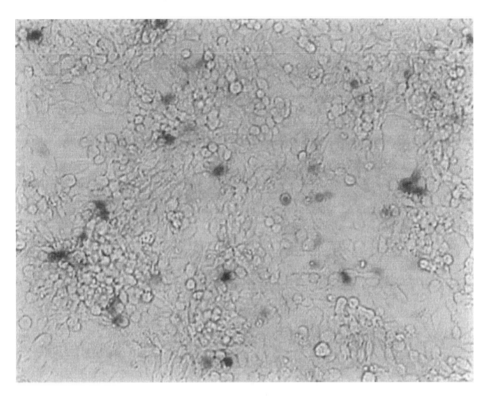

Figure 6.2 Transfection assay: X-Gal expression of the F98 cells after treatment with 10 μg pUT651 complexed with 50 μg DAC-40 and 50 μl magnetosome solution containing 0.06 μg iron determined by atomic adsorption spectroscopy (AAS).

The detection of the GFP expressing cells growing on cover slips was possible without additional staining after different time points using an Olympus IX-70 fluorescence microscope.

6.2.4 Prussian Blue Staining for the Detection of Magnetosome (Iron) Uptake into the Cells

The uptake of magnetosomes and magnetosome complexes is detectable immunocytochemically using a reaction (Prussian blue staining) which gives rise to a blue colouring according to the following reaction:

$$4Fe^{3+} + 3K_4Fe(CN)_6 \rightarrow Fe_4[Fe(CN)_6]_3 + 12K^+$$

Because magnetosomes contain [Fe(II,III)-Oxid], the particles taken up can be detected in the cells as blue spots using a light microscope. Counter-staining of

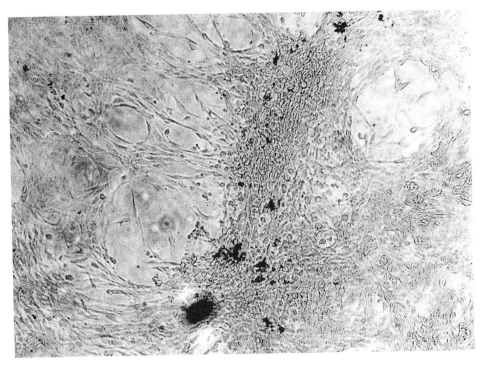

Figure 6.3 Prussian blue staining of F98 cells after treatment with the DAC-40/pUT651/magnetosome complex (see text for composition).

the cells was performed with nuclear fast red, increasing the contrast of the picture (Fig. 6.3).

6.2.5 Electron Microscopy

6.2.5.1 Visualization of Cellular Uptake of Cationic Lipid–Magnetosome–DNA Complexes

The interaction of cationic lipid–magnetosome–DNA complexes with the cells and their fate *in vitro* was visualized by electron microscopy. The cells were fixed, dehydrated and embedded in epoxy resin, and finally ultra-thin sections were prepared. After contrasting with heavy metal salts, the samples were evaluated using transmission electron microscopy [41] (Figs. 6.4–6.6).

6.2.5.2 Immune Electron Microscopy for the Detection of Antibody Binding

To detect the immobilized antibody on the surface of the magnetosomes a second gold-marked goat anti-mouse or the protein A gold technique was used [42] (Figs. 6.7 and 6.8).

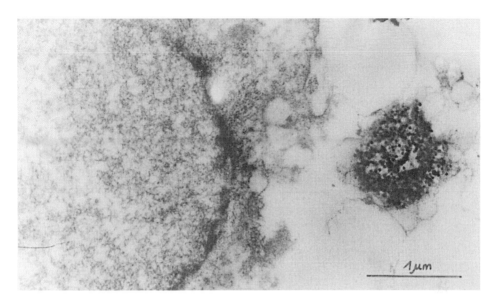

Figure 6.4 Transmission electron micrograph of the DAC-30 liposome/DNA/magnetosome complex in front of the F98 cell surface (bar = 1 μm, magnification × 30 000).

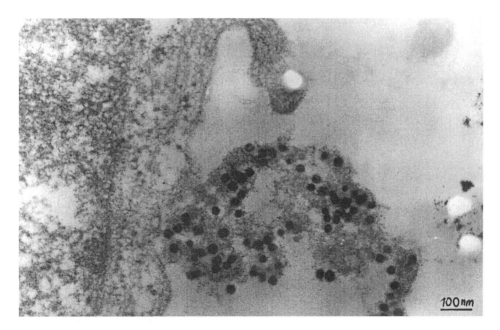

Figure 6.5 Transmission electron micrograph of the uptake of the DAC-30 liposome/DNA/ magnetosome complex by F98 cells (bar = 100 nm, magnification × 100 000).

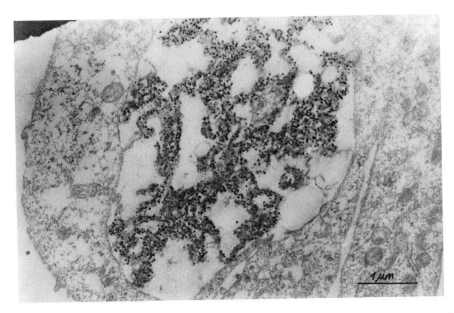

Figure 6.6 Transmission electron micrograph of an internalized DAC-30 liposome/DNA/ magnetosome complex into a lysosomal compartment of F98 cells (bar = 1 μm, magnification × 20 000).

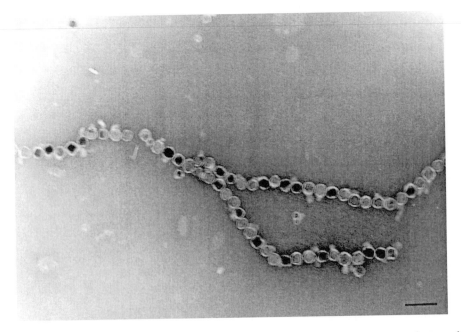

Figure 6.7 Transmission electron micrograph of isolated magnetosomes arranged as a chain. Negative staining using ammonium molybdate, bar = 100 nm, magnification × 100 000).

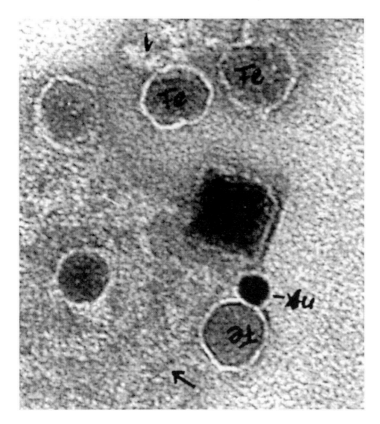

Figure 6.8 Transmission electron micrograph of immuno-magnetosomes. Gold-anti-CEA antibody (Au) coupled on the magnetosome membrane (Fe); magnification × 400 000).

6.3 Future Perspectives

In the future magnetosomes particles in the nanometer range will gain increasing importance in diagnostic and therapeutic approaches. As biocompatible, non-toxic and less immunogenic pharmaceutical ferrofluids with magnetically responsive properties, they can be used in combination with a wide range of biologically active substances such as genes, peptides, hormones, antibiotics, lectins, proteins and radioactive species. A great advantage is the ability to move magnetosomes directly to their target site *in vivo*, driven by a magnetic force. It is therefore possible to enhance local drug concentrations at the target site and reduce the toxicity to normal tissues.

As soon as a production process for magnetosomes is established, ensuring a yield of magnetic particles in sufficient quantity and with validated quality, it is expected that development will simultaneously begin in all the different areas already known today as possible applications for magnetosomes, such as magnetic resonance imaging (MRI) and positron emission tomography (PET) (diagnostic), as well as drug transport, radionuclide delivery and hyperthermia (therapeutic) [7].

In addition, specific genetic manipulation of the magnetosome-producing bacteria offers the possibility of expressing therapeutically relevant molecules such as proteins, peptides and antisense oligonucleotides in the magnetosome membrane. By this means it would be possible to achieve a drug carrier designed on a genetic basis without the use of coupling techniques.

Within the framework of our ongoing research, we will first perform biophysical investigations to evaluate the interaction of DNA–lipid complexes with magnetosomes. The aim of this study is to gain an understanding of the structure and stability parameters of the novel DNA carriers. Animal experiments for the *in vivo* gene therapy of glioblastoma and liver metastases using a rat model will also be carried out under the control of MRI. The systemic and local application, respectively, of apoptosis-inducing and antiangiogenesis-inducing genes provides evidence on the one hand of the circulation half-life time and enrichment in the target tissue in comparison to other organs, and on the other hand to the therapeutic efficiency.

Therefore the novel carrier system offers many opportunities to understand tumor growing parameters such as angiogenesis, tumor necrosis, pressure and pH value of the tumor.

Acknowledgements

I am grateful to Prof. E. Bäuerlein and Dr D. Schüler for collaboration and discussion, Mrs D. Wegner for isolation and purification of the magnetosomes, Mrs C. Lehmann for the electron microscopic pictures, Mrs J. Richter and B. Pohl for their excellent technical assistance and Dr J. E. Diederichs for her help in preparing the manuscript.

References

[1] R. P. Blakemore, *Science*, 1975, *190*, 377–379.
[2] D. Balkwill, D. Maratea, R. P. Blakemore, *J. Bacteriol.* 1980, *141*, 1399–1408.
[3] D. Schüler, *J. Mol. Microbiol. Biotechnol.* 1999, *1*, 79–86.
[4] R. P. Blakemore, R. B. Frankel, *Sci. Am.* 1981, *245*, 42–49.
[5] D. Günther, *Pharmazie in unserer Zeit* 1996, *25*, 130–134.
[6] W. Schütt, C. Grüttner, U. Häfeli, M. Zborowski, J. Teller, H. Putzar, C. Schümichen, *Hybridoma*, 1997, *16*, 109–117.
[7] U. Häfeli, W. Schütt, J. Teller, M. Zborowski (Eds) *Scientific and Clinical Applications of Magnetic Carriers*, Plenum Press, New York and London, 1997.

[8] M. Sarikaya, *Microsc. Res. Tech.* 1994, *27*, 360–375.
[9] D. A. Bazylinski, *ASM News*, 1995, *61*, 337–343.
[10] S. Mann, N. H. C. Sparks, R. G. Board, *Adv. Microbial. Physiol.* 1990, *31*, 125–181.
[11] S. K. Banerje, B. M. Moskowitz, in *Magnetite Biomineralization and Magnetoreception in Organisms* (Eds J. L. Kirschvink, D. S Jones, B. M MacFadden), Plenum Press, New York and London, 1985, pp. 17–41.
[12] T. Matsunaga, S. Kamiya, *Appl. Microbiol. Biotechnol.* 1987, *26*, 328–332.
[13] N. Nakamura, K. Hashimoto, T. Matsunaga, *Anal. Chem.* 1991, *63*, 268–272.
[14] T. Matsunaga, *TIBTECH* 1991, *9*, 91–95.
[15] C. Nakamura, T. Sakaguchi, S. Kudo, J. G. Burgess, K. Sode, T. Matsunaga, *Appl. Biochem. Biotechnol.* 1993, *30/40*, 169–177.
[16] N. Nakamura, J. G. Burgess, K. Yagiuda, S. Kudo, T. Sakaguchi, T. Matsunaga, *Anal. Chem.* 1993, *65*, 2036–2039.
[17] T. Matsunaga, R. Sato, S. Kamiya, T. Tanaka, H. Takeyama, *J. Magn. Matt.* 1999, *1994*, 126–131.
[18] K. Sode, S. Kudo, T. Sakaguchi, N. Nakamura, T. Matsunaga, *Biotechnol. Techniques* 1993, *7*, 688–694.
[19] T. Matsunaga, S. Kamiya, N. Tsujimura, in *Scientific and Clinical Applications of Magnetic Carriers* (Eds U. Häfeli, W. Schütt, J. Teller and M. Zborowski), Plenum Press, New York and London, 1997, pp. 287–294.
[20] T. Matsunaga, Off. Gaz. US Pat. Trademark Off. Pat. 1079 US patent 5 861 285, 1999, 1218: 2181.
[21] H. Takeyama, A. Yamazawa, C. Nakamura, T. Matunaga, *Biotechnol. Techniques*, 1995, *9*, 355–360.
[22] I. M. Verma, N. Somia, *Nature*, 1997, *389*, 239–242.
[23] E. Baeuerlein, D. Schüler, R. Reszka, S. Päuser, German Patent DE 197 16 732 C2; International Application PCT/DE 98/00668, 1998.
[24] R. Reszka, J. H. Zhu, F. Weber, W. Walther, R. Greferath, S. Dyballa, *J. Liposome Res.* 1995, *5*, 149–167.
[25] D. Groth, O. Keil, C. Lehmann, M. Schneider, M. Rudolph, R. Reszka, *Int. J. Pharm.* 1998, *162*, 143–157.
[26] O. Keil, thesis, University of Wuppertal, 1999.
[27] M. Schneider, O. Keil, R. Reszka, D. Groth, Akz.: 196 31 189.6, 1996.
[28] D. Groth, thesis, University of Berlin, 2000.
[29] D. Schüler, M. Köhler, *Zentralbl. Mikrobiol.* 1992, 147, 150–151.
[30] K. H. Schleifer, D. Schüler, S. Spring, M. Weizenegger, R. Amann, W. Ludwig, M. Köhler, *Syst. Appl. Microbiol.* 1991, *180*, 159–162.
[31] D. Schüler, thesis, University of München, 1994.
[32] D. Schüler, E. Baeuerlein, *Bioinorg. Chem.* 1997, Chapter 3, 24–36.
[33] D. Schüler, E. Baeuerlein, *J. Bacteriol.* 1998, *180*, 159–162.
[34] M. Gassmann, thesis, University of München, 1998.
[35] D. Groth, R. Reszka, J. A. Schenk, *Anal. Biochem.* 1996, *240*, 302–304.
[36] H. Madry, thesis, University of Berlin, 1996.
[37] F. J. Martin, D. Papahadjopoulos, *J. Biol. Chem.* 1982, *257*, 286–288.
[38] B. Micheel, G. Scharte, *Hybridoma* 1993, *12*, 227–229.
[39] T. Mosmann, *J. Immunol. Methods* 1983, *65*, 55–63.
[40] D. Groth, O. Keil, M. Schneider, R. Reszka, *Anal. Biochem.* 1998, *258*, 141–143.
[41] T. Watanabe, M. Watanabe, Y. Ishii, H. Matsuba, S. Kimura, *J. Histochem. Cytochem.* 1989, *37*, 347–351.
[42] J. W. Slot, H. J. Geuze, *Eur. J. Cell Biol.* 1985, *38*, 87–93.

7 Enzymes for Magnetite Synthesis in *Magnetospirillum magnetotacticum*

Yoshihiro Fukumori

7.1 Introduction

Magnetospirillum magnetotacticum, which was isolated from microaerobic fresh-water sediments in 1979 by Blakemore *et al.* [1], possesses interesting particles called magnetosomes, with ferrimagnetic iron oxide magnetite crystals (Fig. 7.1) [2]. By using the chained magnetosomes as a magnetic sensor, the bacterium orients itself along the lines of the Earth's magnetic field [3]. The magnetite crystal in the mag-netosome occurs with almost the same size of 50–100 nm in the cytoplasm and enclosed by lipid bilayers with some characteristic proteins [4].

How does the bacterium synthesize the magnetite crystals and regulate the size at room temperature? Mann *et al.* [5] have reported the characterization of magnetic spinels prepared by two independent chemical methods. The first is based on the partial oxidation of Fe(II) solutions in the presence of nitrate at 100 °C and the second on the reaction of hydrated ferric oxide (ferrihydrite) with ferrous ions at room temperature and pH = 7 [5]. Tamaura *et al.* [6] have also reported the chem-ical formation of magnetite by air oxidation of $Fe(OH)_2$ suspensions. The reaction proceeds in three stages: (i) formation of Fe(III) oxides and slower formation of Fe_3O_4; (ii) rapid formation of Fe_3O_4; (iii) linear formation of Fe_3O_4. They also suggested that Fe(II) is oxidized on the surface of the solid phases during the course of air oxidation. On the other hand, Frankel *et al.* [2] proposed that *M. magneto-tacticum* synthesizes magnetites in the following sequence: (i) iron uptake with a reduction of Fe(III) to Fe(II) in the transport process; (ii) formation of low-density hydrous ferric oxide with re-oxidation of Fe(II); (iii) formation of high-density hy-drous ferric oxide (ferryhydrite) through the dehydration of low-density hydrous oxide; (iv) formation of magnetite by the partial reduction of iron and the further dehydration of ferryhydrite. However, these proposals demonstrate no enzymatic mechanisms of iron oxidation and reduction for magnetite synthesis in *M. mag-netotacticum*. This chapter describes the purification and characterization of Fe(II)-nitrite oxidoreductase [7] and NADH-Fe(III) oxidoreductase for magnetite synthesis [8] and discusses the mechanism of enzymatic synthesis of magnetite in *M. magnetotacticum* MS-1 (ATCC31632).

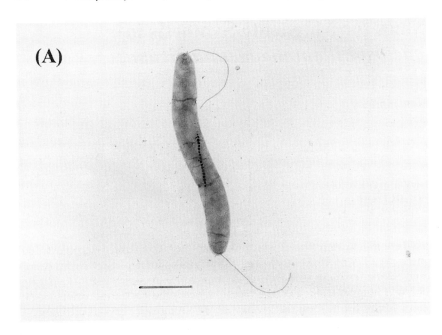

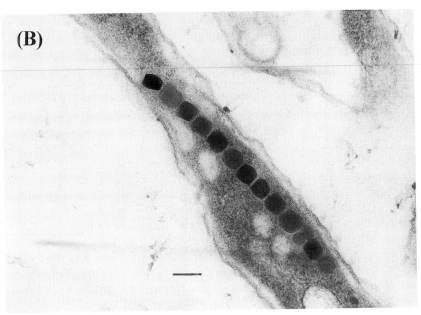

Figure 7.1 (a) Transmission electron micrograph of *Magnetospirillum magnetotacticum* MS-1 (ATCC31632). Electron-opaque spots in the middle of the cell are the magnetosomes. Single bipolar flagella are visible. Bar = 1 μm. (b) Thin section of *M. magnetotacticum*. Bar = 100 nm.

7.2 Ferric Iron Reduction in *M. magnetotacticum*

In general the iron atom exists in its oxidized ferric state under aerobic conditions. A common strategy for bacteria to use in obtaining iron from their environment is to excrete siderophores, which bind and solubilize ferric iron. *M. magnetotacticum* uses a high-affinity siderophore system similar to that used by other Gram-negative organisms for iron acquisition [9]. For the iron held in the iron–siderophore complex to be used by the cell, it first must be removed from the siderophore. Current mechanisms envisioned for the removal of iron from siderophores include *in situ* reduction by ferric iron reductase [10], resulting in release of the ferrous iron. Reduction of Fe(III) to Fe(II) by ferric iron reductase is thought to be an obligatory step in iron uptake as well as the primary factor in making iron available for absorption by bacteria.

On the other hand, Nakamura *et al.* [11] have reported that *Magnetospirillum* sp. AMB-1 produces no siderophores and Schüler and Baeuerlein [12] have also reported that *Magnetospirillum gryphiswaldense* utilizes energy-dependent iron uptake but not siderophore-like compounds. Although the uptake of iron in magnetic bacteria is still poorly understood, iron reduction must occur in the cell because magnetite is an oxide of mixed oxidation state, i.e. Fe(III) and Fe(II).

7.2.1 Localization and Purification of Iron Reductase from *M. magnetotacticum*

After *M. magnetotacticum* had been cultivated microaerobically in the chemically defined growth medium [1], the cells were suspended in 10 mM Tris-HCl buffer (pH 8.0) containing 0.75 M sucrose and incubated with EDTA plus lysozyme at 30 °C for 1 h. The suspension was centrifuged at 104 000 g for 30 min, and the periplasmic fraction was retained as the supernatant. The precipitates obtained were resuspended in water at 4 °C and centrifuged at 104 000 g for 1 h. The supernatant was retained as the cytoplasmic fraction and the pellet was resuspended in 10 mM Tris-HCl buffer (pH 8.0) and utilized as the membrane fraction. The magnetosomes in the membrane fraction were removed with a magnet. Table 7.1 summarizes the localization of ferric iron reductase. Most of the ferric iron reductase is localized in the cytoplasm. The periplasmic fraction did not show any ferric iron reductase activity, although Paoletti and Blakemore [13] reported that the iron reductase activity in the cell-free extracts prepared from *M. magnetotacticum* was localized in the periplasmic space. On the other hand, the membranes retained about 30 % of the total activity detected in the cell-free extract. However, ferric iron reductase activity was not found in the membranes that had been washed with 0.3 M NaCl. Therefore, it seems likely that although the ferric iron reductase of *M. magneto-tacticum* is a soluble enzyme, it may be loosely bound to the cytoplasmic face of the cytoplasmic membrane.

To purify the ferric iron reductase, the cells were broken with two passes through a French pressure cell at 1000 kgf/cm^2 and centrifuged at 10 000 g for 15 min. The

Table 7.1 Localization of ferric iron reductase in *M. magnetotacticum*

Fraction	Ferric iron reductase activity (nmol/min)	Nitrite reductase activity (μmol/min)	Malate dehydro-genase activity (μmol/min)
Periplasm	0	0.43	1.9
Cytoplasm	8.3	0	16.1
Membrane	3.6	N.D.**	N.D.**
Membrane*	0	N.D.**	N.D.**

The periplasm, cytoplasm, membrane and washed membrane were prepared by the method of [14]. The enzymatic activities were measured by the assay described by Noguchi *et al.* [8]
* Membranes were washed with 10 mM Tris-HCl buffer (pH 8.0), containing 0.3 M NaCl and centrifuged at 104 000 g for 1 h. The precipitate obtained was used as the washed membrane fraction.
** N.D., not determined.

supernatant was centrifuged at 104 000 g for 1.5 h and the resulting supernatant was fractionated with ammonium sulfate between 50 % and 65 % saturation. The precipitate was suspended in 100 mM Tris-HCl buffer (pH 8.0) containing ammonium sulfate (30 % saturation) and the suspension was applied to a Butyl-Toyopearl column. The enzyme was eluted with a linear gradient of 30–10 % saturation of ammonium sulfate in 100 mM Tris-HCl buffer (pH 8.0) containing protease inhibitors. The active fractions were saturated with solid ammonium sulfate to 50 % and applied to a Sepharose CL-6B that had been equilibrated with 50 % ammonium sulfate saturated buffer, and eluted with a linear gradient of 50–25 % saturated ammonium sulfate buffer. The fractions showing ferric iron reductase activity were dialyzed against 50 mM sodium phosphate buffer (pH 7.0) containing 300 mM NaCl. The concentrated ferric iron reductase was further applied to the HPLC and purified to an electrophoretically homogenous state. About 0.5 mg enzyme was purified from 40 g (wet weight) cells.

7.2.2 Characterization of *M. magnetotacticum* Ferric Iron Reductase

M. magnetotacticum ferric iron reductase is composed of a single subunit with molecular mass of 36 kDa; the absorption spectrum of the purified enzyme shows an absorption peak at 280 nm, indicating that the enzyme has no prosthetic groups as heme and flavin. However, it should be noted that the enzyme requires essentially FMN as electron mediator from NADH to ferric iron. The activity in the presence of FAD is about 18 % of that in the presence of FMN and almost the same as that in the absence of the substrate, while other bacterial ferric iron reductases have been reported to be able to use FAD as an electron mediator. The K_m values for FMN, NADH and ferric citrate are 0.035, 1.3 and 15.5 μM, respectively. The V_{max} is about 0.87 s^{-1}.

The effects of $ZnSO_4$, $CaCl_2$, $MgSO_4$ and $MnCl_2$ on ferric iron reductase activity were examined. Zn^{2+} strongly inhibited the activity of the ferric iron reductase. The

Table 7.2 Ferric iron reductase activity of the soluble fraction and the magnetosome numbers in a cell cultivated at various concentrations of Fe(III)-quinate

Fe(III)-quinate in the medium (μM)	Magnetosome numbers*	Iron reductase activity** (nmol Fe(II)-ferrozine formed/min/mg)
0	4.2	4.3
1	8.9	6.9
2.5	12	5.7
5	14	11.7
20	15	10.5

*Average magnetosome numbers were determined by counting electron-dense particles in the micrographs from a total of about 100 cells.
** The enzymatic activity was determined by the method described by Noguchi *et al.* [8].

K_i values for Zn^{2+} were approximately 19.2 μM and approximately 23.9 μM with respect to NADH and FMN, respectively. Other divalent cations had no effects on the enzymatic activity.

7.2.3 Function of Ferric Iron Reductase in *M. magnetotacticum*

Ferric iron reductases have been found in several other bacteria and are thought to be involved in many intracellular iron metabolisms. The ferric iron reductase of *M. magnetotacticum* has similar enzymatic and biochemical properties to those of *Rhodopseudomonas sphaeroides* [15] and *Azotobacter vinelandii* [16], suggesting that the enzyme has the same functions *in vivo*. However, *M. magnetotacticum* requires many more ferrous irons to synthesize magnetite than other heterotrophic bacteria. In fact, the magnetic cells of *M. magnetotacticum* contain 100 times as much iron as the heterotrophic bacteria [1]. Table 7.2 summarizes the effects of extracellular iron on ferric iron reductase activity of the soluble fraction and magnetite synthesis. The average number of particles per cell and ferric iron reductase activity of the soluble fraction decrease in parallel with the concentrations of ferric quinate in the medium. These results suggest that the iron reductase of *M. magnetotacticum* may participate in magnetite synthesis *in vivo*.

To demonstrate the participation of ferric iron reductase in magnetite synthesis, the effects of Zn^{2+} in the medium on magnetite synthesis were investigated by comparing the ferric iron reductase activity of the soluble fractions and the average numbers of magnetosomes in the cells, which were cultivated in the presence of various concentration of $ZnSO_4$. As shown in Fig. 7.2, the bacterial growth was not affected by $ZnSO_4$, while the ferric iron reductase activity and the average number of magnetosomes decreased in parallel with the concentration of $ZnSO_4$ in the medium. Furthermore, the non-magnetic cells increased in parallel with the concentration of $ZnSO_4$ in the medium. In the cells grown in the presence of 75 μM $ZnSO_4$, about 48 % of cells did not have magnetic particles. These results strongly

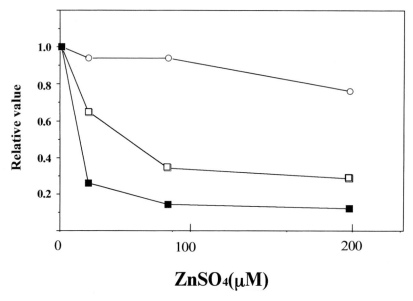

Figure 7.2 Effects of $ZnSO_4$ on growth, magnetosome numbers per cell and ferric iron reductase activity. *M. magnetotacticum* was cultivated in the medium supplemented with various concentrations of $ZnSO_4$. The growth ($\bigcirc$) was determined by measuring the absorbance at 600 nm. The magnetosome numbers ($\square$) were obtained by counting electron-dense particles in micrographs from a total of about 50 cells in each sample. The ferric iron reductase activity in the presence of Zn^{2+} ($\blacksquare$) was determined by the method of [8].

suggest that the ferric iron reductase plays a role in supplying ferrous iron to magnetite *in vivo*.

7.3 Ferrous Iron Oxidation in *M. magnetotacticum*

Blakemore *et al.* investigated the optimal growth conditions for magnetite synthesis by *M. magnetotacticum* and found that the bacterium produces much magnetite under microaerobic denitrifying conditions [17, 18]. The bacterium scarcely synthesizes magnetites under aerobic conditions and, furthermore, does not grow using nitrate respiration (denitrification) under strictly anaerobic conditions. Therefore, the bacterium can obtain the energy for life processes and synthesize magnetite by respiring with nitrate and oxygen as terminal electron acceptors simultaneously. Recently, Tamegai and Fukumori [19] have purified the *ccb*-type cytochrome *c* oxidase from *M. magnetotacticum*. The enzyme is very similar to those found in some microaerobic bacteria such as *Bradyrhizobium japonicum* [20] and *Rhodobacter sphaeroides* [21] and constitutively synthesized in both magnetic cells and

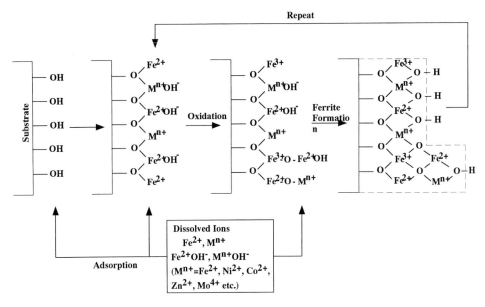

Figure 7.3 Principle of ferrite plating. A substrate with OH groups on its surface is immersed in a reaction solution containing Fe(II) and other metal ions. These ions are then adsorbed onto the surface mediated by the OH groups. When an oxidizing reagent such as $NaNO_2$ is introduced, some of the Fe(II) ions are oxidized to Fe(III), which causes a ferrite formation reaction [24].

non-magnetic cells [19]. Therefore, *M. magnetotacticum* appears to produce ATP with microaerobic respiration using the *ccb*-type cytochrome *c* oxidase but not denitrification.

What is the function of denitrification in *M. magnetotacticum*? Denitrification involves four distinct enzyme systems – nitrate reductase, nitrite reductase, nitiric oxide reductase and nitrous oxide reductase – which catalyze the reaction, $NO_3 \Rightarrow NO_2 \Rightarrow NO \Rightarrow N_2O \Rightarrow N_2$ [22]. Abe and Tamaura [23, 24] have reported that nitrite can be used as an effective oxidizing reagent for the chemical synthesis of spinel-type ferrites in aqueous solution. Figure 7.3 shows the principles of magnetite synthesis by the ferrite-plating method developed by Egusa *et al.* [24]. A substrate with OH groups on its surface is immersed in a reaction solution containing ferrous ions. The ferrous ions are then adsorbed on the surface mediated by the OH groups. When an oxidizing reagent such as sodium nitrite is introduced at 70 °C, some of the ferrous ions are oxidized to ferric ions, which causes a magnetite formation reaction. In this model, nitrite functions as an oxidizing reagent for the chemical synthesis of magnetite, suggesting that a similar phenomenon, i.e. the enzymatic oxidation of ferrous iron by nitrite, might be occurring in *M. magnetotacticum*. Recently, we have purified cytochrome cd_1 (nitrite reductase) from *M. magneto-tacticum* and found that the enzyme has high Fe(II)-nitrite oxidoreductase activity [7]. Blakemore *et al.* analyzed spectrophotometrically the cytochrome composition in the soluble fractions and membranes prepared from *M. magnetotacticum* and found that cytochrome cd_1 was highly expressed in the denitrifying cells, which have large quantities of magnetites [25]. In this section, we describe the purification and novel enzymatic properties of cytochrome cd_1 from *M. magnetotacticum* and discuss the function of cytochrome cd_1 in magnetite synthesis.

7.3.1 Purification of *M. magnetotacticum* Cytochrome cd_1

The soluble fraction prepared from *M. magnetotacticum* was applied to a DEAE-Toyopearl column equilibrated with 10 mM Tris-HCl buffer (pH 8.0). The flow-through fraction was dialyzed against 10 mM sodium phosphate buffer (pH 6.5) for 12 h and the desalted solution was applied to a CM-Toyopearl column equilibrated with the same buffer used for dialysis. Cytochrome cd_1 was eluted with a linear gradient of 0–0.2 M NaCl in the same buffer. The fraction containing cytochrome cd_1 was subjected to gel filtration on a Sephacryl S-200 column equilibrated with 10 mM sodium phosphate buffer (pH 6.5) containing 0.2 M NaCl. The fraction containing cytochrome cd_1 was concentrated by ultrafiltration in an Amicon unit and used as the purified enzyme.

7.3.2 Spectral Properties and Molecular Features of *M. magnetotacticum* Cytochrome cd_1

The oxidized form of the purified cytochrome cd_1 had absorption peaks at 643, 409 and 280 nm (Fig. 7.4). After reduction with sodium dithionite, the absorption peaks were observed at 663, 551, 522 and 418 nm. Although these spectral properties resemble those of *P. aeruginosa* cytochrome cd_1, *M. magnetotacticum* cytochrome cd_1 showed some novel spectral features, different from those of other cytochrome cd_1. First, the α-peak attributed to heme *c* of *M. magnetotacticum* cytochrome cd_1 is symmetrical. The α-peak of the enzyme is not split at room temperature. Secondly, a prominent shoulder around 460 nm in the reduced form is not observed in *M. magnetotacticum* cytochrome cd_1. These spectral features suggest that the heme-binding environments of *M. magnetotacticum* cytochrome cd_1 may be different from those of other bacterial cytochrome cd_1.

The molecular mass of cytochrome cd_1 was determined to be about 133 000 on gel filtration. However, on SDS-PAGE, the value was estimated to be about 54 000. These results suggest that the cytochrome cd_1 of *M. magnetotacticum* exists as a dimer.

7.3.3 Enzymatic Properties and Function of *M. magnetotacticum* Cytochrome cd_1

M. magnetotacticum cytochrome cd_1 showed N,N,N′,N′-tetramethyl-p-phenyl-enediamine (TMPD)-nitrite oxidoreductase activity. The V_{max} is 6.3 s^{-1} and the K_m value for nitrite is 1.47 µM, while the V_{max} of *P. aeruginosa* cytochrome cd_1

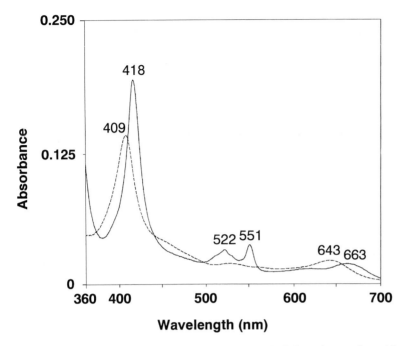

Figure 7.4 Absorption spectra of pure cytochrome cd_1 nitrite reductase from *M. magnetotacticum*. The oxidized form of the purified cytochrome cd_1 had absorption peaks at 643, 409 and 280 nm. After reduction with sodium dithionite, the absorption peaks were observed at 663, 551, 522 and 418 nm. Broken line, oxidized; full line, reduced with $Na_2S_2O_4$.

is about 80 s^{-1}. Cytochrome cd_1 usually utilizes ferrocytochrome c as an electron donor *in vivo*. Although *M. magnetotacticum* has cytochrome c-550, which is homologous to cytochrome c_2 [26], the cytochrome cd_1 could not oxidize ferrocytochrome c-550 in the presence of nitrite. Furthermore, *M. magnetotacticum* cytochrome cd_1 was not equally induced in the magnetic cells and non-magnetic cells, even though both cells were grown under the same culture conditions. These results indicate that although *M. magnetotacticum* cytochrome cd_1 reduces nitrite to NO, the physiological function may not be closely related to the denitrifying respiratory chain.

On the other hand, Fe(II) is chemically oxidized with nitrite at pH 8, as previously reported by Moraghan and Buresh [27]. As shown in Fig. 7.5, Fe(II) is chemically oxidized by nitrite under anaerobic conditions. However, Fe(II) is more rapidly oxidized by nitrite in the presence of *M. magnetotacticum* cytochrome cd_1. *P. aeruginosa* cytochrome cd_1, which showed much higher TMPD-nitrite oxidoreductase activity than *M. magnetotacticum* cytochrome cd_1, did not activate the oxidation of Fe(II).

Cytochromes cd_1 of denitrifying bacteria are considered to be localized in the periplasmic space or loosely bound to the cytoplasmic membrane. Recently, we have found that most *M. magnetotacticum* cytochrome cd_1 is localized in the peri-

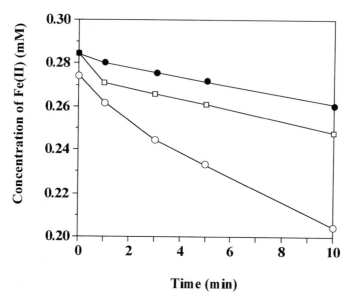

Figure 7.5 Enzymatic and chemical oxidation of Fe(II) in the presence of nitrite. The reaction mixture was assayed in 10 mM Tris-HCl buffer (pH 8.0) containing 0.6 μM cytochrome cd_1, D-glucose oxidase (2.4 U), catalase (60 U), 0.1 M glucose and 0.5 mM $FeSO_4$ in a total volume of 3 ml [7]. After the oxygen was removed by the action of the glucose/glucose oxidase/catalase system, 15 μl of 20 mM $NaNO_2$ was added at time zero to the reaction mixture. The following are indicated: the rate of Fe(II) oxidation in the absence of *M. magnetotacticum* and *P. aeruginosa* cytochrome cd_1 (●); the rate of Fe(II) oxidation in the presence of *M. magnetotacticum* cytochrome cd_1 (○); the rate of Fe(II) oxidation in the presence of *P. aeruginosa* cytochrome cd_1 (□).

plasmic fraction [7]. Furthermore, some of these cytochromes were identified on the cytoplasmic membranes. Therefore, *M. magnetotacticum* cytochrome cd_1 seems to be situated on the periplasmic face of the cytoplasmic membrane. However, magnetosome vesicles are present in the cytoplasm, and the magnetosome membranes do not appear to be contiguous with the cytoplasmic membranes [4]. Therefore, to elucidate the involvement of cytochrome cd_1 in magnetite synthesis *in vivo*, the mechanism of the transport system of Fe(II) to the periplasmic space across the cytoplasmic membrane and the formation of magnetosome vesicles should be studied in future.

7.4 Structure and Function of the 22 kDa Protein Localized in the Magnetosome Membrane

The bacterium possesses a 'magnetosomes chain', consisting of 10–30 magnetosomes. The magnetosome is composed of a single crystal of magnetite (Fe_3O_4)

which occurs in the cell with a diameter of 50–100 nm and enclosed by lipid membranes. Recently, we have found that a 22 kDa protein (MAM22) is localized in the magnetosomes prepared from *M. magnetotacticum* and is a new member of the tetratricopeptide repeat (TPR) protein family [28]. Based on the consensus amino acid sequence pattern, MAM22 contains six tandemly arranged TPR units, although the first, second and last sequence have lower similarity to the consensus sequence pattern (Fig. 7.6). In the case of MAM22, the TPR motifs occupy 90 % of the whole of MAM22.

The TPR motif was first identified as a tandemly repeated degenerate 34 amino acid sequence in the cell division cycle genes [29, 30]. It is now realized that over 25 proteins are present, and organisms as diverse as bacteria and humans contain TPR motifs. In addition to cell cycle regulation, biological processes such as transcription control, mitochondrial and peroxisomal protein transport, neurogenesis, protein kinase inhibition, Rac-mediated activation of NADPH oxidase and protein folding involve TPR motifs [31].

Recently, the crystal structure of the N-terminal TPR domain of a human Ser/Thr protein phosphatase, PP5, was reported [32]. The authors proposed a structural model for the tandemly arranged TPR motifs. According to this structural model, it was proposed that six TPR motifs containing MAM22 form the 6/7 turn of a superhelix. Furthermore, the top and outside of the superhelix consisted of negatively-charged residues, the bottom of the superhelix consisted of positively-charged residues and helix B of TPR1′ exposed to the outside and inside of the superhelix consisted of hydrophobic residues. Therefore, the inside of the superhelix interacts with a helix of other magnetosome proteins and MAM22 is localized on the magnetosome surface. Furthermore, the opposite charges of the top and bottom of MAM22 and/or hydrophobic residues of TPR1′ cause the interaction of MAM22 themselves.

The finding of a protein with TPR motifs in magnetosomes would provide important clues concerning the mechanism of magnetite synthesis and/or maintenance of the chain structure of magnetosomes in magnetotactic bacteria, although the elucidation of the mechanism would require detailed information about the structures of MAM22 and other magnetosome proteins such as MAM12 and MAM28. It should be studied in future, combining the construction of an MAM22-deficient mutant.

7.5 Proposed Mechanism of Magnetite Synthesis in *M. magnetotacticum*

Figure 7.7 shows a proposed mechanism for the enzymatic synthesis of magnetite in *M. magnetotacticum*. First, the bacterium takes up ferric quinate with a very poorly characterized system and transports it across the membrane into the cytoplasm. Once in the cell, Fe(III) is reduced by iron reductase, using NADH as the reducing power in the presence of FMN in the cytoplasm. The iron reductase of *M. magne-*

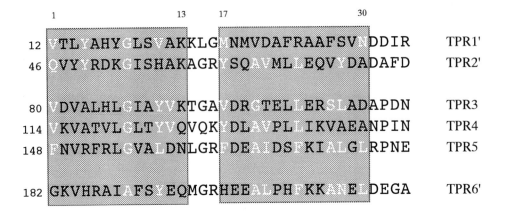

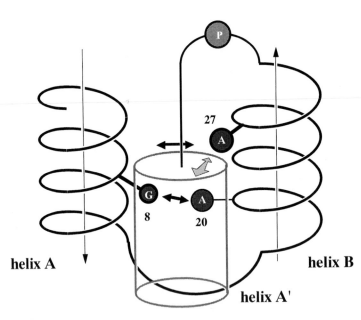

Figure 7.6 (a) Consensus sequence pattern of TPR motif and six amino acid sequences of the TPR motif found in MAM22. The highly conserved large and small hydrophobic residues positioned among the TPR proteins are emphasized by white letters on the shaded background. The residues in the shaded box construct α-helix A and B, respectively. The number on the left of each sequence is the position of the first residue in the MAM22 sequence (GenBank accession No. D82942). (b) The structural model of the TPR motif. Residues 8 and 20 are located at the position of closest contact between the A and B α-helices of a TPR, whereas residue 27 on helix B is located at the interface of three helices [A, B and A′ (helix A of next TPR motif)] within a 3-helix bundle.

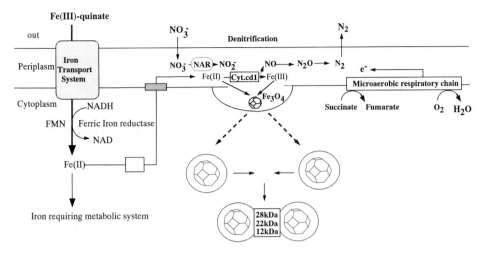

Figure 7.7 Proposed mechanism of magnetite synthesis and magnetosome formation in *M. magnetotacticum*. See the text for discussion.

totacticum appears to be an assimilatory enzyme, which has similar properties to those found in *R. sphaeroides* [15] and *A. vinelandii* [16]. The iron reductase is strongly inhibited by Zn^{2+} and the amounts of magnetite in the cell decrease proportionally to increasing concentrations of Zn^{2+} in the culture medium. Although the mechanisms of these inhibitory effects are still not clear, the iron reductase seems to be participating in magnetite synthesis in *M. magnetotacticum*.

M. magnetotacticum has a microaerobic respiratory chain [19]. Oxygen is essential for growth and is rapidly reduced to H_2O by cytochrome *c* oxidase [19]. It is therefore possible that most of the $Fe(II)$ may be present as free ions at neutral pH even under microaerobic conditions. However, it is not clear whether or not $Fe(II)$ is stored as intracellular Fe-binding components.

In 1983, Abe *et al.* invented the ferrite plating technique [23]. The method enables crystalline ferrite films to be grown from an aqueous solution at $100\,°C$. The formation of Fe_3O_4 from an $Fe(II)$ aqueous solution has two important steps: (i) oxidation of $Fe(II)$ to $Fe(III)$, forming γ-FeOOH; (ii) adsorption of $Fe(II)$ onto γ-FeOOH, which accompanies the spinel formation. Therefore, $Fe(II)$ oxidation plays an essential role in the ferrite plating, for which $NaNO_2$ has been found to be the most expedient oxidizing reagent. As described in Section 7.3, the respiratory enzyme in denitrification, cytochrome cd_1, shows high $Fe(II)$-nitrite oxidoreductase activity. Furthermore, the enzyme is weakly bound to the cytoplasmic membrane in the periplasmic space. These results suggest that $Fe(II)$ oxidation may occur on the cytoplasmic membrane surface, which might mimic the plate.

On the basis of HRTEM and Mossbauer spectroscopy, many researchers have reported that the magnetite produced by *M. magnetotacticum* is a well ordered single-domain octahedral crystal [2, 33]. Therefore, the nucleation of magnetite occurs at one primary nucleation site, suggesting the possibility that protein molecules, active in nucleation, are spatially organized at a unique site in the magneto-some membrane or cytoplasmic membrane. Furthermore, the nucleating proteins seem to have high affinity for Fe(II)/Fe(III). Recently, we have found that the 38 kDa protein in the cytoplasmic membrane has high affinity for ferric ions (unpublished data). The N-terminal amino acid sequence is homologous to those of ADP-L-glycero-D-manno-heptose-6-epimerase, which has the fingerprint sequence Gly-X1-Gly-X2-X3-Gly. The sequence is characteristic of the ADP-binding babefold in FAD-binding and NAD-binding proteins [34]. Therefore, although the 38 kDa protein has the same ADP-binding motif at the N-terminus as ADP-L-glycero-D-manno-heptose-6-epimerase, the function may be different to that of the *E. coli* enzyme. The molecular mechanism of single magnetite crystal formation at room temperature will be a problem of central interest in *M. magnetotacticum*. Genetic systems in the bacterium have only recently begun, and little is known at the enzyme level. Improvement of genetic systems will be critical in learning about the regulatory mechanism of magnetite crystallization.

References

[1] R. P. Blakemore, D. Maratea, R. S. Wolfe, *J. Bacteriol.* 1979, *140*, 720–729.
[2] R. B. Frankel, G. C. Papaefthymiou, R. P. Blakemore, W. O'Brien, *Biochim. Biophys. Acta* 1983, *763*, 147–159.
[3] R. P. Blakemore, *Annu. Rev. Microbiol.* 1982, *36*, 217–238.
[4] Y. A. Gorby, T. J. Beveridge, R. P. Blakemore, *J. Bacteriol.* 1988, *170*, 834–841.
[5] S. Mann, N. H. C. Sparks, S. B. Couling, M. C. Larcombe, R. B. Frankel, *J. Chem. Soc., Faraday Trans.* 1989, *85*, 3033–3044.
[6] Y. Tamaura, K. Ito, T. Katsura, *J. Chem. Soc., Dalton Trans.* 1983, 189–194.
[7] T. Yamazaki, H. Oyanagi, T. Fujiwara, Y. Fukumori, *Eur. J. Biochem.* 1995, *233*, 665–671.
[8] Y. Noguchi, T. Fujiwara, K. Yoshimatsu, Y. Fukumori, *J. Bacteriol.* 1999, *181*, 2142–2147.
[9] L. C. Paoletti, R. P. Blakemore, *J. Bacteriol.* 1986, *167*, 73–76.
[10] F. Halle, J.-M. Meyer, *Eur. J. Biochem.* 1992, *209*, 621–627.
[11] C. Nakamura, T. Sakaguchi, S. Kudo, J. G. Burgess, K. Sode, T. Matsunaga, *Appl. Biochem. Biotechnol.* 1993, *39/40*, 169–176.
[12] D. Schüler, E. Baeuerlein, *J. Bacteriol.* 1998, *180*, 159–162.
[13] L. C. Paoletti, R. P. Blakemore, *Curr. Microbiol.* 1988, *17*, 339–342.
[14] P. R. Alefounder, S. J. Ferguson, *Biochem. J.* 1980, *192*, 231–240.
[15] M. D. Moody, H. A. Dailey, *J. Bacteriol.* 1985, *163*, 1120–1125.
[16] M. Huyer, W. J. Page, *J. Bacteriol.* 1989, *171*, 4031–4037.
[17] J. C. Escalante-Semerena, R. P. Blakemore, R. S. Wolfe, *Appl. Environ. Microbiol.* 1980, *40*, 429–430.
[18] R. P. Blakemore, K. A. Short, D. A. Bazylinski, C. Rosenblatt, R. B. Frankel, *Geomicrobiol. J.* 1985, *4*, 53–71.
[19] H. Tamegai, Y. Fukumori, *FEBS Lett.* 1994, *347*, 22–26.
[20] O. Preisig, R. Zufferey, L. Thony-Meyer, C. A. Appleby, H. Hennecke, *J. Bacteriol.* 1996, *178*, 1532–1538.
[21] J. A. Garcia-Horsman, E. Berry, J. P. Shapleigh, J. O. Alben, R. B. Gennis, *Biochemistry* 1994, *33*, 3113–3119.
[22] W. G. Zumft, *Microbiol. Mol. Biol. Rev.* 1997, *61*, 533–616.
[23] M. Abe, Y. Tamaura, *Jpn. J. Appl. Phys.* 1983, *22*, L511.

[24] K. Egusa, K. Marugame, M. Abe, T. Itoh, in *Proceedings of the 6th International Conference on Ferrites (ICF-6)* (Eds T. Yamaguchi and M. Abe), The Japan Society of Powder and Powder Metallurgy, Tokyo-Kyoto, 1992, pp. 11–14.
[25] W. O'Brien, L. C. Paoletti, R. B. Blakemore, *Curr. Microbiol.* 1987, *15*, 121–127.
[26] K. Yoshimatsu, T. Fujiwara, Y. Fukumori, *Arch. Microbiol.* 1995, *163*, 400–406.
[27] J. T. Moraghan, R. J. Buresh, *Soil Sci. Soc. Am. J.* 1977, *41*, 47–50.
[28] Y. Okuda, K. Denda, Y. Fukumori, *Gene* 1996, *171*, 99–102.
[29] R. S. Sikorski, M. S. Boguski, M. Goebl, P. Hieter, *Cell* 1990, *60*, 307–317.
[30] T. Hirano, N. Kinoshita, K. Morikawa, M. Yamagata, *Cell* 1990, *60*, 319–328.
[31] G. L. Blatch, M. Lassle, *Bioessays* 1999, *21*, 932–939.
[32] A. K. Das, P. T. W. Cohen, D. Barford, *EMBO J.* 1998, *17*, 1192–1199.
[33] S. Mann, R. B. Frankel, R. P. Blakemore, *Nature* 1984, *310*, 405–407.
[34] R. K. Wierenga, P. Terpstra, W. G. Hol, *J. Mol. Biol.* 1986, *187*, 101–107.

8 Characterization of the Magnetosome Membrane in *Magnetospirillum gryphiswaldense*

Dirk Schüler

8.1 Introduction

Despite the considerable diversity of biomineralization processes, there are several fundamental principles governing the formation of inorganic materials in organisms [1]. One of these principles is the spatial control of nucleation and growth of inorganic crystals. In many organisms, this is achieved by the intracellular compartmentalization of biomineralization processes [2]. Compartmentalization in intracellular vesicles allows the biomineralization process to be strictly controlled through the chemical and spatial partitioning of the mineralization environment. One of the best studied examples of intracellular mineral formation is the iron storage protein ferritin. Here, a protein shell forms a hollow sphere of defined size and shape, in the cavity of which iron is stored in the compact form of the iron mineral ferryhydrite [3]. The protein shell acquires iron(II) through hydrophilic channels, catalyzes its oxidation at ferroxidase centers, and induces mineralization within its cavity by specific nucleation sites [4]. Thus, the several functions leading to iron mineralization within the ferritin shell are accomplished by a single protein.

In magnetotactic bacteria (MTB), the biomineralization of magnetosomes is also linked to the accumulation of substantial amounts of iron and the intracellular formation of an iron mineral within a special intracytoplasmic compartment [5]. However, unlike the iron core in ferritin, the inorganic phase of magnetosomes consists of well ordered, highly crystalline particles of a magnetic iron mineral, which in most MTB is magnetite (Fe_3O_4). Moreover, compared to the about 4500 iron atoms housed within the 8 nm cavity of the protein shell in ferritin, the formation of magnetosomes, which are 40–120 nm in diameter, requires the uptake, transport and mineralization of iron on a different scale. Therefore, while following similar strategies in the spatial control of the physicochemical conditions within a particular compartment, the synthesis of magnetite in MTB obviously requires a more complex structure – this is provided by the magnetosome membrane (MM).

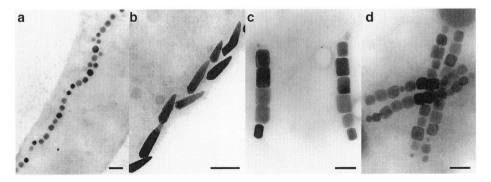

Figure 8.1 Crystal shapes and intracellular organization of magnetosomes found in various magnetotactic bacteria. Shapes include cubo-octahedral (a), bullet-shaped (b) and elongated-prismatic (c, d) morphologies. The particles are arranged within the cell in one (a, b) or multiple (c, d) chains. Bar = 0.1 μm.

8.2 The Magnetosome Membrane Is an Unique Structure in Magnetotactic Bacteria

Most magnetotactic bacteria studied so far are able to synthesize intracytoplasmic crystals of the ferromagnetic iron mineral magnetite (Fe_3O_4). The magnetite particles are characterized by narrow size distributions and uniform, species-specific crystal habits (Fig. 8.1).

Balkwill *et al.* [6] examined the ultrastructure of the magnetic spirillum *Magnetospirillum* (formerly *Aquaspirillum*) *magnetotacticum* strain MS-1. In transmission electron micrographs, each particle appeared to be surrounded by an electron-dense layer, separated from the particle surface by an electron-transparent region. The term "magnetosomes" was proposed for the inorganic particles together with their enveloping layer [6].

In another study, isolated magnetosomes from the same organism were analyzed [7]. The magnetite particles of *M. magnetotacticum* were each enclosed by a biological membrane consisting of a lipid bilayer incorporating proteins. The overall composition of the magnetosome membrane did not appear significantly different from other cell membranes. SDS-gel electrophoresis revealed a complex protein pattern of the magnetosome membrane, sharing numerous proteins bands in various quantities with either the outer or inner membrane. Two proteins of 15 kDa and 33 kDa appeared restricted to the magnetosome membrane in this organism. Neutral lipids, free fatty acids, glycolipids, sulfolipids and phospholipids were all detected in the MM.

A more distinctive protein pattern of the magnetosome membrane of the same organism (*M. magnetotacticum*) was detected by gel electrophoresis, following a slightly modified protocol for the isolation of the magnetosomes [8]. Although several proteins were shared between the MM and cell membranes, three proteins of 12, 22 and 28 kDa appeared to be restricted to the MM. Cloning and sequence analysis of a gene encoding the 22 kDa protein (MAM-22) revealed homology to members of the tetratricopeptide repeat protein family [8].

A similar structure of a lipid bilayer membrane associated with proteins was found in magnetosomes from other cultivated magnetic spirilla, including *M. gryphiswaldense* [9, 10] and *Magnetospirillum* spec. strain AMB-1 [11]. Although a detailed biochemical analysis of the MM in magnetotactic bacteria other then Magnetospirilla has not yet been accomplished, ultrastructural studies in uncultivated MTB, including magnetic vibrios, cocci and rods, indicate similar structures surrounding the magnetite crystals [12–14]. Thus, the presence of a magnetosome membrane seems to be a trait common to all MTB investigated so far.

In electron micrographs of magnetic spirilla, the magnetosome particles were mostly found arranged in a chain-like structure adjacent or in close proximity to the cytoplasmic membrane (CM) [6, 9]. However, connections between the MM and the CM have never been observed and the MM does not appear to be continuous with the cell membrane. Thus, it is not clear how the magnetosome chain is positioned within the cell, and how the MM vesicles are synthesized during growth. At room temperature, no relative motion of magnetosome particles within *M. magnetotacticum* cells was detectable [15]. Therefore, it seems unlikely that the individual magnetosome particles are free to rotate within the cytoplasm. Mechanical anchoring of the magnetosome to the cell envelope would also be required to provide the magnetic torque for aligning the cell in a magnetic field. A membranous "superstructure" may also account for the integrity of the more complex intracellular arrangements of magnetosomes found in some MTB with two or multiple chains of magnetosomes (Fig. 8.1c, d), which would otherwise collapse due to the magnetic energy of the particles. It therefore seems likely that some sort of connection or association with the cell membrane exists at some stage of intracellular development, which would also explain the biosynthetic origin of the MM.

Empty and partially filled vesicles have been observed in iron-starved cells of *M. magnetotacticum* and *M. gryphiswaldense* [7, 9], so the MM probably pre-exists as an "empty" vesicle prior to the synthesis of the mineral phase.

Compartmentalization through the formation of membrane vesicles enables the processes of magnetite mineralization to be regulated by biochemical pathways. The MM is probably the crucial component in the control of crystal growth, thereby providing spatial constraints for the shaping of species-specific crystal morphologies. Magnetite formation requires the presence of mixed valence complexes in solution. Biomineralization of this material therefore requires a precise regulation of both the redox potential and the pH. The growth of magnetite crystals is ultimately regulated by the uptake mechanisms and depends on a controlled flux of iron ions over the MM to provide a supersaturating iron concentration within the vesicle [16]. Although the exact role of the magnetosome-associated proteins has not been elucidated, it has been speculated that these have specific functions in the transport and accumulation of iron, nucleation of crystallization, and redox and pH control [7, 17].

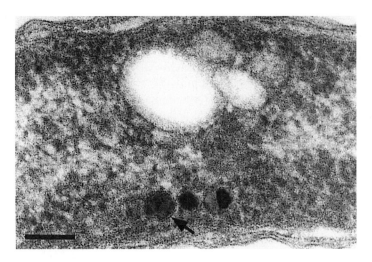

Figure 8.2 Electron micrograph of an ultrathin section of a *M. gryphiswaldense* cell. The arrow indicates the membrane surrounding each electron-dense particle. Bar = 0.1 μm.

8.3 Analysis of the Magnetosome Membrane in *M. gryphiswaldense*

8.3.1 Isolation of Magnetosomes

The freshwater magnetotactic bacterium *M. gryphiswaldense* [18, 19] forms up to 60 cubo-octahedral magnetite crystals that are surrounded by the MM (Fig. 8.2). The strain can be readily cultivated to obtain ample material for biochemical and molecular genetic analysis [20].

Magnetosomes were isolated by a method combining centrifugation and a separation method in a high-gradient magnetic field. After cell disruption using a French press, the cell-free extracts were passed through magnetic separation columns filled with a ferromagnetic matrix. In the presence of a strong magnetic field generated by a Sm–Co magnet, magnetic particles were bound to the matrix. After removal from the magnet, magnetic particles were eluted from the column. Residual contaminating material was removed by loading the magnetosome suspension on top of a sucrose cushion and ultracentrifugation. The purified magnetosome particles formed a pellet at the bottom of the centrifugation tube whereas any contaminating material was retained in the supernatant. Transmission electron microscopy indicated that individual magnetite crystals were enveloped by an intact membrane and were apparently free of contaminating material (Fig. 8.3). About 5 mg of highly purified magnetite were obtained from 1 g magnetic cells (wet weight) by this technique.

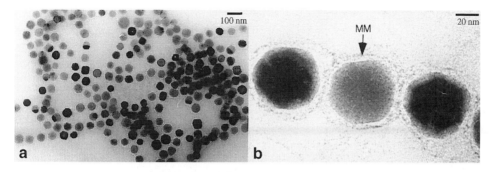

Figure 8.3 Transmission electron micrographs of purified magnetosomes from *M. gryphiswaldense*. (a) Isolated magnetosomes form stable suspensions due to the presence of the enveloping membrane, which prevents agglomeration. Bar = 100 nm (b) High magnification of magnetite crystals enveloped by a membrane (MM) of 8–12 nm thickness. Bar = 20 nm.

8.3.2 Biochemical Analysis

Due to the presence of the magnetosome membrane, isolated intact magnetosomes formed stable, well-dispersed suspensions with a ferrofluid-like behavior, while solubilization of the membrane by treatment with detergents resulted in an immediate agglomeration of the remaining magnetite cores. Besides a specific phospholipid composition [21], the analysis of the SDS-extracted MM from *M. gryphiswaldense* revealed that the MM is associated with specific proteins. By SDS-polyacrylamide gel electrophoresis, at least nine MM-specific polypeptides in various quantities were identified (Fig. 8.4), whereas only a few faint bands were shared by the MM and non-magnetic cellular fractions. In contrast, Gorby *et al.* [7] and Okuda *et al.* [8] could detect only two and three MM-specific polypeptides in *M. magnetotacticum*, respectively, and numerous bands were shared by the magnetic and non-magnetic fractions. Thus, either the composition of the MM is more distinctive in *M. gryphiswaldense*, or the different technique used for the isolation of magnetosomes from this organism yields preparations which are less contaminated. Antisera raised against peptide fragments of the 15 kDa and 24 kDa MM-polypeptide recognized their respective antigens in Western blots of MM preparations and crude extracts, but not in the soluble fraction and in the cytoplasmic and outer membrane [22], indicating a specific localization of these polypeptides in the MM.

8.3.3 Cloning and Sequence Analysis of Genes Encoding Magnetosome Proteins

N-terminal amino acid sequences were determined from two MM-specific polypeptides and used for cloning and subsequent sequence analysis of their respective genes from a genomic library of *M. gryphiswaldense* [23].

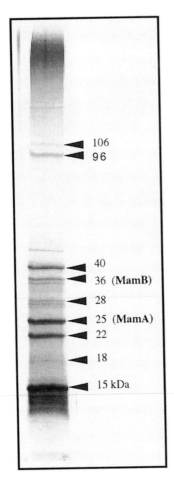

Figure 8.4 Polyacrylamide-gel electrophoresis of the magnetosome membrane (MM)-associated proteins from *M. gryphiswaldense*. The bands were visualized by staining with Comassie blue. Nine major MM-specific proteins were identified (arrows) and assigned MM-15 to MM-106, based on their apparent molecular weights.

8.3.3.1 MamA

An open reading frame that was assigned *mamA* encodes the second most abundant MM protein (MM-25). It is composed of 217 amino acids (aa) with a calculated molecular mass of 24 kDa, which is consistent with an apparent mass of 25 kDa observed in gel electrophoresis. The amino acid sequence of *mamA* was found to be 97% similar to an MM-associated protein from the closely related *M. magneto-tacticum* [8]. Although the overall protein pattern of the MM from different *Magnetospirillum* species appears to be different, the previous analysis of isolated magnetosomes by SDS-PAGE and Western blots suggested that this protein is common to the MM in several different *Magnetospirillum* strains [22]. In addition, *mamA* has significant sequence homology with a number of proteins belonging to the tetra-tricopeptide repeat (TPR) family and contains four repeats of a degenerate 34-aa consensus sequence of the TPR motif. TPR proteins perform a variety of functions as diverse as transcriptional repression, signal transduction, stress response, mitochondrial and peroxisomal protein transport, protein secretion, DNA replication and cell division [24]. The repeated tetratricopeptide motif is thought to be involved in protein–protein interactions. Accordingly, a functional role as a receptor and in the interaction with cytoplasmic proteins was suggested for the homologous MAM22-protein from *M. magnetotacticum* [8]. Unlike typical membrane proteins, *mamA* is lacking the characteristic membrane spanning segments, and the overall hydrophobicity is relatively low, suggesting that *mamA* may be electrostatically bound to the cytoplasmic surface of the MM.

8.3.3.2 MamB

MamB was identified as the gene for another principal MM polypeptide (MM-36), encoding a protein of 297 amino acids with a calculated molecular mass of 31 kDa. It exhibits significant sequence homology to members of the ubiquitous cation-diffusion facilitator (CDF) family, which are exclusively involved in the transport of several different heavy metals and in some cases known to confer heavy metal resistance by export or vesicular sequestration [25]. Based on sequence analysis, *mamB* exhibits the characteristic topology of bacterial CDF members and possesses the family-specific signature sequence [26]. The significant similarity to proteins specifically involved in metal transport and the specific localization in the MM imply that *mamB* may be involved in iron transport into the magnetosome vesicles.

8.3.4 Arrangement of Magnetosome Genes

MamB is located about 2 kb downstream of *mamA*, indicating a clustering of several MM-related genes in the chromosome of *M. gryphiswaldense*. Sequence analysis revealed that *mamA* and *mamB* are separated by a DNA region containing two complete open reading frames (ORFs) encoding two putative proteins of 270 and 72 aa-residues, respectively (Fig. 8.5). The aa-sequences of the ORF-derived proteins are not similar to any sequence from databases. All identified ORFs including *mamA* and *mamB* are arranged in a collinear manner, closely followed by one another and spaced by only a few nucleotides. This organization indicates that *mamA* and *mamB* may be parts of a larger operon, which in bacteria is often characteristic of genes of related function. Thus, the identified ORFs between *mamA* and *mamB*, as well as the two flanking truncated ORFs, may code for proteins of so far unknown functions which are putatively involved in magnetite biomineralization.

8.4 Conclusions

The structural, biochemical and molecular genetic analysis of the magnetosome membrane is key to understanding bacterial magnetite biomineralization at a molecular level. Sequence analysis of the structural genes and their genetic character-

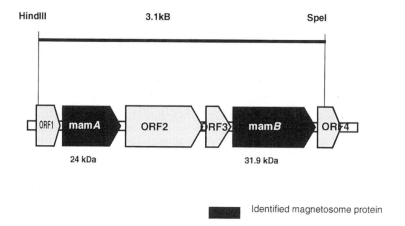

Figure 8.5 Molecular organization of ORFs on a chromosomal *HindIII/SpeI* fragment from *M. gryphiswaldense*. Arrows indicate the direction of gene transcription. Two truncated (ORF1, ORF4) and four complete ORFs were detected on the 3.1 kB fragment. Two ORFs (*mamA*, *mamB*) were identified as the genes for two major MM proteins (MM-25, MM-36).

ization will provide insights into the structural and functional basis of magnetite synthesis.

The formation of intracytoplasmic magnetosome vesicles is the result of an intracellular differentiation process that is poorly understood. It is not known, for example, if the MM vesicles originate from the cytoplasmic membrane by invagination, or if they represent an independent compartment that is discontinuous with the CM. More research is required to study the intracellular membrane topography in magnetic and non-magnetic cells during growth. These experiments should be accompanied by analysis of the expression of magnetosome-specific genes at the transcriptional and translational level and the intracellular localization of their gene products during the growth of cells.

Iron-starved cells of *Magnetospirillum* strains were observed to contain empty MM vesicles. The isolation and analysis of magnetite-free, premature MM vesicles from non-magnetic cells may provide substantial information about the differentiation and biosynthesis of the MM. Isolated, functional MM vesicles may also be used in experiments to reconstitute the mineral iron core *in vitro*. In similar studies using isolated ferritin molecules, the native ferrihydrite core could be reconstituted [27]. By careful variation of the reaction conditions, various other minerals including Fe_3O_4 could substitute for ferrihydrite [28].

Finally, the emerging question is how does the MM structure and composition affect the characteristic size, structure and shape of the diversity of magnetite crystals found in MTB different from the cubo-octahedral magnetosome type? To answer this question, increased effort should be put into research involving the great natural diversity of MTB and their magnetosome crystals. The understanding of how different bacteria achieve species-specific control over the biomineralization process could be used to tailor different crystal morphologies with desired properties.

Because of their unique magnetic and crystalline properties, the biomineralization of magnetic nanoparticles is currently attracting growing interest from material scientists and biotechnologists for their potential use in numerous biotechnological applications [29]. Because of their defined composition, MMs can be used for the immobilization of bioactive substances to the magnetite crystals as magnetic carriers [14, 30]. In addition, the presence of the MM stabilizes suspensions of isolated magnetosomes by preventing the agglomeration of nano-scaled particles. A biochemical understanding of the MM may therefore be ultimately useful for potential biotechnological applications of magnetite biomineralization.

Acknowledgements

I would like to acknowledge the continued collaboration and invaluable discussions with E. Baeuerlein, D. A. Bazylinski, R. B. Frankel, M. Hildebrand, S. Spring and B. Tebo.

This work was supported by grants from the Deutsche Forschungsgemeinschaft and the Max-Planck-Gesellschaft.

References

[1] H. A. Lowenstam, *Science* 1981, *211*, 1126–1131.
[2] S. Mann, *Nature* 1993, *365*, 499–505.
[3] A. Treffry, P. M. Harrison, M. I. Cleton, W. deBruijn, S. Mann, *J. Inorg. Biochem.* 1987, *31*, 1–6.
[4] N. D. Chasteen, P. M. Harrison, *J. Struct. Biol.* 1999, *126*, 182–194.
[5] D. Schüler, *J. Mol. Microbiol. Biotechnol.* 1999, *1*, 79–86.
[6] D. Balkwill, D. Maratea, R. P. Blakemore, *J. Bacteriol.* 1980, *141*, 1399–1408.
[7] Y. A. Gorby, T. J. Beveridge, R. P. Blakemore, *J. Bacteriol.* 1988, *170*, 834–841.
[8] Y. Okuda, K. Denda, Y. Fukumori, *Gene* 1996, *171*, 99–102.
[9] D. Schüler, PhD thesis, Technical University, Munich, 1994.
[10] D. Schüler, E. Baeuerlein, *J. Phys. IV*, 1997, *7*, 647–650.
[11] C. Nakamura, J. G. Burgess, K. Sode, T. Matsunaga, *J. Biol. Chem.* 1995, *270*, 28392–28396.
[12] D. A. Bazylinski, A. Garratt-Reed, R. B. Frankel, *Microsc. Res. Tech.* 1994, *27*, 389–401.
[13] H. Vali, J. L. Kirschvink, in *Iron Biominerals* (Eds R. B. Frankel and R. P. Blakemore), Plenum Press, New York and London, 1991, pp. 97–116.
[14] T. Matsunaga, *TIBTECH* 1991, *9*, 91–95.
[15] S. Ofer, I. Novik, E. R. Bauminger, G. C. Papaefthymiou, R. B. Frankel, R. P. Blakemore, *Biophys. J.* 1984, *46*, 57–64.
[16] D. Schüler, E. Baeuerlein, *Arch. Microbiol.* 1996, *166*, 301–307.
[17] S. Mann, N. H. C. Sparks, R. G. Board, *Adv. Microb. Physiol.* 1990, *31*, 125–181.
[18] D. Schüler, M. Köhler, *Zentralbl. Mikrobiol.* 1992, 147, 150–151.

[19] K. H. Schleifer, D. Schüler, S. Spring, M. Weizenegger, R. Amann, W. Ludwig, M. Köhler, *Syst. Appl. Microbiol.* 1991, *14*, 379–385.
[20] D. Schüler, E. Baeuerlein, *J. Bacteriol.* 1998, *180*, 159–162.
[21] M. Gassmann, Diploma thesis, University of Tübingen/München, 1996.
[22] D. Schüler, E. Baeuerlein, D. A. Bazylinski, *Abstracts of the General Meeting of the ASM, Miami, USA*, 1997, p. 332.
[23] D. Schüler, B. Tebo, submitted for publication.
[24] N. C. Kyrpides, C. R. Woese, *TIBS* 1998, *23*, 245–247.
[25] D. H. Nies, S. Silver, *J. Ind. Microbiol.* 1995, *14*, 186–199.
[26] I. T. Paulsen, M. H. Saier, *Membrane Biol.* 1997, *156*, 99–103.
[27] S. Mann, J. M. Williams, A. Treffry, P. M. Harrison, *J. Mol. Biol.* 1987, *198*, 405–416.
[28] F. C. Meldrum, B. R. Heywood, S. Mann, *Science* 1992, *257*, 522–523.
[29] D. Schüler, R. B. Frankel, *Appl. Microbiol. Biotechnol.* 1999, *52*, 464–473.
[30] E. Baeuerlein, D. Schüler, R. Reszka, S. Päuser, Patent PCT/DE 98/00668, 1998.

9 Molecular and Biotechnological Aspects of Bacterial Magnetite

Tadashi Matsunaga, Toshifumi Sakaguchi

9.1 Introduction

Magnetic bacteria synthesize ferromagnetic crystalline particles intracellularly, which consist of magnetite (Fe_3O_4) and occur within a specific size range (50–100 nm). Bacterial magnetic particles (BMPs) can be distinguished from abiologically formed magnetite by their regular morphology and the presence of an organic membrane enveloping the crystals (Fig. 9.1) [1]. In addition, the particle is the smallest magnetic crystal that has a regular morphology within single-domain size. Therefore, BMPs have a enormous potential value in various technological applications and are not just of scientific interest. However, the molecular and genetic mechanisms of magnetite biomineralization are poorly understood, even though iron oxide formation occurs widely in many organisms such as algae [2], chitons [3], honey bees [4, 5], yellowfin tuna [6], sockeye salmon [7, 8], etc. In order to elucidate the molecular and genetic mechanisms of magnetite biomineralization, a magnetic bacterium *Magnetospirillum* sp. AMB-1 [9], for which gene transfer techniques have recently been developed [10], was used as a model organism. Several findings and advanced applications worthy of note were achieved during the 1990s by means of studies on this model organism and another related organism. In this chapter, we describe the current molecular and biotechnological knowledge relating to magnetic bacteria and bacterial magnetic particles (BMPs), obtained from our recent studies of the genetic engineering and analysis of magnetic bacteria, and also the advanced applications of bacterial magnetite.

9.2 Isolation and Cultivation of Magnetic Bacteria

9.2.1 Pure Cultivation of Magnetic Bacteria

Since the achievement of a pure culture of a magnetic bacterium in 1979, *Magnetospirillum (Aquaspirillum) magnetotacticum* strain MS-1 [11], only eight pure cultures of magnetic bacteria have since been achieved. The described strains of magnetic bacteria are an obligate microaerophilic spirillum, *Magnetospirillum*

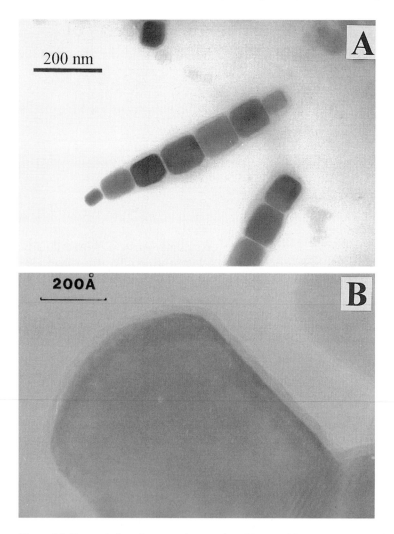

Figure 9.1 Transmission electron micrographs of BMPs. (a) BMPs extracted from magnetic bacteria by a French press, arranged in a chain. Average size of BMPs was 50–100 nm within single-domain region. Bar = 200 nm. (b) Magnified picture of single BMP. A thin organic membrane was observed covering the particle, with a thickness of about 3.5 nm. Bar = 200 Å.

(Aquaspirillum) magnetotacticum strain MS-1 [11], an aerotolerant microaerophilic spirillum *Magnetospirillum gryphiswaldense* strain MSR-1 [12], three facultatively anaerobic vibrios, strains MV-1 [13], MV-2 and MV-4 [14], a microaerophilic coccus, strain MC-1 [15], two facultatively anaerobic spirilla, *Magnetospirillum* sp. strains AMB-1 [9, 16] and MGT-1 [1, 16, 17], and an obligate anaerobe, strain RS-1 [18]. Many morphological types of magnetic bacteria can be found in natural

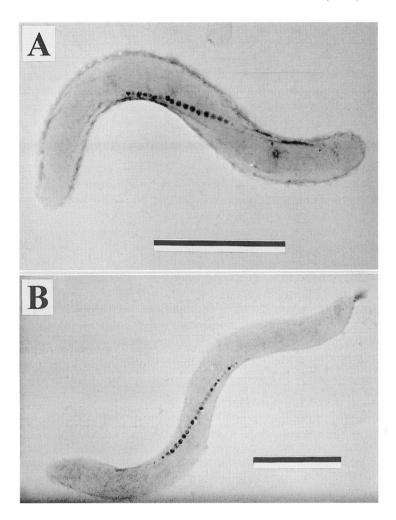

Figure 9.2 Transmission electron micrographs of magnetic bacteria, *Magnetospirillum* spp. strains AMB-1 (a) and MGT-1 (b). Cells were grown anaerobically in the growth medium [9]. Bar = 1.0 μm.

aquatic environments [19, 20] but they cannot be cultured in artificial conditions. Therefore, new methods are necessary to obtain physiological information about non-cultured magnetic bacteria. Recent progress has been made in pure culture. Our isolated magnetic spirilla from freshwater sediment, *Magnetospirillum* sp. strains AMB-1 and MGT-1 (Fig. 9.2) are facultatively anaerobic denitrifiers that can grow well either aerobically or anaerobically by dissimilatory nitrate reduction. Cells of both strains have oxidase activity when they are grown in aerobic conditions. Although they are catalase-negative, they are aerotolerants capable of forming colonies on solid agar medium without the addition of catalase. AMB-1 and MGT-1 can grow in free gaseous exchange with an air atmosphere under shaking at 120 rpm. These characteristics against oxygen in AMB-1and MGT-1 are different from that of an obligate microaerophile, *Magnetospirillum magneto-tacticum* MS-1 [9, 16]. Additionally, a respiratory inhibition assay using potassium cyanide, 2-heptyl-4-hydroxyquinoline N-oxide (HQNO) and dicumarol suggest that iron reduction via quinone is directly related to BMP formation in AMB-1. Growth, BMP formation, and iron reduction of AMB-1 are reduced or inhibited by an electron transferring inhibitor of quinone, dicumarol, in anaerobic conditions [21]. This differs from the iron reduction of MS-1 via cytochrome oxidase, which is affected by the addition of a cytochrome oxidase inhibitor, sodium azide [22]. These results suggest a metabolic differentiation between AMB-1 and MS-1 in the respiratory chains.

AMB-1 and MGT-1 both grow aerobically, and both can form colonies. In addition, the spiral morphology, cell size, ability to produce BMPs and the crystal morphology of synthesized BMPs are similar for both AMB-1 and MGT-1. However, they are regarded as different strains because of their differences in carbon source utilization. AMB-1 can use 12 carboxylic acids, and MGT-1 can use 15 carboxylic acids when they are grown in anaerobic conditions [16]. Our recent study on the evolutionary relationship among *Magnetospirillum* spp. based on 16SrDNA sequence indicates that AMB-1 and MGT-1 can be distinguished phylogenetically as different species in the genus [23].

9.2.2 Mass Cultivation of Magnetic Bacteria

In our initial studies on the application of BMPs [24–26], fine biogenic magnetite was obtained from non-cultured magnetic bacteria that were enriched by the addition of nutrients such as nitrate and succinate into plastic containers containing sediments [26] or which were directly collected from natural sediments by the harvesting apparatus [24]. Today, the achievement of pure cultiures of magnetic bacteria in artificial media has enabled great progress in the application of BMPs. Mass cultivation of magnetic bacteria for BMP production is one of the most important biotechnological processes in the application of BMPs. Most of the magnetic bacteria isolated so far are difficult to grow on a large scale due to their fastidious culture requirements. BMP production from the aerotolerant magnetic bacteria, AMB-1 and MGT-1, has recently been achieved in large-scale batch culture (1000 l using MGT-1 and AMB-1) [17] and by using a fed-batch culturing system (4 l fermenter using AMB-1) [27]. BMPs were purified magnetically after the disruption of harvested cells by ultrasonication or using a French press. In the large-scale batch culture, about 2.6 mg of BMPs was yielded per liter of culture (in total 2.6 g BMP). In the fed-batch culturing system by feeding nitric acid and succinate, 1 l of AMB-1 culture could produce 4.5 mg BMPs from 0.34 g dry cells [27]. Mass cultivation of magnetic bacteria will be an important technique in the future for the production of genetically modified functional BMPs and biomass supply for the elucidation of magnetite biomineralization.

9.2.3 Obligately Anaerobic Magnetic Bacteria

Magnetic bacteria were discovered in 1975 [28], and since then they have been observed in many different habitats such as soil, marine and freshwater sediments, and sulphide-rich habitats [29]. We have recently isolated and purely cultured a sulphate-reducing bacterium, designated RS-1, that can synthesize intracellular magnetite particles [18] (Fig. 9.3). A novel isolation procedure without magnetic collection using magnetotaxis was performed for the pure cultivation of this bacterium [30]. Basic characteristics of RS-1 were investigated. RS-1 is also an obligate anaerobe capable of producing magnetic iron sulphides extracellularly. The intracellular BMPs have a different morphology (irregular bullet-shaped) (Fig. 9.4) to previously cultured magnetic bacteria. This isolate illustrates the wider metabolic diversity of magnetic bacteria and suggests the presence of a novel mechanism of magnetite biomineralization. Phylogenetic analysis of RS-1 based on its 16S rDNA sequence shows it is a new bacterial species ascribing to a member of δ-Proteobacteria [31]. The genetic analysis of RS-1 has now been started, to elucidate the mechanism of magnetite formation.

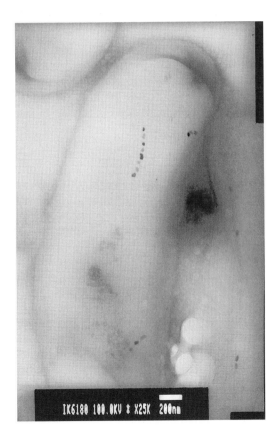

Figure 9.3 Transmission electron micrograph of a sulphate-reducing magnetic bacterium, strain RS-1 (bar = 200 nm).

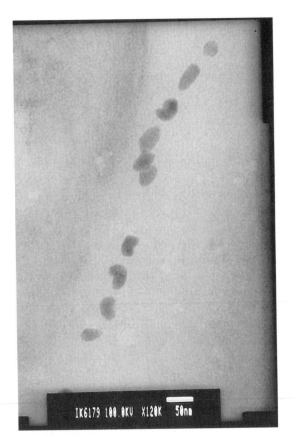

Figure 9.4 Intracellular BMPs in RS-1 cells grown anaerobically under sulphate-reducing conditions, showing irregular (randomly bullet-shaped) morphology. The single cell has ∼ 10–15 BMPs. Bar = 50 nm.

9.3 Genetic Analysis of Magnetic Bacteria

9.3.1 Gene Transfer and Transposon Mutagenesis of Magnetic Bacteria

Gene transfer in the magnetic bacterium *Magnetospirillum* sp. AMB-1 has been achieved for the first time by employing a conjugation method using broad-host-range IncP and IncQ plasmids, pRK415 [32] and pKT230 [33] respectively [10]. In addition, a mobilizable plasmid containing transposon Tn5, pSUP1021 [34], was transferred to AMB-1. Five non-magnetic kanamycin-resistant mutants (NM1, NM2, NM3, NM5 and NM7) of AMB-1 were obtained, in which Tn5 was shown to be integrated into the chromosome (Fig. 9.5). Southern hybridization analysis of *Eco*RI-digested genomic DNA from these mutants with Tn5 as a probe showed the presence of several chromosomal regions involved in synthesis of BMP. Different genomic DNA fragments from the non-magnetic mutants containing the muta-genized regions were cloned into *E. coli*. After determination of the site of Tn5

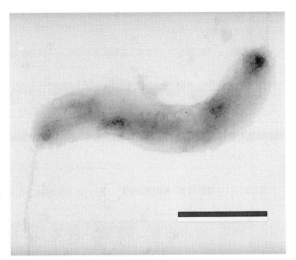

Figure 9.5 Non-magnetic kanamycin-resistant mutant (strain NM 2) of *Magnetospirillum* sp. AMB-1. The mutant cells were grown anaerobically in the presence of 33 μM ferric quinate, in which the wild type cells synthesize intracellular BMPs. This figure shows that the mutant is deficient in BMP production at the genetic level by transposon insertion into the chromosome. Bar = 1.0 μm.

insertion, one of these fragments, isolated from non-magnetic mutant NM5, was a 2.6 kbp *Eco*RI fragment of genomic DNA interrupted by Tn5. The corresponding DNA fragment without Tn5 (NM5 fragment) was isolated from λZAP gene bank of the wild type AMB-1. Physical map of the Tn5 insertion site and open reading frame in the mutagenized region of the strain NM5 was ascertained (Fig. 9.6). Nucleotide sequence and expression analysis of the NM5 fragment were carried out. Three open reading frames including the *magA* gene which had been interrupted by Tn5 were found. Gene products of two open reading frames designated as *magA* and *magB* were inferred from the homology search using protein databanks. The predicated amino acid sequence of the *magA* gene product has high homology with cation transport proteins, in particular KefC, a potassium ion translocating protein in *E. coli* [35] (Fig. 9.7) and NapA, the putative Na^+/H^+ antiporter in

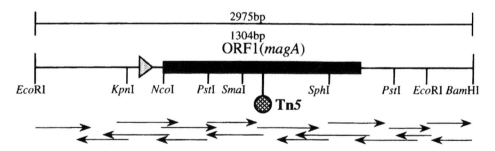

Figure 9.6 Physical map of *Eco*RI–*Bam*HI DNA fragment (∼ 3.0 kb) from the non-magnetic kanamycin-resistant mutant strain NM5, which contains the Tn5 insertion site and open reading frame (*magA*) in the mutagenized region. The ORF (*magA*) is shown by the boxed line. The putative promoter region is indicated by a triangle. Arrows represent sequenced regions and the direction of sequencing.

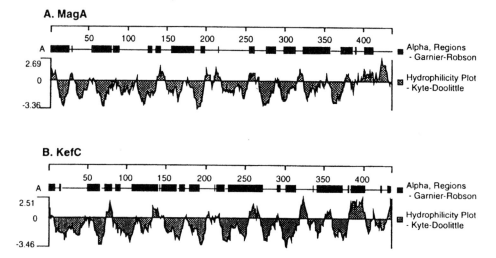

Figure 9.7 Comparison of hydrophilicity plot and alpha region in amino acid sequence between MagA and KefC.

Enterococcus hirae [36]. Additionally, the *magA* protein shows high hydrophobicity. Therefore, the gene product from *magA* was probably located in the membrane fraction and may function as a cation transporter protein in AMB-1 [37]. Meanwhile, *magB* codes for a protein that is highly homologous to the *E. coli* RNase HII protein. The gene product could complement RNase HII mutants of *E. coli* [38].

9.3.2 Function Analysis of *magA* for Molecular Architecture

To investigate the original function of the MagA protein *in vivo*, a 2094 bp DNA fragment from *Nco*I-*Bam*HI digestion containing the *magA* gene was connected to the multi-cloning site of pTrc99A [39] to express the *magA* gene in *E. coli*. This plasmid (pTMG5) was transferred into *E. coli* DH5α, and its transformant was used to prepare inverted membrane vesicles. Iron transport by the MagA protein was verified by the direct measurements of iron in membrane vesicles. When ATP was added to the transformant, iron uptake could be observed (Fig. 9.8). However, when ATP was omitted, iron uptake was not observed. This result indicates that the gene product of *magA* functions as an iron transporter in the cytoplasmic membrane in *E. coli* and that the energy of transport is coupled with ATP hydrolysis. The function of the MagA protein in magnetic bacteria could be predicted by using *E. coli* cells [37]. In addition, the Northern blot analysis of gene expression under iron-limited conditions demonstrates that *magA* is an iron-regulated gene. The transcription of *magA* is found to be enhanced by low iron concentrations in the wild-type AMB-1 cells [37].

To determine the intracellular localization of the MagA protein, a *magA-luc* fusion gene was constructed (Fig. 9.9) [40]. About 1.7 kbp *Bam*HI-*Sac*I DNA

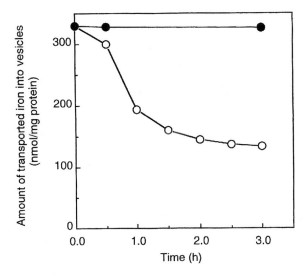

Figure 9.8 Iron accumulation ability of the membrane vesicles isolated from *E. coli* in which the MagA protein was expressed. Membrane vesicles were prepared as described in [36] and assayed in the TMSM buffer [36] at a final protein concentration of 10 mg ml^{-1}, supplemented with 330 μM iron and 5 mM ATP. The curves indicate change in iron concentration due to vesicles from the pTrc99A transformant of *E. coli* DH5α (full circles) and the pTMG5 transformant (open circles).

fragment containing whole *luc* gene was ligated into the *Bam*HI-*Sac*I sites of a gene transfer for AMB-1, pRK415 [32]. This plasmid, pKLC, was digested with *Bam*HI and blunted, and then a 1402 bp *Kpn*I-*Sph*I restriction fragment containing about 90% whole *magA* gene was ligated into the blunted site. The firefly luciferase gene, *luc*, was cloned downstream of the *magA* hydrophilic C terminal domain (pKML). Moreover, for the promoter assay of *magA* gene, a 265 bp *Kpn*I-*Nco*I restriction fragment containing the putative promoter region of *magA* gene but excluding its open reading frame was connected with the *luc* gene in pKLC. Luciferase activity in transconjugants of AMB-1 was investigated for determinations of the localization and the promoter assay. As the results, the luciferase activity was observed in the fraction of BMPs membrane. Approximately 7-fold luciferase activity was detected for the transconjugant with pKML (containing *magA-luc* gene) in the BPM fraction compared with that of pKPL (containing only *luc* gene). MagA-Luc fusion protein was confirmed to be localized in magnetic particle and cell membranes [41]. Moreover, the promoter activity was regulated by iron concentration [41]. This phenomenon supports the result of the Northern blot analysis [37] of the gene expression under iron-limited conditions.

These results indicated that the MagA protein was a membrane-bound protein and that the *magA* gene fusion was applicable to the biotechnological application in making directly modified BMPs. In addition, they showed that direct display of functional enzymes onto BMPs could be achieved using MagA as an anchor protein.

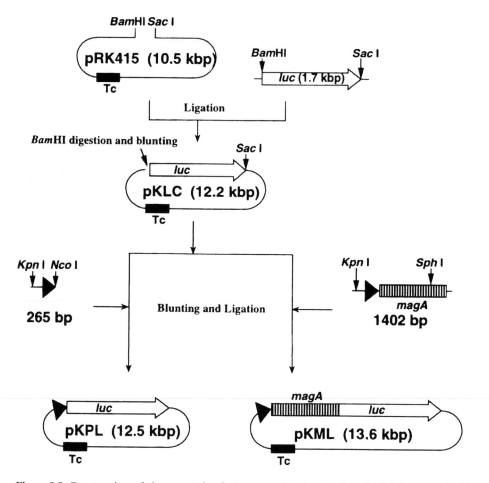

Figure 9.9 Construction of the *magA–luc* fusion gene. Black triangles, shaded boxes and white boxes show the *magA* promoter, the *magA* gene and the *luc* gene respectively.

9.4 Advanced Applications of Bacterial Magnetite

9.4.1 Immunoassay Using BMPs Modified by Chemical Crosslinking Reagents

Ultrafine hexahedral BMP purified by ultrasonication and magnetic separation from AMB-1 has a large surface-to-volume ratio due to its small size; aggregation of BMPs is prevented by the presence of intact membranes enveloping the particles.

Chemical reagent for cross-linking

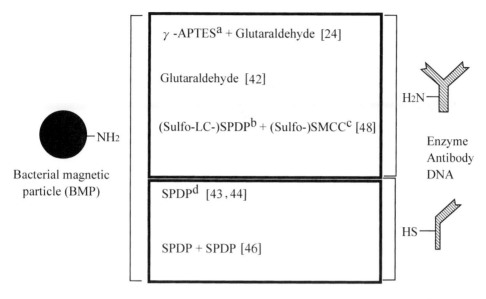

Figure 9.10 Chemical crosslinking reagents used in our studies for immobilization of enzymes, antibodies or DNA onto the BMPs: (a) γ-aminopropyltriethoxysilane; (b) sulfosuccinimidyl 6-[3′-(2-pyridyldithio)propionamido]hexanoate; (c) sulfosuccinimidyl 4-(N-maleimidomethyl)cyclohexane-1-carboxylate; (d) N-hydroxy-succinimidyl 3-(2-pyridyldithio)-propionate.

So far, we have carried out immobilizations of enzyme [24], antibody [42–44] for the sensing devices in biosensors, and DNA for the genetic carrier [45], and magnetic recovery of specific mRNA [46, 47]. In our previous study, crosslinking reagents (e.g. glutaraldehyde, sulfo-LC-SPDP, sulfo-SMCC) had been used to immobilize antibody onto BMPs for various immunoassays [42–4, 48]. Chemical reagents for crosslinking between BMPs and functional biomaterials (enzyme, antibody and DNA), which have been used in our studies, are summarized in Fig. 9.10. In our latest study on BMPs immunoassay using chemical crosslinkers, antibody was immobilized onto BMPs using the heterobifunctional reagents sulfosuccinimidyl 6-[3′-(2-pyridyldithio)propionamido]hexanoate (sulfo-LC-SPDP) and sulfosuccinimidyl 4-(N-maleimidomethyl)cyclohexane-1-carboxylate (sulfo-SMCC) (Fig. 9.11) [48]. A highly sensitive immunoassay has been achieved with these BMPs. A good correlation is obtained between the luminescence intensity and mouse IgG concentration in the range of $1–10^5$ fg/ml (Fig. 9.12). The minimum detectable concentration of IgG is 1 fg/ml (corresponding to 6.7 zmol = 4000 molecules as calculated from Avogadro's number). This high sensitivity is due to the specific immobilization onto the BMPs and its high dispersion in the reaction solution. The sensitivity (the minimum detectable concentration of IgG) was 5×10^8 times higher than that

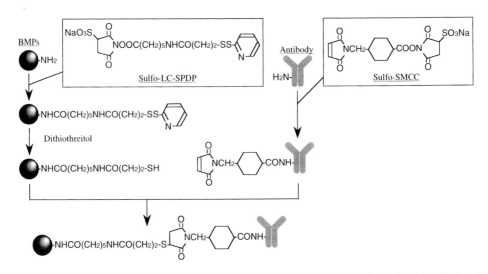

Figure 9.11 Schematic diagram of antibody immobilization onto BMP using sulfo-LC-SPDP and sulfo-SMCC (sulfo-LC-SPDP: sulfosuccinimidyl 6-[3′-(2-pyridyldithio)propionamido]hexanoate; sulfo-SMCC: sulfosuccinimidyl 4-(*N*-maleimidomethyl)cyclohexane-1-carboxylate).

of fluoroimmunoassay using antibody-immobilized BMPs [42]. Furthermore, the amount of BMPs and the reaction time was optimized in the measurement. A rapid chemiluminescence enzyme immunoassay method has been developed based on the sandwich immunoassay using alkaline phosphatase labeled anti-IgG antibody and anti-mouse IgG antibody immobilized BMPs. This assay can be completed within 10 minutes. A linear relationship was obtained between the luminescence and mouse IgG concentration in the range of 10–1000 ng/ml.

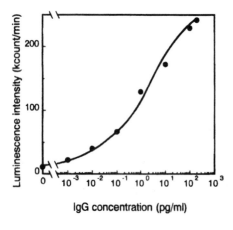

Figure 9.12 Relationship between luminescence intensity and mouse IgG concentration. Alkaline phosphatase labeled anti-IgG antibody and anti-mouse IgG antibody immobilized BMPs were used for the measurement of various mouse IgG concentrations. 50 μg of the BMPs was used in each immunoreaction.

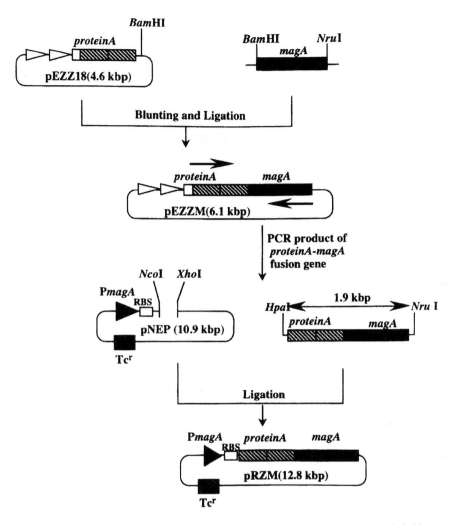

Figure 9.13 Plasmid construction procedure for expressing the *proteinA–magA* hybrid gene.

9.4.2 Immunoassay Using ProteinA Displayed on BMPs

The Z-domain of proteinA in *Staphylococcus aureus* has the ability to tightly bind to the Fc region of an immunoglobulin (IgG) [40]. Recently, we have constructed BMPs that have ProteinA expressed on the surface using *magA* gene fusion (Fig. 9.13). Direct conjugation of antibody (IgG) onto BMPs has been carried out for enzyme immunoassay. A plasmid, pEZZ 18 [40], was used as a *proteinA* gene fusion vector. *Bam*HI–*Nru*I genomic fragment involving the *magA* structural gene was blunted and ligated into the multi-cloning site (3′ end of the *proteinA* gene encoded the ZZ region) in pEZZ 18. The plasmid, pEZZM containing *proteinA–magA*

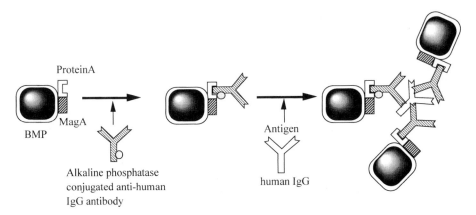

Figure 9.14 Homogenous chemiluminescence enzyme immunoassay using Protein A–BMP complexes. Alkaline phosphatase conjugated anti-human IgG antibody was bound onto the BMPs (produced from pRZM containing AMB-1 cells) based on the specific molecular recognition to the Z-region of ProteinA. These functional BMPs were used for the detection of human IgG at various concentrations. Luminescence intensity was measured in the supernatant after magnetic aggregation, due to the immunoreaction between the functional BMPs and human IgG.

hybrid gene was constructed. The *proteinA–magA* hybrid gene was amplified using two primers (26 mers DNA containing *Hap*I and *Nru*I restriction sites, respectively) based on the upstream and downstream regions of the hybrid gene respectively. The amplified hybrid gene fragment was cloned into pNEP consisting of the *magA* promoter and pRK415 [32]. The constructed plasmid, pNEP, was transferred into AMB-1 by transconjugation [10]. BMPs that were displayed on the ProteinA–MagA fusion protein were produced for immunoassay and purified magnetically after cell disruption with ultrasonication. A homogenous chemiluminescence enzyme immunoassay using anti-human IgG antibody (conjugated alkaline phosphatase) bound ProteinA–BMP complexes has been developed for detection of human IgG (Fig. 9.14) [49]. We showed in this study that the minimum and maximum detectable concentrations of IgG were 1 and 1000 ng/ml respectively [49]. Construction of functional BMPs without chemical crosslinking reagents has been achieved in the conjugation of antibodies.

9.4.3 Fully Automatic Immunoassay Using ProteinA–BMPs

We have also developed a fully automated immunoassay for the determination of human insulin in blood serum using antibody–Protein A–BMP complexes. A ProteinA–BMP complex is obtained from *Magnetospirillum* sp. AMB-1 transformed by a plasmid, pRZM, containing the *ProteinA–magA* fusion gene [49]. The particle size distributions of antibody–ProteinA–BMP complexes are no different before and after the introduction of antibody. On the other hand, the weight percentage of 80–100 nm decreased to less than 40%. The luminescence intensity (kcount/s/µg antibody) from antibody–ProteinA–BMP complexes after immuno-reaction in the presence of human insulin and ALP-antibody was higher than that from BMPs chemically conjugated to antibody. This is explained by a difference in dispersion between chemically modified BMPs and antibody–ProteinA–BMP complexes. An automated immunoassay system containing a reaction station cor-responding to a 96-well microtiter plate, tip track, multi-pipettor consisting of eight tips and magnetic separation apparatus using a magnet, was developed. Dose–response curves of human insulin concentration in PBS and blood serum were obtained by using the fully automated chemiluminescence enzyme immunoassay system. Linear regression analyses yielded a correlation coefficient (r) of 0.999, with a slope of 0.99 between methods in the automatic procedure, $r = 0.989$ and slope $= 1.01$ in the manual procedure. The fully automated immunoassay system allows a precise assay of human insulin in serum.

9.4.4 High-Throughput Genotyping Using BMPs

A DNA microarray system has been developed for rapid and species-specific (high-throughput) identification of fish meat, using BMPs which was conjugated to species-specific oligonucleotide probes based on the sequence of ATPase and cyto-chrome oxidase subunit III genes. There are seven nominal species and one sub-species in large tunas of the genus *Thunnus*. Species identification at adult stage is important for stock assessment and management, but it may be more difficult when a few diagnostic external morphological characters are removed and/or when they are filleted. PCR-RFLP analysis of the mitochondrial DNA segment flanking ATPase and cytochrome oxidase subunit III genes (ATCO region) is reported to be reliable for tuna species [50] but this method takes a long time for the identification. A higher throughput method is required. Thus, we have designed specific DNA probes for the identification of each *Thunnus* spp. and its sub. spp. The DNA probes were immobilized onto BMPs using chemical crosslinking reagents, and its BMPs were applied to a DNA microarray system. Our system could identify all five species of tuna using an array pattern due to the hybridization with seven specific DNA probes which were labeled on BMPs.

References

[1] T. Matsunaga, *Trend Biotechnol.* 1991, *9*, 91–95.
[2] F. F. Torres de Araujo, M. A. Pires, R. B. Frankel, C. E. M. Bicudo, *Biophys. J.* 1986, *50*, 375–378.
[3] H. Lowenstam, *Geol. Soc. Am. Bull.* 1962, *73*, 435–438.
[4] D. A. Kuterbach, B. Walcott, R. J. Reeder, R. B. Frankel, *Science* 1982, *218*, 695–697.
[5] C.-Y. Hsu, C.-W. Li, *Science* 1994, *265*, 95–96.
[6] M. M. Walker, J. L. Kirschvink, S.-B. R. Chang, A. E. Dizon, *Science* 1984, *224*, 751–753.
[7] Y. Sakai, T. Tomiyama, M. Kato, M. Ogura, *IEEE Trans. Magnet.* 1990, *26*, 1554–1556.

[8] S. Mann, N. H. C. Sparks, M. M. Walker, J. L. Kirschvink, *J. Exp. Biol.* 1988, 35–49.

[9] T. Matsunaga, T. Sakaguchi, F. Tadokoro, *Appl. Microbiol. Biotechnol.* 1991, *35*, 651–655.

[10] T. Matsunaga, C. Nakamura, J. G. Burgess, K. Sode, *J. Bacteriol.* 1992, *174*, 2748–2753.

[11] R. P. Blakemore, D. Maratea, R. S. Wolfe, *J. Bacteriol.* 1979, *140*, 720–729.

[12] K. H. Schleifer, D. Schüler, S. Spring, M. Weizenegger, R. Amann, W. Ludwig, M. Köhler, *Syst. Appl. Microbiol.* 1991, *14*, 379–385.

[13] D. A. Bazylinski, R. B. Frankel, H. W. Jannasch, *Nature* 1988, *334*, 518–519.

[14] F. C. Meldrum, S. Mann, B. R. Heywood, R. B. Frankel, D. A. Bazylinski, *Proc. R. Soc. Lond. B* 1993, *251*, 237–242.

[15] F. C. Meldrum, S. Mann, B. R. Heywood, R. B. Frankel, D. A. Bazylinski, *Proc. R. Soc. Lond. B* 1993, *251*, 231–236.

[16] T. Matsunaga, T. Sakaguchi, in *Proceedings of the 6th International Conference on Ferrites (ICF 6)*, (Eds T. Yamaguchi and M. Abe), The Japan Society of Powder and Powder Metallurgy, Tokyo-Kyoto, 1992, pp. 262–267.

[17] T. Matsunaga, F. Tadokoro, N. Nakamura, *IEEE Trans. Magnet.* 1990, *26*, 1557–1559.

[18] T. Sakaguchi, J. G. Burgess T. Matsunaga, *Nature* 1993, *365*, 47–49.

[19] R. H. Thornhill, J. G. Burgess, T. Sakaguchi, T. Matsunaga, *FEMS Microbiol. Lett.* 1994, *115*, 169–176.

[20] R. H. Thornhill, J. G. Burgess, T. Matsunaga, *Appl. Environ. Microbiol.* 1995, *61*, 495–500.

[21] T. Matsunaga, N. Tsujimura, *Appl. Microbiol. Biotechnol.* 1993, *39*, 368–371.

[22] K. A. Short, R. P. Blakemore, *J. Bacteriol.* 1986, *167*, 729–731.

[23] J. G. Burgess, R. Kawaguchi, T. Sakaguchi, R. H. Thornhill, T. Matsunaga, *J. Bacteriol.* 1993, *175*, 6689–6694.

[24] T. Matsunaga, S. Kamiya, *Appl. Microbiol. Biotechnol.* 1987, *26*, 328–332.

[25] T. Matsunaga, S. Kamiya, in *Biomagnetism '87* (Eds K. Atsumi, M. Kotani, S. Ueno, T. Kalita and S. J. Williamsen), Tokyo Denki University Press, Tokyo, 1988, pp. 410–413.

[26] T. Matsunaga, K. Hashimoto, N. Nakamura, K. Nakamura, S. Hashimoto, *Appl. Microbiol. Biotechnol.* 1989, *31*, 401–405.

[27] T. Matsunaga, N. Tsujimura, S. Kamiya, *Biotechnol. Techniques* 1996, *10*, 495–500.

[28] R. P. Blakemore, *Science* 1975, *190*, 377–379.

[29] S. Spring, K.-H. Schleifer, *Syst. Appl. Microbiol.* 1995, *18*, 147–153.

[30] T. Sakaguchi, N. Tsujimura, T. Matsunaga, *J. Microbiol. Methods* 1996, *26*, 139–145.

[31] R. Kawaguchi, J. G. Burgess, T. Sakaguchi, H. Takeyama, R. H. Thornhill, T. Matsunaga, *FEMS Microbiol. Lett.* 1995, *126*, 277–282.

[32] N. T. Keen, S. Tamami, D. Kobayashi, D. Trollinger, *Gene* 1988, *70*, 191–197.

[33] M. Bagdasarian, R. Lurz, B. Rückert, F. C. Franklin, M. M. Bagdasarian, J. Frey, K. N. Timmis, *Gene* 1981, *16*, 237–247.

[34] R. Simon, U. Preifer, A. Pühler, *Bio/Technology* 1986, *1*, 784–791.

[35] A. W. Munro, G. Y. Ritchie, A. J. Lamb, R. M. Douglas, I. R. Booth, *Mol. Microbiol.* 1991, *5*, 607–616.

[36] M. Waser, D. Hess-Bienz, K. Davies, M. Solioz, *J. Biol. Chem.* 1992, *267*, 5396–5400.

[37] C. Nakamura, J. G. Burgess, K. Sode, T. Matsunaga, *J. Biol. Chem.* 1995, *270*, 28392–28396.

[38] C. Nakamura, Y. Hotta, R. H. Thornhill, T. Matsunaga, *J. Mar. Biotechnol.* 1995, *3*, 97–100.

[39] E. Amann, B. Ochs, K. J. Abel, *Gene* 1988, *69*, 301–315.

[40] B. Nilsson, T. Moks, B. Jansson, L. Abrahamsen, A. Elmblad, E. Holmgren, C. Henrichson, T. A. Jones, M. Uhlen, *Protein Eng.* 1987, *1*, 107–113.

[41] C. Nakamura, T. Kikuchi, J. G. Burgess, T. Matsunaga, *J. Biochem.* 1995, *118*, 23–27.

[42] N. Nakamura, K. Hashimoto, T. Matsunaga, *Anal. Chem.* 1991, *63*, 268–272.

[43] N. Nakamura, J. G. Burgess, K. Yagiuda, S, Kudo, T. Sakaguchi, T. Matsunaga, *Anal. Chem.* 1993, *65*, 2036–2039.

[44] N. Nakamura, T. Matsunaga, *Anal. Chim. Acta* 1993, *281*, 585–589.

[45] H. Takeyama, A. Yamazawa, C. Nakamura, T. Matsunaga, *Biotechnol. Techniques* 1995, *9*, 355–360.

[46] K. Sode, S. Kudo, T. Sakaguchi, N. Nakamura, T. Matsunaga, *Biotechnol. Techniques* 1993, *7*, 688–694.

[47] T. Matsunaga, T. Sakaguchi, K. Sode, S. Kudo, N. Nakamura, in Advanced Materials '93 II/ A: Biomaterials, Organic and Intelligent Materials (Proceedings of the 3rd IUMRS International Conference on Advanced Materials), Vol. 15A (Eds H. Aoki et al.), Trans. Mat. Res. Soc. Jpn., Tokyo, 1993, pp. 449–454.

[48] T. Matsunaga, M. Kawasaki, X. Yu, N. Tsujimura, N. Nakamura, *Anal. Chem.* 1996, *68*, 3551–3554.

[49] T. Matsunaga, R. Sato, S. Kamiya, T. Tanaka, H. Takeyama, *J. Magn. Magn. Mater.* 1999, *194*, 126–131.

[50] S. Chow, H. Kishino, *J. Mol. Evol.* 1995, *41*, 741–748.

Eukaryotes

10 A Grand Unified Theory of Biomineralization

Joseph L. Kirschvink, James W. Hagadorn

10.1 Introduction

The geological record indicates that the major animal phyla began biomineralizing in a relatively short interval of time during the Cambrian evolutionary explosion, about 525 Myr ago. Because these phyla diverged well before this biomineralization event, it was triggered either by an unprecedented lateral genetic transfer, or was the result of parallel exaptation of an ancestral biomineral system in many separate lineages. As magnetite (Fe_3O_4) biomineralization is the most ancient matrix-mediated system, and is present in most animal groups, it may have served as this ancestral template for exaptation. Complete sequencing of the genome of a magnetotactic bacterium, and identifying the magnetite operon, might provide a "road map" for unraveling the genetics of biomineralization in higher organisms, including humans.

One of the most sobering things a modern biologist can do is to examine the results of a 2-D gel taken from biomineral-forming tissue. The complexity and the number of protein products involved in what looks like a rather simple biological process is daunting. Years of work can go into unraveling the identity and function of a single product, such as the sea urchin SM50 protein [1].

Perhaps the despair is premature, as the existing complexity observed today had to evolve from a pre-existing, presumably simpler system. One major process by which complex biological systems evolve is by taking an existing genetic pattern that evolved for one function, and then duplicating it, linking it up differently, and adapting it for a new role. The nascent system is then gradually debugged and improved through the process of random mutation and natural selection. This evolutionary pattern has been called "exaptation" [2]. If the biomineralization systems present in the major animal phyla today evolved in this fashion, then it makes sense to examine that ancient system first, and to use it to provide an evolutionary road map for the modern processes. The paleontological record of biomineralization supports the idea that many animal groups experienced an evolutionary "trigger" during the Cambrian Explosion, which, as noted below, is compatible with repeated exaptation of an ancestral process. Certainly, there appears to be some fundamental underlying immunological similarity between the macromolecules involved in hydroxyapatite formation in vertebrates, and those involved in the aragonite present in molluscan nacre [3, 4]; freshly ground nacre fails to elicit an immune response in humans, and in fact stimulates bone regeneration. This would be highly unlikely if both biominerals had evolved via separate pathways, and argues for a common ancestor.

10.2 Geological Record of the Cambrian Explosion

The geological record provides an important clue to the origin of biomineralization in the animal phyla, as has been noted by many previous authors [5–12]. Prior to the Cambrian (~ 544 Myr ago), almost the only evidence for animal life was the soft-bodied Ediacaran fauna (Fig. 10.1). Together with related bed-parallel tracks

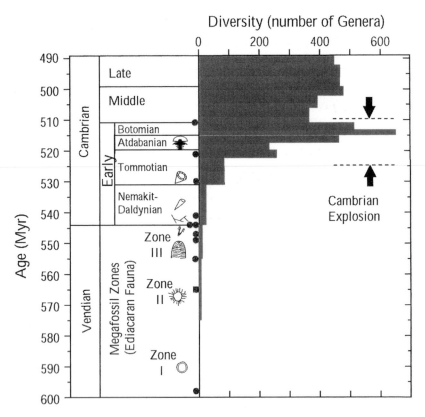

Figure 10.1 Radiometric age constraints and generic diversity for Vendian and Cambrian time. Position of the actual U/Pb constraints are indicated by solid blue dots. Data for the generic diversity have been compiled from Narbonne *et al.* [27] and Sepkoski [28], and are plotted as the number of taxa reported in each reference time interval. The black arrows indicate the approximate time of the Cambrian Explosion/true polar wander event [29]. The original figure was adapted from that of Kirschvink *et al.* [31].

and trails of this interval, these extraordinarily rare fossils are typically preserved as casts/molds at bed interfaces, and display no evidence of biomineralization. Molecular clock studies of protein divergence times for the major animal phyla have indicated consistently that major phyla separated up to several hundred million years prior to their first appearance in the fossil record [13–18]. The first clear evidence of matrix-mediated calcium biomineralization is the latest Precambrian invertebrate *Cloudina* [19] (Fig. 10.2). Extensive work on the numerical calibration of the geological time scale for this interval illustrates that *Cloudina* first appeared about 550 Myr ago, with other mineralized forms appearing during the next 40 Myr [20–26]. However, the largest burst of new biomineralization activity is clearly in the Tommotian/Atdabanian interval, where the diversity of fossil organisms (mostly new biomineralizing groups) increases almost exponentially over a ~ 10 Myr interval.

The trigger for this Cambrian Explosion has been the subject of extensive debate in the geological literature. It seems to have been a time of general climatic instability, as reflected by intense oscillations in the stable isotope record of carbon [17, 29, 30], probably driven by tectonic events of global magnitude [31]. Thus, it was a good time for new evolutionary innovations, because the climatic instability provided opportunities for novel forms to become fixed within small populations. But the mineralization drive itself may have been triggered by a separate development: the evolution of an animal predator, as suggested by Stanley [5]. Although there is no evidence of animal predation among Ediacaran paleocommunities, some of the earliest trace fossils associated with the Ediacarans appear to be scratch marks made by the hardened radula of an unidentified mollusc [32]. Similarly, the earliest skeletonized fossils (*Cloudina*) contain clear evidence of predatorial borings – borings likely to have been made by the rasping of a mineralized feeding apparatus [12]. It is therefore likely that the adaptive advantages conferred by skeletal and tooth biomineralization were an important factor influencing the intensity of the Cambrian explosion, and they may have been amplified by the co-evolution of predator/prey systems.

The mere presence of a good evolutionary driving force is not enough to completely reinvent a complex biochemical system multiple times within a geologically short interval. Lowenstam and Margulis [6] noted that the vast majority of new biomineral products observed in the Early Cambrian were based on calcium – either some form of $CaCO_3$, or Ca phosphate minerals, as shown in Fig. 10.2. Noting that precise control of intracellular calcium is necessary for the formation of the microtubules needed by all eukaryotic cells, they suggested that these calcium regulation and transport systems provided the evolutionary prerequisites for their eventual use in biomineralization (in effect, exaptation, although the word had not been coined in 1980). Calcium, carbonate and phosphate ions are abundant in the world's oceans, and hence would be favored for use in skeletons over much rarer materials such as Fe, Sr, Mn, etc.

However, gathering the components is only a small part of the biologically controlled mineralization process. They need to be brought together in a confined volume, in a controlled fashion, and induced to crystallize. The growing crystallites need to be properly tended, fed, and confined to the desired size, shape and crystallographic orientation. Hence, it seems that another biochemical/genetic system,

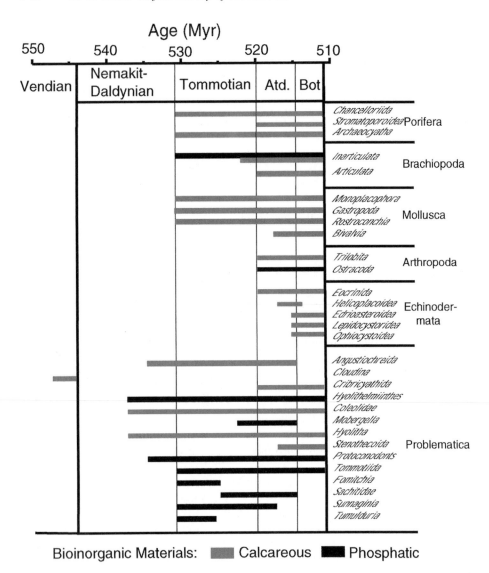

Figure 10.2 Stratigraphic ranges and first appearances of major fossil taxa that employ calcareous and phosphatic biomineralization. Not all families and problematic taxa are listed, nor are silica biomineralizing groups; for summaries of the stratigraphic range of these groups, see Bengtson [11]. Note that the mineralogy of *Cloudina* is poorly constrained; Grant [19] inferred a primary mineralogy of high-magnesian calcite based on preferential dolomitization of shell layers. Data from Lowenstam and Margolis [6] and Bengtson [11].

in addition to the ion transport system, was involved in the exaptation that led to widespread biomineralization in the Early Cambrian. In all of these respects, the magnetite (Fe_3O_4) biomineralization system present in extant magnetotactic bacteria appears to be this missing link. As noted below, it is ancient, appears to be present in most of the animal phyla, and has all of the essential aspects of biologically-controlled mineralization processes present in higher organisms. With apologies to physicists, it seems appropriate to dub this concept the "Grand Unified Theory of Biomineralization". Unlike some physical "GUT" theories, this one can be tested easily.

10.3 Magnetite Biomineralization

Heinz A. Lowenstam of the California Institute of Technology first discovered biochemically precipitated magnetite as a capping material in the radula (tongue plate) teeth of chitons (marine molluscs of the class *Polyplacophora* [33]). He and his students were able to demonstrate the biological origin of this material through a variety of radioisotope tracing studies and by detailed examination of the tooth ultrastructure [34–36]. Prior to this discovery, magnetite was thought to form only in igneous or metamorphic rocks under high temperatures and pressures. In the chitons, the magnetite serves to harden the tooth caps, enabling the chitons to extract and eat endolithic algae from within the outer few millimeters of rock substrates. Nesson and Lowenstam [36] reported the results of detailed histological and ultrastructural examinations of magnetite formation within the radula, and noted that the process begins with an initial transport of metabolic iron to the posterior end of the radula sac. This iron is deposited as the mineral ferrihydrite within a preformed proteinaceous mesh [34], forming one or two distinct rows of reddish teeth. This ferrihydrite is converted rapidly to magnetite via an unknown process.

Magnetotactic bacteria were the second organisms found to contain biogenic magnetite [37, 38], a typical example of which is shown in Fig. 10.3. They precipitate individual sub-micron sized magnetite crystals within an intracellular phospholipid membrane vacuole, forming structures called "magnetosomes" [39, 40]. Chains of these magnetosomes act as simple compass needles which passively torque the bacterial cells into alignment with the Earth's magnetic field, and allow them to seek the microaerophilic zone at the mud/water interface of most natural aqueous environments. These bacteria swim to the magnetic north in the northern hemisphere [37], to the magnetic south in the southern hemisphere [41, 42], and both ways on the geomagnetic equator [43, 44]. Magnetite-bearing magnetosomes have also been found in eukaryotic magnetotactic algae, with each cell containing several thousand crystals [45]. The magnetite formation process in bacteria has an overall similarity to that in chiton teeth, as both involve deposition of a ferrihydrite-like mineral precursor prior to magnetite formation [35, 46].

Magnetite crystals formed within these magnetosome vesicles have five main features that distinguish them from magnetites formed through geological pro-

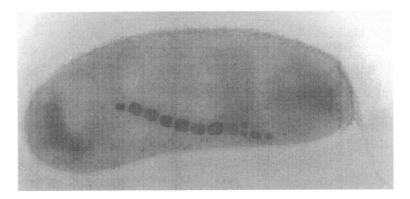

Figure 10.3 TEM image of a typical magnetotactic bacterium. The bacterium is 3 μm in size, with typical magnetite crystals on the order of 30–50 nm in length.

cesses: (1) High-resolution TEM studies reveal that bacterial magnetites are almost perfect crystals, which (2) often violate the cubic crystal symmetry of magnetite. They (3) are usually elongate in the [111] direction [40, 47–49], (4) are chemically quite pure Fe_3O_4, and (5) are restricted in size and shape so as to be uniformly magnetized (single-magnetic-domains). Inorganic magnetites are usually small octahedral crystals, often with lattice dislocations, chemical impurities, and other crystal defects. The elongation of biogenic crystals in the [111] direction serves to stabilize the magnetic moment of the particle, and presumably is the result of natural selection for their magnetic properties [40, 50]. Bacterial magnetite crystals are restricted to a size range of about 35–500 nm, with shapes that confine them to the single-domain magnetic stability field [51, 52]. Inorganic magnetites tend to have log–normal size distributions that often extend up into the multi-domain size region. Bacterial magnetites tend to be rather pure iron oxide, with no detectable titanium, chromium or aluminum, which are often present in geologically-produced magnetite. An additional feature is the alignment of the crystals into linear chains, which can be preserved in the fossil record [53, 54]. These characteristic features have enabled bacterially-precipitated magnetites to be identified in Earth sediments up to 2 billion years old [44], and possibly in 4 billion year old carbonate inclusions in the ALH84001 meteorite from Mars [55, 56].

As shown in Figs. 10.4 and 10.5, many of these same features are shared by the magnetite crystals extracted from salmon [57] and from the human brain [58, 59]. The simplest interpretation of these results is that many higher organisms, including humans, possess the biochemical ability to form magnetite.

In higher animals, an obvious function of magnetite biomineralization is its role in magnetoreception [61–63]. Magnetoreception is now well established in virtually all major groups of animals [64], and specialized cells containing single-domain chains of magnetite are the best candidates for the receptor cells [60, 65]. In the brown trout, Walker *et al.* [60] have shown elegantly that magnetically-sensitive nerves in the ophthalmic branch of the trigeminal nerve connect to specialized, tri-lobed cells in the olfactory laminae which contain magnetite crystals. Similarly, behavioral work with honey bees and birds has shown that brief magnetic pulses are

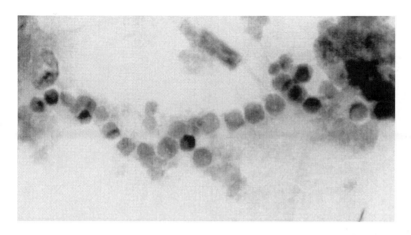

Figure 10.4 Single-domain magnetite crystals extracted from the frontal tissues of the sockeye salmon [57]. These particles are structurally nearly identical to those present in magnetotactic bacteria. Recent studies have shown that these are indeed present in the magnetosensory cells in fish [60].

able to alter the magnetic responses, confirming that a ferromagnetic material like magnetite is indeed part of the magnetic sensory system [66–72].

From an evolutionary perspective, it now seems clear that magnetite-based magnetoreception, and hence magnetite biomineralization, date back at least to the last common ancestor of Chordata, Mollusca and Arthropoda (~ 600–900 Myr ago). The existence of magnetotactic protists argues that this genetic ability for magnetite biomineralization may go back even further, to the evolution of the first eukaryotes nearly 2 billion years ago. Indeed, Vali and Kirschvink [40] argue that the ancestral eukaryotes probably inherited the ability to make magnetite from magnetotactic bacteria during the endosymbiotic events that formed this cell type. Therefore, the magnetite system must have been present in most of the animal phyla in the massive biomineralization episode during the Cambrian Explosion, and hence available for exaptation to form other biomineral systems. This may account for the apparent lack of a human immune response to molluscan nacre noted earlier. The fact that two of the most primitive molluscan groups, the *Archaeogastropods* and the *Polyplacophorans*, both use iron minerals to harden their radular teeth (goethite and magnetite, respectively [33, 73]) indicates that they adapted this from a pre-existing iron biomineral system. Hence, magnetite biomineralization is a prime candidate for this missing "evolutionary precursor".

10.4 Discussion

The magnetotactic bacteria are the most primitive organisms known which use a vacuole-based system to form their biomineral products. In mammals, much of the

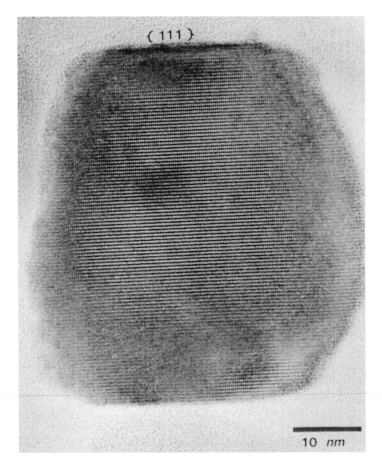

Figure 10.5 Magnetite crystal extracted from tissues of the human brain [58].

hydroxyapatite in bone and teeth is formed via a similar process, in which the chemical precursors are transported first to a vacuole storage system, and then discrete peptides are added to nucleate the desired crystal forms [9]. Understanding how this process evolved in bacteria ought to provide insights into how the more complex biomineralization systems in eukaryotes operate.

Most of the published genetic analyses of magnetotactic bacteria to date have focused on using 16s RNA to infer phylogenetic relationships [74–77]. As of this writing, only one protein directly involved in the magnetite biomineralization process has been found using transposon mutagenesis [78], in an as yet unnamed magnetotactic bacterium dubbed AMB-1. The protein coded by this open reading frame (called MagA) is known to reside both in the cell membrane and in the magnetosome membrane [79], and is involved in transporting and accumulating iron within the magnetosomes [80]. It demonstrates strong homology with known Ca^{2+} transmembrane transport proteins in other bacteria.

In an attempt to attack the magnetosome problem from the standpoint of the iron mediating enzymes, Bertani *et al.* [81] focused on the gene coding for bacterioferritin (bfr) in *Magnetospirillum magnetotacticum*. In contrast to *E. coli*, which has only one bfr gene, *M. magnetotacticum* has two genes, bfr1 and bfr2, that are strongly homologous to bacterioferritins. The subunit encoded by bfr1 is more similar to *E. coli* bacterioferritin than it is to the subunit encoded by bfr2. These genes are strange in two other ways. First, the open reading frames overlap by one base pair, with the last base pair of the stop codon of the first gene serving as the first codon of the second protein. Second, the amino acid residues of the region in the putative bfr2 subunit, which is thought to be involved in the binding and nucleating of the iron oxide mineral at the core of the ferritin protein (the mineral ferrihydrite), are completely different from the other bacterioferritins, which are otherwise highly conserved. This characteristic indiciates that the proteins are *not* acting to nucleate ferrihydrite deposition, and Bertani *et al.* [81] speculate that this peculiar feature may have something to do with the mineralization process.

In summary, the genetic basis of biomineralization – for all mineral systems – is still a mystery. The "Grand Unified Theory of Biomineralization" presented here suggests that an understanding of magnetite biomineralization in the magnetotactic bacteria might provide a template for unraveling, or at least understanding, portions of vacuolar-based biomineral systems in higher animals, including humans. The first major step in understanding the bacterial system would be, of course, determining the complete genome sequence for a magnetotactic bacterium.

It is fitting to close this article with the first stanza of Rudyard Kipling's famous poem, *Cold Iron:*

> Gold is for the mistress – silver for the maid –
> Copper for the craftsman cunning at his trade.
> "Good!" said the Baron, sitting in his hall,
> "But Iron – Cold Iron – is master of them all."

Acknowledgments

We thank the NASA Astrobiology Institute, the Electric Power Research Institute WO-4307-03, and an anonymous donor for support. This paper is dedicated to the memory of Heinz A. Lowenstam, who initiated the systematic study of the diversity of biomineral products.

References

[1] K. W. Makabe, C. V. Kirchhamer, R. J. Britten, E. H. Davidson, *Development* 1995, *121*, 1957.

[2] S. J. Gould, E. S. Vrba, *Paleobiology* 1982, *8*, 4.

[3] P. Westbroek, F. Marin, *Nature* 1998, *392*, 861.
[4] G. Atlan, N. Balmain, S. Berland, B. Vidal, E. Lopez, *Comptes Rendus De L Academie Des Sciences Serie Iii-Sciences De 320* 1997, 253.
[5] S. M. Stanley, *Proc. Natl Acad. Sci. USA* 1973, *70*, 1486.
[6] H. A. Lowenstam, L. Margulis, *BioSystems* 1980, *12*, 27.
[7] H. A. Lowenstam, *Science* 1981, *211*, 1126.
[8] H. A. Lowenstam, J. L. Kirschvink, in *Magnetite Biomineralization and Magnetoreception in Organisms: A New Biomagnetism*, Vol. 5 (Eds J. L. Kirschvink, D. S. Jones, B. McFadden), Plenum Press, New York, 1985, p. 3.
[9] H. A. Lowenstam, S. Weiner, *On Biomineralization*, Oxford University Press, Oxford, 1989.
[10] H. A. Lowenstam, *Precambrian Research* 1980, *11*, 89.
[11] S. Bengtson, J. D. Farmer, M. A. Fedonkin, J. H. Lipps, B. N. Runnegar, in *The Proterozoic Biosphere: A Multidisciplinary Study* (Eds J. W. Schopf, C. Klein and D. Des Marais), Cambridge University Press, Cambridge, 1992, p. 425.
[12] S. Bengtson, Y. Zhao, *Science* 1992, *257*, 367.
[13] B. Runnegar, *Lethaia* 1982, *15*, 199.
[14] S. J. Peroutka, T. A. Howell, *Neuropharmacology* 1994, *33*, 319.
[15] E. H. Davidson, K. J. Peterson, R. A. Cameron, *Science* 1995, *270*, 1319.
[16] G. A. Wray, J. S. Levinton, L. H. Shapiro, *Science* 1996, *274*, 568.
[17] A. H. Knoll, S. B. Carroll, *Science* 1999, *284*, 2129.
[18] F. J. Ayala, A. Rzhetsky, *Proc. Natl Acad. Sci. USA* 1998, *95*, 606.
[19] S. W. F. Grant, *Am. J. Sci.* 1990, *290A*, 261.
[20] W. Compston, I. Williams, J. L. Kirschvink, Z. Zichao, M. Guogan, *J. Geol. Soc. Lond.* 1992, *149*, 171.
[21] W. Compston, M. S. Sambridge, R. F. Reinfrank, M. Moczydlowska, V. G. S. Claesson, *J. Geol. Soc. Lond.* 1995, *152*, 599.
[22] J. A. Cooper, R. J. F. Jenkins, W. Compston, I. S. Williams, *J. Geol. Soc. Lond.* 1992, *149*, 185.
[23] S. A. Bowring, J. P. Grotzinger, C. E. Isachsen, A. H. Knoll, S. M. Pelechaty, P. Kolosov, *Science* 1993, *261*, 1293.
[24] C. E. Isachsen, S. A. Bowring, E. Landing, S. D. Samson, *Geology* 1994, *22*, 496.
[25] J. P. Grotzinger, S. A. Bowring, B. Z. Saylor, A. J. Kaufman, *Science* 1995, *270*, 598.
[26] R. D. Tucker, W. S. Mckerrow, *Can. J. Earth Sci.* 1995, *32*, 368.
[27] G. M. Narbonne, A. J. Kaufman, A. H. Knoll, *Geol. Soc. Am. Bull.* 1994, *106*, 1281.
[28] J. J. Sepkoski, in *The Proterozoic Biosphere: A Multidisciplinary Study* (Eds J. W. Schopf, C. Klein and D. Des Maris), Cambridge University Press, Cambridge, 1992, p. 1171.
[29] M. Magaritz, W. T. Holser, J. L. Kirschvink, *Nature* 1986, *320*, 258.
[30] M. D. Brasier, R. M. Corfield, L. A. Derry, A. Y. Rozanov, A. Y. Zhuravlev, *Geology* 1994, *22*, 455.
[31] J. L. Kirschvink, R. L. Ripperdan, D. A. Evans, *Science* 1997, *277*, 541.
[32] J. G. Gehling, PhD thesis, University of California, Los Angeles, 1996.
[33] H. A. Lowenstam, *Bull. Geol. Soc. Am.* 1962, *73*, 435.
[34] K. M. Towe, H. A. Lowenstam, *J. Ultrastruct. Res.* 1967, *17*, 1.
[35] J. L. Kirschvink, H. A. Lowenstam, *Earth Planet. Sci. Lett.* 1979, *44*, 193.
[36] M. H. Nesson, H. A. Lowenstam, in *Magnetite Biomineralization and Magnetoreception in Organisms: A New Biomagnetism*, Vol. 5 (Eds J. L. Kirschvink, D. S. Jones, B. McFadden), Plenum Press, New York, 1985, p. 333.
[37] R. P. Blakemore, *Science* 1975, *190*, 377.
[38] R. B. Frankel, R. P. Blakemore, R. S. Wolfe, *Science* 1979, *203*, 1355.
[39] Y. A. Gorby, T. J. Beveridge, R. P. Blakemore, *J. Bacteriol.* 1988, *170*, 834.
[40] H. Vali, J. L. Kirschvink, in *Iron Biomineralization* (Eds R. P. Frankel and R. P. Blakemore), Plenum Press, New York, 1991, p. 97.
[41] J. L. Kirschvink, *J. Exp. Biol.* 1980, *86*, 345.
[42] R. P. Blakemore, R. B. Frankel, A. J. Kalmijn, *Nature* 1980, *286*, 384.
[43] R. B. Frankel, R. P. Blakemore, F. F. Torres de Araujo, E. M. S. Esquivel, J. Danon, *Science* 1981, *212*, 1269.

[44] S.-B. R. Chang, J. L. Kirschvink, *Annu. Rev. Earth Planet. Sci.* 1989, *17*, 169.
[45] F. F. Torres de Araujo, M. A. Pires, R. B. Frankel, C. E. M. Bicudo, *Biophys. J.* 1985, *50*, 375.
[46] R. B. Frankel, R. P. Blakemore, *Phil. Trans. R. Soc. Lond. B* 1984, *304*, 567.
[47] S. Mann, R. B. Frankel, R. P. Blakemore, *Nature* 1984, *310*, 405.
[48] S. Mann, T. T. Moench, R. J. P. Williams, *Proc. R. Soc. Lond. B* 1984, *221*, 385.
[49] S. Mann, in *Magnetite Biomineralization and Magnetoreception in Organisms: A New Biomagnetism*, Vol. 5 (Eds J. L. Kirschvink, D. S. Jones, B. McFadden), Plenum Press, New York, 1985, p. 311.
[50] J. L. Kirschvink, *Bioelectromagnetics* 1989, *10*, 239.
[51] R. F. Butler, S. K. Banerjee, *J. Geophys. Res.* 1975, *80*, 4049.
[52] J. C. Diaz-Ricci, J. L. Kirschvink, *J. Geophys. Res.* 1992, *97*, 17309.
[53] J. L. Kirschvink, S.-B. R. Chang, *Geology* 1984, *12*, 559.
[54] N. Petersen, T. von Dobeneck, H. Vali, *Nature* 1986, *320*, 611.
[55] D. McKay, E. Gibson, K. Thomaskeprta, H. Vali, C. Romanek, S. Clemett, X. Chillier, C. Maechling, R. Zare, *Science* 1996, *273*, 924.
[56] K. L. Thomas-Keprta, D. A. Bazylinski, J. L. Kirschvink, S. J. Clemett, D. S. McKoy, S. J. Wentworth, J. Voli, E. K. Gibson, Jr., *Geochimica Cosmochiunica Acta*, in press.
[57] S. Mann, N. H. C. Sparks, M. M. Walker, J. L. Kirschvink, *J. Exp. Biol.* 1988, *140*, 35.
[58] J. L. Kirschvink, A. Kobayashi-Kirschvink, B. J. Woodford, *Proc. Natl Acad. Sci.* 1992, *89*, 7683.
[59] J. Dobson, P. Grass, *Brain Res. Bull.* 1996, *39*, 255.
[60] M. M. Walker, C. E. Diebel, C. V. Haugh, P. M. Pankhurst, J. C. Montgomery, C. R. Green, *Nature* 1997, *390*, 371.
[61] J. L. Gould, J. L. Kirschvink, K. S. Deffeyes, *Science* 1978, *201*, 1026.
[62] C. Walcott, J. L. Gould, J. L. Kirschvink, *Science* 1979, *205*, 1027.
[63] J. L. Kirschvink, J. L. Gould, *BioSystems* 1981, *13*, 181.
[64] R. Wiltschko, W. Wiltschko, *Magnetic Orientation in Animals*, Vol. 33, Springer, Berlin, 1995.
[65] J. L. Kirschvink, *Nature* 1997, *390*, 339.
[66] J. L. Kirschvink, A. Kobayashi-Kirschvink, *Am. Zool.* 1991, *31*, 169.
[67] J. L. Kirschvink, S. Padmanabha, C. K. Boyce, J. Oglesby, *J. Exp. Biol.* 1997, *200*, 1363.
[68] R. C. Beason, R. Wiltschko, W. Wiltschko, *The Auk* 1997, *114*, 405.
[69] U. Munro, J. A. Munro, J. B. Phillips, W. Wiltschko, *Aust. J. Zool.* 1997, *45*, 189.
[70] U. Munro, J. A. Munro, J. B. Phillips, R. Wiltschko, W. Wiltschko, *Naturwissenschaften* 1997, *84*, 26.
[71] W. Wiltschko, U. Munro, R. C. Beason, H. Ford, R. Wiltschko, *Experimentia* 1994, *50*, 697.
[72] W. Wiltschko, R. Wiltschko, *J. Comp. Physiol.* 1995, *177*, 363.
[73] H. A. Lowenstam, *Science* 1962, *137*, 279.
[74] J. G. Burgess, R. Kawaguchi, T. Sakaguchi, R. H. Thornhill, T. Matsunaga, *J. Bacteriol.* 1993, *175*, 6689.
[75] R. H. Thornhill, J. G. Burgess, T. Matsunaga, *Appl. Environ. Microbiol.* 1995, *61*, 495.
[76] R. Kawaguchi, J. G. Burgess, T. Sakaguchi, H. Takeyama, R. H. Thornhill, T. Matsunaga, *FEMS Microbiol. Lett.* 1995, *126*, 277.
[77] S. Spring, R. Amann, W. Ludwig, K.-H. Schleifer, H. van Gemerden, H. Petersen, *Appl. Environ. Microbiol.* 1993, *59*, 2397.
[78] T. Matsunaga, C. Nakamura, J. G. Burgess, K. Sode, *J. Bacteriol.* 1992, *174*, 2748.
[79] C. Nakamura, T. Kikuchi, J. G. Burgess, T. Matsunaga, *J. Biochem.* 1995, *118*, 23.
[80] T. Matsunaga, N. Tsujimura, S. Kamiya, *J. Phys. IV* 1997, *7*, 651.
[81] L. E. Bertani, J. Huang, B. Weir, J. L. Kirschvink, *Gene* 1997, *201*, 31.

11 The Biochemistry of Silica Formation in Diatoms

Nils Kröger, Manfred Sumper

11.1 Introduction

Diatoms (order *Bacillariophyceae*) are eukaryotic, unicellular algae that are ubiquitously present in almost every water habitat on Earth. In the oceans diatoms dominate phytoplankton populations and algal blooms [1] and they are responsible for a considerable proportion of the world's net primary production [2]. Apart from this ecological significance, diatoms are mainly known for the intricate geometries and spectacular patterns of their silica-based cell walls (Fig. 11.1). These patterns are species-specific and are precisely reproduced during each cell division cycle, documenting a genetic control of this biomineralization process. Therefore, biogenesis of the diatom cell wall has been regarded as a paradigm for controlled production of nanostructured silica [3]. In the past, investigations into the biogenesis of diatom cell walls were mainly based on ultrastructural (electron microscopy) and physiological (silicon uptake studies) experiments. These investigations provided a first insight into the basic cell biological processes involved in biosilica formation [4] and revealed that silicic acid is actively taken up from the environment by specific transporter proteins located in the diatom's plasma membrane [5]. However, the mechanism that controls the formation of the species-specific morphology of biosilica has remained enigmatic.

Biochemical studies provided evidence that diatom biosilica may be a composite material containing (glyco-)proteins in addition to the inorganic phase [6–8]. It has been speculated that these organic macromolecules are involved in controlling biosilica formation. However, further analysis has been hampered by the lack of both structural data on these molecules and a defined *in vitro* system of studying their interaction with silica. In the past few years, a number of proteins associated with diatom biosilica have been purified to homogeneity and structurally characterized by both chemical and molecular genetic techniques. Functional studies of their role in silica precipitation have also been carried out. Data from these studies are summarized in this chapter and discussed with respect to the role these proteins may have in biosilica morphogenesis.

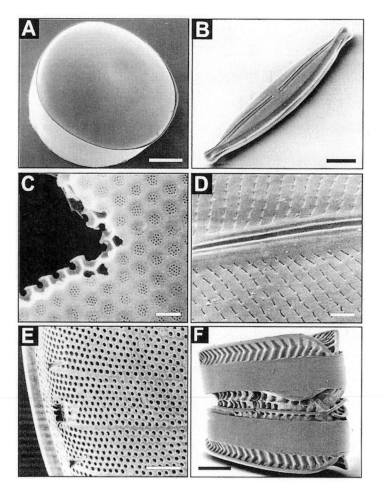

Figure 11.1 Morphology of diatom cell walls. Scanning electron microscopy (SEM) images of isolated cell walls from different diatom species. (a) *Coscinodiscus granii* (bar: 50 μm) (b) *Craticula sp.* (bar: 10 μm). (c) Detail of fragmented valve (outside surface) from *C. granii* (bar: 500 nm. (d) Detail of valve (outside surface) from *Nitzschia sp.* (bar: 1 μm). (e) Detail of valve (inside surface) from *C. granii* (bar: 5 μm). (f) *Surirella sp.* (a pennate diatom) shortly after cell division. The two daughter cells are still attached to each other (bar: 20 μm).

11.2 The Diatom Cell Wall

The protoplast of a diatom cell is tightly enclosed by a cell wall (also called a frustule) that is constructed in a petri dish-like fashion (Fig. 11.2a). It contains a top half (epitheca) that overlaps the slightly smaller bottom part (hypotheca). In most diatom species the construction and morphology of hypotheca and epitheca are

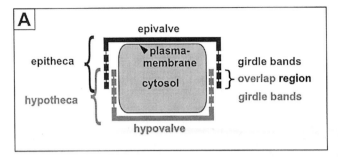

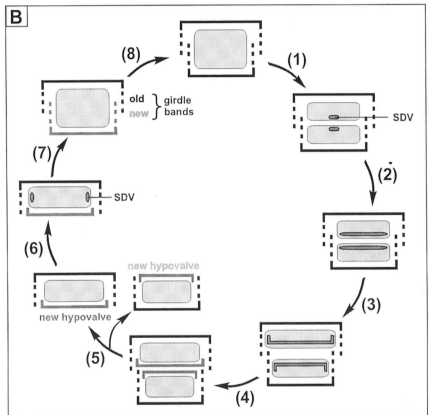

Figure 11.2 Structure and biogenesis of the diatom cell wall. Diatom cells are shown in cross-section (schematic). (a) Diatom cell wall structure. The cell wall is made up of two half shells, called epitheca and hypotheca, which together fully enclose the protoplast. Each theca consists of a valve and one or more girdle bands that run laterally along the outline of the cell. The terminal girdle bands of each theca constitute the overlap region of the cell wall in which the slightly larger epitheca overlaps the hypotheca. (b) Diatom cell cycle: (1) cytokinesis and formation of a valve SDV in each daughter protoplast; (2) and (3) expansion of the SDV and formation of a new hypovalve within each SDV; (4) exocytosis of SDV contents; (5) separation of daughter cells; (6) formation of the first girdle band SDV; (7) consecutive formation and secretion of girdle bands; (8) DNA replication.

identical. Each theca is made up of a valve and several girdle bands that span the circumference of the cell. Both these theca elements are siliceous, but it is the valve that displays the ornamental morphological features. The girdle band surface is either completely smooth or uniformly perforated. With respect to the symmetry of their frustules diatoms are divided into two groups, the centrics and the pennates. Pennate diatoms are bipolar, with the longitudinal axis in most species running parallel to the plane of symmetry (Fig. 11.1b). In contrast, the frustules of centric diatoms are – in most species – radially symmetrical about an axis that passes through the center of the cell (Fig. 11.1a). However, some centric species exhibit lower symmetries, displaying triradiate or multiradiate morphologies. With respect to diatom cell wall evolution, the centric cell wall morphology is thought to be the more ancient one, because it appears first in the fossil record (about 200 million years ago). Thus, it is hypothesized that the pennate morphology has evolved from a centric ancestor.

11.3 Diatom Cell Wall Biogenesis

The cellular events initiating biogenesis of the diatom cell wall have been extensively investigated by electron microscopic techniques (see Fig.11.2b for a summary). These studies led to the important discovery that diatoms use a specialized intracellular compartment, called a silica deposition vesicle (SDV), for the production of silica [9]. Subsequent studies on other silicifying protists have shown that the SDV is in fact not a speciality of diatoms but rather is a general organelle for biosilica formation [10]. Consequently, the SDV is regarded as a cellular "reaction vessel" in which silica precipitation and patterning is controlled to yield the ornamentations characteristic of the respective diatom species. After completion of the silica forming/patterning processes, the content of the SDV (the new valve) is transferred to the extracellular space in a remarkable exocytosic process that involves almost half of the plasma membrane area of the cell. Thus, the characteristic ornamentations of a diatom's cell wall must be a direct consequence of species-specific properties of the diatom SDV. Therefore, it appears reasonable to postulate species-specific differences with respect to the molecular composition of the SDV.

11.3.1 The Silica Deposition Vesicle

A straightforward approach to studying the biochemistry of biosilica formation would be a detailed analysis of all the SDV components. Unfortunately, no method has been developed up to now that allows isolation of a pure SDV preparation. Therefore, the scarce information available on the SDV stems from rather indirect methods. In electron microscopic images of thin sections from diatom cells, the

SDV membrane (silicalemma) appears as a typical lipid bilayer of 5–7 nm thickness. Results from fluorescence and electron microscopy show that Rhodamine 123 and DAMP [3-(2,4-dinitroanilino)-3′-amino-*N*-methyl-propylamine] accumulate within the SDV, indicating that an electrochemical potential exists across the silicalemma and that the SDV lumen exhibits an acidic pH [11, 12]. With respect to these properties the SDV resembles the lytic compartments of plant and animal cells, namely vacuoles and lysosomes, respectively. These compartments are connected to the Golgi apparatus of the cell by transport vesicles that deliver membrane and protein material. Likewise it has been suggested that formation and expansion of the SDV is achieved by the fusion of Golgi-derived vesicles. There are numerous reports claiming the close association of small vesicles (16–100 nm) with the rim of the growing SDV [13] but the origin of these vesicles is still unclear. Therefore, the SDV's origin and relation to other endomembrane compartments remains the subject of debate.

11.3.2 Silicic Acid Accumulation

Chemical analysis of diatom cell walls has shown that almost pure hydrated silicium dioxide (silica) doped with small amounts of aluminium and iron represents the inorganic component of biosilica [14]. The silica of diatom cell walls is amorphous rather than crystalline, because no long-range order of the atomic lattice is detected by high-resolution TEM and X-ray diffraction analysis [15]. It is not possible to ascribe an exact chemical formula to diatom silica because it is a nonstoichiometric inorganic polymer made up of a varying ratio of SiO_2 and H_2O.

Silica is formed by a polycondensation process from monosilicic acid, $Si(OH)_4$, which is the soluble silicon supply in aqueous systems. In natural water habitats dissolved silicon is available in concentrations of 1–100 µM [16]. It has been demonstrated that uptake of silicic acid by diatoms requires active transport across the plasma membrane and is mediated by Na^+-dependent silicon transporter proteins [17]. Recently, a gene family encoding seven different silicon transporter proteins (SIT1–SIT7) has been characterized from the diatom *Cylindrotheca fusiformis* [18]. One member of this family, SIT1, was indeed shown to mediate Na^+-dependent silicic acid uptake [19]. Thus, it is reasonable to assume that the transporter protein SIT1 is responsible for monosilicic acid uptake into the diatom cell. A large quantity of silicic acid needs to be transported and accumulated for diatom cell wall formation. The amount of silica present in the cell walls of the diatoms *Thalassiosira weissflogii* [20] and *Ethmodiscus* Castr. [21] has been determined. These results indicate that shortly before cytokinesis the intracellular silicic acid concentration ranges between 10 and 100 mM to account for the subsequent production of the cell wall silica. In this concentration range silicic acid is no longer stable under cellular pH conditions and undergoes spontaneous polycondensation (see Section 11.5.1). Therefore, the precise control of biosilica formation in diatoms requires the management of silicic acid uptake, intracellular storage and delivery to the SDV. It has been postulated that diatoms might have developed stable silicon complexes as intracellular storage and transport forms for silicic acid. However, the chemical nature of these hypothetical derivatives of silicic acid is still completely unknown.

11.3.3 Silica Deposition

Because the transport form of intracellular silicon is unknown, the monomeric precursor for biosilica formation must remain elusive. However, regardless of the molecular nature of intracellular silicon it is possible that monosilicic acid is liberated from a putative silicon transport compound when it has reached the SDV. Indeed, Shimizu *et al.* [22] have characterized a protein (called silicatein-α) from sponge biosilica that catalyzes the hydrolysis of synthetic monosilicic acid ester compounds. Silicatein-α-like proteins could also be present inside the diatom SDV, releasing monosilicic acid from a putative precursor molecule. If so, silica formation inside the SDV would simply be ruled by the inorganic chemistry of silicic acid and its condensation reactions. The central question of biosilica morphogenesis is then obvious: is there a pre-pattern within the SDV that determines where silica is precipitated or does the pattern of biosilica arise by a physicochemical process of self-organization? Both possibilities have been discussed in the literature. The theory of an organic matrix inside the SDV has been debated [4, 8, 23, 24] – this postulates that the diatom SDV contains organic macromolecules (proteins, polysaccharides), arranged in a regular pattern (pre-pattern) and serving as a template onto which silica is precipitated. According to this model the diatom cell wall pattern would reflect the spatial arrangement of this putative matrix. In contrast, Gordon and Drum [25] have developed an alternative theory. These authors postulate that pattern formation is achieved by instabilities during diffusion-limited silica precipitation inside the SDV. Their scenario relinquishes the need for templating by an organic matrix. Using computer simulations the authors were able to demonstrate the formation of basic patterns of diatom cell walls simply by the diffusion-limited and irreversible adsorption of colloidal silica particles onto a nucleating center within the SDV. However, the authors also admit that there must be present non-siliceous components controlling the kinetic parameters of biosilica formation, in order to explain both the creation of species-specific patterns and to account for the extremely high rate of biosilica formation.

It is reasonable to assume that some or all of the SDV components putatively involved in the control of silica deposition will remain tightly associated with the cell wall after exocytosis of the silica skeleton. Therefore, in searching for these components the organic constituents of a diatom cell wall have been analyzed [24, 26]. During the past years *Cylindrotheca fusiformis* has become a model organism for molecular studies of diatom cell wall biogenesis. A variety of cell wall proteins from this diatom have been characterized with respect to structure and localization [27–29]. Furthermore, a genetic transformation system has been established that will allow alteration of the protein composition of the cell wall [30] and thus may in future prove a useful tool in analyzing the influence of a particular protein on biosilica structure. In Section 11.4 we summarize what is known about the structure and function of protein components isolated from *C. fusiformis* cell walls.

11.4 Diatom Cell Wall Proteins

Early studies investigating the amino acid composition of isolated diatom cell walls suggested that proteins rich in hydroxy amino acids and glycine are enriched in cell wall preparations [7, 8]. Most remarkably, Nakajima and Volcani found that three unusual amino acids, namely 3,4-dihydroxyproline, ε-*N*-trimethyl-δ-hydroxylysine and its phosphorylated derivative, are present within the cell walls of different diatom species including *C. fusiformis* [6]. Given this unusual amino acid composition the authors postulated that a matrix composed of proteins, which carry a high proportion of hydroxy amino acids, glycine and the modified lysine and proline residues, may be involved in diatom biosilica formation by constituting an organic matrix within the SDV. However, the authors did not succeed in isolating the corresponding protein molecules and so were unable to carry out functional studies.

Recently, three families of diatom cell wall protein from the diatom *C. fusiformis* have been characterized and named as frustulins, HEPs (HF-extractable proteins; now pleuralins [32]) and silaffins [26, 31]. These proteins are highly hydrophilic and some indeed contain a high proportion of serine and glycine residues as well as post-translationally modified amino acids. However, neither 3,4-dihydroxyproline nor ε-*N*-trimethyl-δ-hydroxylysine have so far been found in these proteins.

Frustulins were shown to be distributed throughout the cell wall, constituting an outer protein coat bound within the cell wall via Ca^{2+} bridges. They become completely extracted from the cell wall by EDTA treatment [27]. In contrast, HEPs and silaffins are much more tightly bound to the cell wall and can only be extracted after dissolution of all the silica by treatment with anhydrous hydrogen fluoride (HF) [28, 31]. The tight incorporation of HEPs and silaffins within the cell wall could be explained by at least three different modes of interactions:

Physical entrapment within the silica.
Covalent linkage to the silanol groups of the silica surface by Si–O–C bonds.
Covalent crosslinking of cell wall components by HF-labile glycosidic bonds.

At present, no unequivocal decision can be made from these possibilities.

11.4.1 Structure and Localization of HEPs

SDS-PAGE of an HF-treated cell wall preparation reveals three high molecular mass proteins, collectively called HEPs (Fig. 11.3a), in addition to components with low molecular masses (Section 11.4.2). These proteins display apparent molecular masses of 200 kDa (HEP200), 180 kDa (hepBp) and 150 kDa (hepCp). HEP200 and hepCp are almost equally abundant whereas hepBp is a minor component. By cloning of the corresponding genes, the primary structures of these HEPs were deduced and are shown in Fig. 11.3b. HEPs are highly acidic proteins with predicted isoelectric points below pI 4. They are made up of modular arrangements of repetitive as well as unique domains. Each N-terminus contains a typical signal peptide sequence that is required for ER translocation, thus introducing HEPs into the

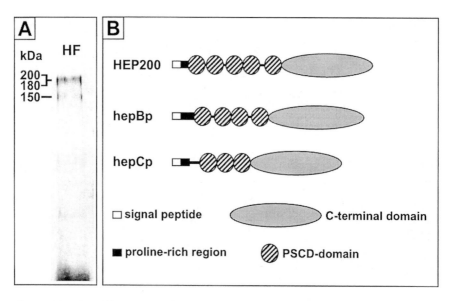

Figure 11.3 HEPs: (a) SDS-PAG (12% acrylamide) of the HF-extract from *C. fusiformis* cell walls (HF). (b) Schematic primary structures of HEPs.

secretory pathway of the cell. Adjacent to the signal peptide, HEPs contain an extremely proline-rich stretch (> 65% proline) that is followed by different numbers of PSCD-domains. Each PSCD-domain contains 87 or 89 amino acid residues, of which more than 50% are constituted by proline (22%), serine (11%), cysteine (11%) and aspartate (9%). The C-terminal parts of the HEPs are not related to the PSCD sequences and do not contain any internal sequence repetitions. HEPs show no significant sequence homology to any other protein in the data bank. To obtain information about their function, the localization of HEPs within the cell wall (Fig. 11.4a) was investigated. Antibodies were raised against the PSCD-domains of HEP200 and used for immunolocalization. Remarkably, HEP200 has a highly restricted distribution within the *C. fusiformis* cell wall. Using immunoelectron microscopy it was shown that HEP200 is specifically associated with the girdle bands of the cell wall. More precisely, HEP200 is localized only on those girdle bands that make up the region in which the epitheca overlaps the hypotheca (Fig. 11.4b, c). This specific localization led to the suggestion that HEPs might be involved in formation of the overlapping girdle bands [29]. However, in post-embedding immunolocalization experiments using thin sections of *C. fusiformis* cells, anti-HEP200 antibodies fail to label girdle band SDVs at any developmental stage. This result strongly suggests that HEPs are not involved in formation of the girdle band silica.

Why then are HEPs so precisely confined to the overlapping girdle bands? It should be emphasized that the overlap region is probably the most critical part of the cell wall, because integrity of the cell wall is only disrupted in this region. There-

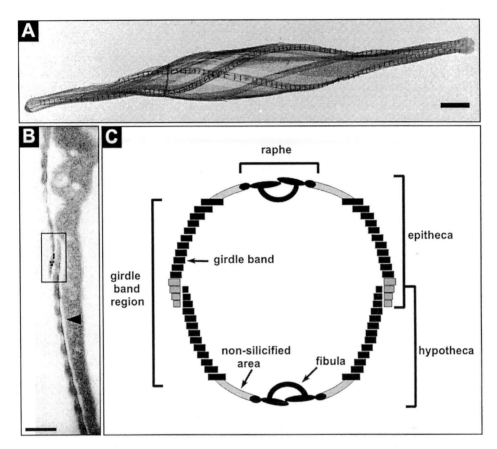

Figure 11.4 Immunolocalization of HEPs. (a) Transmission electron microscopy (TEM) image of *C. fusiformis* cell wall. Bar = 2.5 μm. (b) Post-embedding immuno-gold labeling of sectioned *C. fusiformis* cells. A TEM image of a detail from the girdle band region is shown. The black arrow indicates the position of the plasma membrane and points towards the extracellular space. The rows of oval-shaped structural elements outside the protoplast represent girdle bands from hypotheca (coming from below) and epitheca (coming from the top), respectively. The overlap region is boxed. For indirect immunolabeling anti-HEP200 antibody was used. Bar = 200 nm. (c) Cross-section through *C. fusiformis* cell wall (schematic). The girdle bands that become labeled by the anti-HEP200 antibodies are shaded.

fore, machinery should exist that controls the reversible opening of the overlap region: at cell division the contact between epitheca and hypotheca must be eliminated in order to allow dissociation of the daughter cells (Fig. 11.2b). Subsequently the tight connection between epitheca and hypotheca has to be re-established in each daughter cell to ensure mechanical stability of the cell wall. It could indeed be demonstrated that HEP200 becomes associated with the girdle bands of the overlap region during cell division, suggesting that these proteins may be involved in establishing the reversible connection of epitheca and hypotheca [32].

11.4.2 Structure and Properties of Silaffins

The most prominent components within the HF-extract of the *C. fusiformis* cell wall are three low molecular mass polypeptides exhibiting apparent masses of 17, 8 and 4 kDa in Tricine-SDS-PAGE (Fig. 11.5a). For reasons that will be described below, these polypeptides were called silaffin-1A (4 kDa band), silaffin-1B (8 kDa band) and silaffin-2 (17 kDa band). Silaffins are strongly basic and contain post-translationally modified amino acids. N-terminal amino acid sequencing of each of the silaffins revealed a high degree of sequence homology between silaffin-1A and silaffin-1B (Fig. 11.5a). In both sequences modified amino acid derivatives are present at posi-

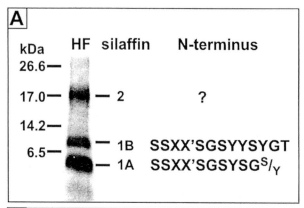

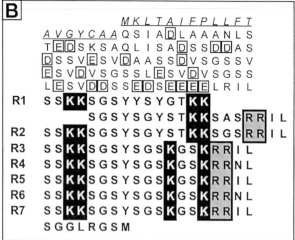

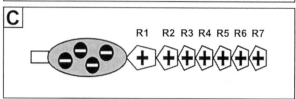

Figure 11.5 Silaffins. (a) Tricine/SDS-PAG (16 % acrylamide) of the HF-extract from *C. fusiformis* cell walls (HF). The positions of the silaffins and the corresponding N-terminal sequences are indicated on the right side of the lane. (b) Amino acid sequence of sil1p. The signal peptide is shown in italics and underlined. The N-terminal acidic domain is shown in normal letters. Aspartate and glutamate residues are boxed. The silaffin generating C-terminal domain is shown in bold letters. Lysine and arginine residues are highlighted. R1 to R7 indicate the repeating sequence elements within the C-terminal domain. (c) Schematic primary structure of sil1p. The white box represents the signal peptide. The shaded oval represents the negatively-charged N-terminal domain. The silaffin-bearing part of sil1p is shown as a row of pentagons. Each pentagon (R1 to R7) corresponds to a sequence repeat as indicated in (b).

tions 3 and 4. Little sequence information was obtained from N-terminal sequencing of silaffin-2 because most Edman degradation cycles yielded unidentified amino acid derivatives.

11.4.2.1 Polypeptide Sequences of Silaffins

The sequence information obtained from the N-terminus of silaffin-1B allowed cloning of the corresponding sil-1 gene. As is the case with all other diatom cell wall proteins characterized so far, the sil-1 encoded polypeptide (designated sil1p) displays a modular primary structure that is composed of unique as well as repetitive sequence elements (Fig. 11.5b). The N-terminus represents a typical signal peptide sequence (amino acids 1 to 19) that is followed by a highly negatively-charged domain (amino acids 20 to 107). Starting from amino acid 108 the C-terminal part of sil1p is composed of seven strongly basic repeat units. The first two repeat units R1 (amino acids 108 to 140) and R2 (amino acids 141 to 162) contain 33 and 22 amino acids, respectively. The remaining five repeat units (R3 to R7) are slightly shorter, each being composed of 19 amino acid residues. The characteristic feature of each repeat unit is the presence of clusters of lysine and arginine residues that are connected by stretches containing hydroxy amino acids (serine, tyrosine) and glycine. Comparison of the sil1p sequence with the N-terminal sequences of silaffin-1A and 1B revealed two interesing features:

The C-terminal part of sil1p (repeat units R1 to R7) becomes extensively proteolytically processed *in vivo*, releasing each repeat unit as an individual peptide. According to the results of N-terminal sequencing, silaffin-1B is derived from repeat unit R1, whereas silaffin-1A represents a mixture of peptides derived from repeat units R2 to R7. So far, no peptide sequence has been found in cell wall extracts that corresponds to the acidic N-terminal half of sil1p.

The lysine residues are the targets for post-translational modification in silaffin-1A and 1B. Amino acid analysis has shown that in fact all of the lysine residues are modified.

For further characterization of silaffin-1A and 1B the molecular structures of the lysine modifications were determined.

11.4.2.2 Lysine Modifications

The N-terminal sequence SSXX'SGSY (X and X' represent differently modified lysine residues) is a common sequence motif in silaffin-1A and 1B. Combination of NMR spectroscopy and fragmentation analysis by ESI–MS revealed the structures of these modifications. X' was found to be a ε-N-dimethyllysine residue (Fig. 11.6). The other type of modification, X, also involves the ε-amino group of lysine. However, only one substituent is a methyl group – the other substituent is an oligomeric structure consisting of N-methylpropylamine units (Fig. 11.6). This modification is heterogenous with respect to chain length because between 5 and 10 N-methylpropylamine units have been found to be attached to lysine residue X (Fig. 11.6). This lysine modification found in silaffin-1A and 1B is unique for two reasons. First, oligo-N-methylpropylamine is the only oligomeric alkyl modification of a protein described so far. Second, silaffins are the first extracellular proteins to be

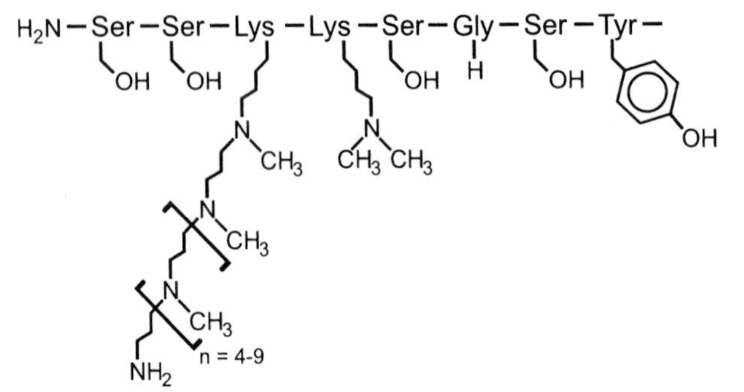

Figure 11.6 Chemical structure of lysine modifications. The modified lysine residues X and X′ are shown in the context of the peptide sequence SSXX′SGSY which is contained in silaffin-1A and 1B, respectively. The peptide backbone is indicated as three-letter amino acid symbols connected by horizontal lines. The structures of the side chains are shown below the respective amino acid symbol.

identified that carry alkylated lysine modifications. In this context it should again be noted that Nakajima and Volcani [6] have isolated ε-N-trimethyl-δ-hydroxylysine from diatom cell walls but so far the protein carrying this amino acid residue has not been identified.

11.4.2.3 Silica Precipitation by Silaffins

As was described in the previous section, silaffins are a unique component of diatom cell walls suggesting that they may be required for diatom cell wall development. In fact, silaffins have a strong impact on silica formation *in vitro*: each of the silaffins, as well as the mixture of silaffins, is able to precipitate silica within seconds when added to a freshly prepared solution of metastable silicic acid. In the absence of silaffins the silicic acid solution remains homogenous for at least a few hours. Thus, silaffins greatly accelerate silica formation. The amount of precipitated silica is proportional to the amount of silaffin applied (Fig. 11.7a). At any protein concentration silaffins completely co-precipitate with the silica as long as silicic acid is present in excess.

The structure of the precipitate formed was analysed by scanning electron microscopy (SEM). The silaffin-1A-induced precipitate was found to be composed of a network of nearly spherical silica particles with diameters of 500–700 nm. Within the networks neighbouring silica spheres are closely attached to each other or are partly fused (Fig. 11.7b). When the unfractionated mixture of silaffins (containing silaffin-1A, 1B and 2) was used for precipitation, aggregates of much smaller silica particles (diameters < 50 nm) were obtained (Fig. 11.7c). This result suggests that silaffins may also have an influence on the morphology of the precipitating silica.

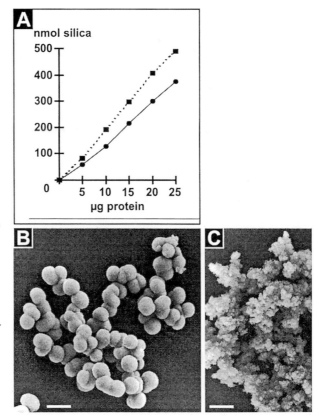

Figure 11.7 Silica precipitation induced by silaffins. (a) Correlation between silaffin concentration and the amount of silica precipitated from a silicic acid solution. The dotted line represents the result obtained for the silaffin mixture, the solid line shows the result for pure silaffin-1A. (b, c) SEM micrographs of silica precipitated by silaffin-1A (b) and the mixture of silaffins (c). The diameter of the silica particles is 500–700 nm (b) and <50 nm (c). Bar = 1 μm.

11.5 Silica Formation *in Vitro*

In order to propose a model for the action of silaffins in the production of silica particles, knowledge of the inorganic chemistry of silica formation is required. Formation of silica from silicic acid is an extraordinarily complex process. However, a general picture of the basic mechanisms of silica formation has emerged [33] which is briefly summarized below.

11.5.1 The Chemistry of Silica Formation

Within the pH range 2–9 monosilicic acid $Si(OH)_4$ is stable in water, providing its concentration remains below 2 mM. At higher concentrations monosilicic acid spontaneously polymerizes by condensation to form dimers (Fig. 11.8a), trimers and higher molecular weight species of silicic acid. Thus, after an initial poly-

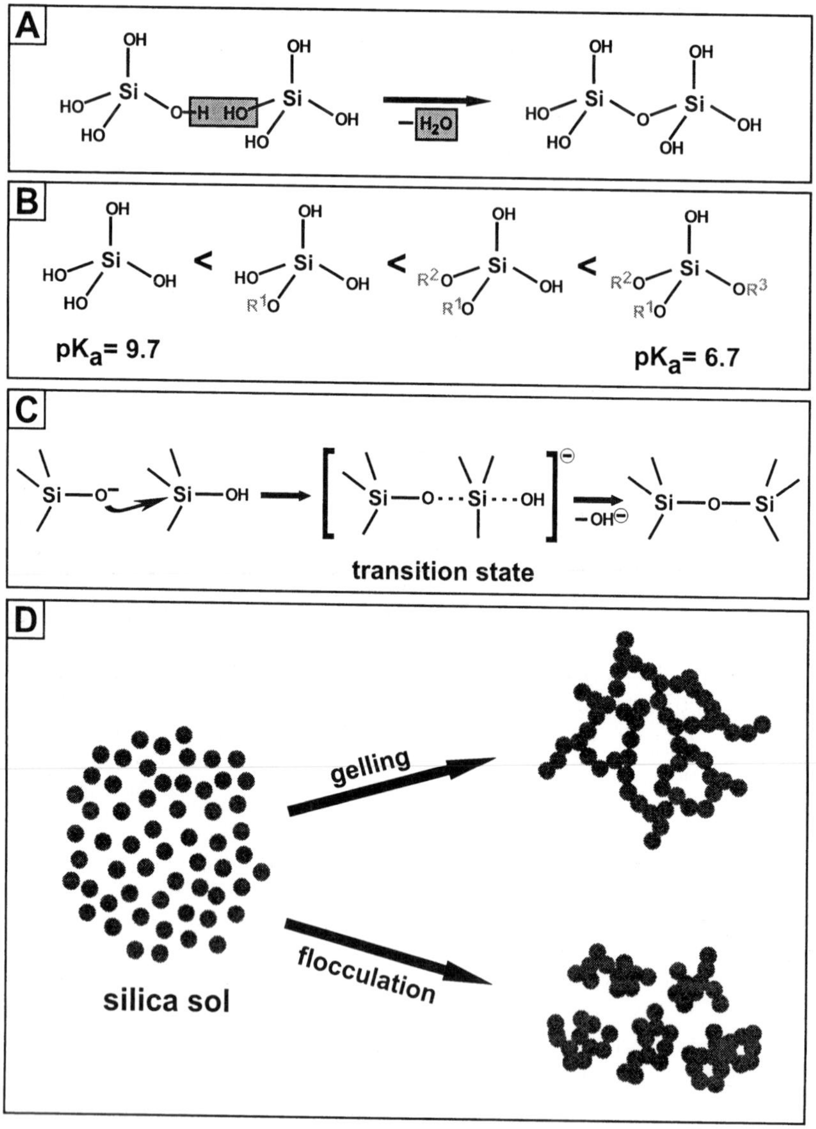

Figure 11.8 Silicic acid condensation and silica formation. (a) Condensation of two monosilicic acid molecules. The Si–OH group is called a silanol group and the Si–O–Si group is called a siloxane group. (b) Acidity of silanol groups. The residues R^1 to R^3 indicate different silicic acid substituents. (c) Mechanism of siloxane bond formation. At pH > 2 the formation of the silanolate anion Si–O$^-$ is the rate-limiting step in siloxane bond formation. (d) Formation of hard silica from a silica sol. Each sphere represents a polysilicic acid particle.

condensation phase, different oligosilicic acid species are present that have a strong tendency to further polymerize in such a way that a maximum of siloxane bonds (Si–O–Si) is produced with a minimum of uncondensed silanol groups (Si–OH) remaining. This peculiar polycondensation behaviour of silicic acids is explained by the following facts:

The acidity of a silanol group increases with the number of siloxane linkages present on the Si atom. Thus, in the sequence shown in Fig. 11.8b, acidity increases from left to right, with monosilicic acid having the lowest acidity. The acidity of trisubstituted silicic acids increases by three orders of magnitude.

Above pH 2 the rate-limiting step of condensation between two silicic acid molecules is the attack of a silanolate ion towards a silanol species (Fig. 11.8c).

Intramolecular siloxane bond formation is kinetically favored over intermolecular condensation reactions. Thus, adjacent silanol groups on an oligomeric silicic acid will undergo ring closure, thereby forming highly branched species.

According to these properties, branched oligmers are formed in a supersaturated silicic acid solution and further polycondense at a higher rate than linear oligomers. Therefore compact, highly branched species of polysilicic acid predominate. When monosilicic acid is used up, the polysilicic acids develop into larger spherical particles by Ostwald ripening. In this process larger particles grow at the expense of smaller particles, because the latter have a higher solubility. The final particle size depends on temperature and pH of the solution but typically ranges from 1 to 10 nm diameter. These nanoscale polysilicic acid particles contain between 50 Si units in a 1 nm particle (molecular mass 3000 Da) and 50 000 Si units in a 10 nm particle (molecular mass 3.2×10^6 Da). Although such a polysilicic acid solution has a homogenous appearance, it is a colloidal system (called a sol) in which two different phases, namely water and particulate polysilicic acids, are present. These polysilicic acid particles (silica particles) are the nuclei for silica formation and thus the solution is denoted a silica sol. However, it depends on the pH of the solution and the presence of additional components if hard silica can form from a silica sol. Above pH 7 the silanol groups of highly substituted polysilicic acids ($pK_s = 6.7$) are considerably dissociated. Thus, the polysilicic acid particles of the sol bear a negative surface charge and repel each other. As a result, alkaline silica sols form stable solutions from which silica does not precipitate. In contrast, below pH 6 the sol particles bear only little negative surface charge and thus can collide with each other. This results in the formation of branched particle chains that are crosslinked by siloxane bonds creating a wide-meshed three-dimensional network throughout the solution. As a consequence the sol becomes viscous and ultimately hardens as a transparent gel (gelling). This process is rather slow and may take several hours or days. A different effect is observed when electrolytes (e.g. NaCl, tetramethylammonium chloride) or polycationic polymers (e.g. polyethyleneimine) are added to silica sols. When either of these substances is present, flocculation occurs and a silica precipitate is formed rather than a silica gel. The reason for this effect is that in the presence of the flocculating agents the polysilicic acid particles become intimately connected. In particular, polycationic polymers interconnect the sol particles by simultanous interaction with the negative surface charges on adjacent particles. This creates localized, narrow-meshed networks of silica particles, which are very much different from the wide-meshed network that develops during gelling (Fig. 11.8d). Thus, flocculation leads to the formation of a silica-enriched hard phase that rapidly (seconds to several minutes) precipitates from the less dense, silica-depleted solution.

Apparently, diatom biosilica formation corresponds to silica flocculation rather than gelling, because numerous observations have been made in different diatom species that silica is initially laid down as tightly associated spheres of up to 100 nm in diameter [13]. Subsequently, the interspace between the spheres appears to be filled with silica and thus the mature silica skeleton displays a rather smooth surface. However, the particulate nature of diatom biosilica is revealed when partial dissolution of the silica skeleton has taken place [14]. There have also been reports of the presence of fibrillar silica strands within developing SDVs of two diatom species [13, 34]. Such fibrils might actually be composed of linear rows of partially fused silica spheres.

11.5.2 Silaffin Mediated Silica Formation

In a diatom cell silica formation probably takes place at a pH below 7 (see Section 11.3.1). As was outlined in the previous section, silica formation under these conditions is a very slow process that results in gelling unless flocculating agents are present. In fact, silica formation in diatoms is a rather rapid process that usually takes less than an hour (Section 11.3.3) and leads to silica flocculation. This demands for the presence of molecules in diatoms that accelerate silicic acid polycondensation and mediate flocculation of the resulting silica sol. We propose that these *in vivo* processes are mediated by polycationic organic molecules like the silaffins that have been identified in *C. fusiformis*. Silaffins appear to be ideally designed to promote formation of hard silica for several reasons. As was pointed out by Iler [33], cationic and hydrogen bonding polymers are highly active flocculating agents because they can interact with both the silanolate groups and the silanol groups of the silica sol particles, respectively. Exactly these structural elements are present in silaffins because they are polycationic molecules containing a high proportion of hydroxy amino acids (serine and tyrosine). The *in vitro* experiments have shown that in the presence of silaffins precipitates of crosslinked silica particles are indeed formed (Fig. 11.7). The effect of silaffin-1A on silicic acid solutions is remarkable because even between pH 5 and 6 extremely large particles (diameter 500–700 nm) are formed. In acidic solutions at ambient temperature silica particles do not normally grow larger than 2–3 nm in diameter, because particle growth by Ostwald ripening is much slower than gelling or precipitation. It should be noted that the silica elements in the *C. fusiformis* cell wall have cross-sectional diameters of about 10–100 nm, indicating that the formation of silica spheres of several hundred nanometers in diameter is not required *in vivo*. However, investigating the ultrastructure of the large silica spheres formed *in vitro* may provide clues to the mechanism of silaffin-1A-induced silica precipitation and could reveal whether or not silaffin-1A has a templating activity and whether it may even exert control over silica morphogenesis.

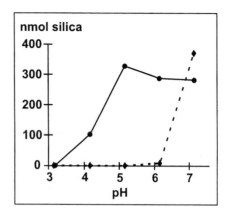

Figure 11.9 pH dependence of peptide-induced silica precipitation. The solid line shows the result for silaffin-1A, the dotted line represents the result for synthetic peptide pR5 (sequence: SSKKSGSYSGSKGSKRRIL). At each pH value the amount of respective peptide applied was 28 nmol.

The *in vitro* silica precipitation assay was used to study the influence of the lysine modifications on the function of silaffin-1A. The silica precipitation activities of silaffin-1A and synthetic peptide pR5 (without any modifications) were compared at different pH values. Synthetic peptide pR5, containing 19 amino acid residues, corresponds to the amino acid sequence of repeat R5 of the silaffin precursor polypeptide (Fig. 11.5b). Due to the high content of basic amino acid residues, pR5 is polycationic and can act as a silica flocculating agent. At pH 7 the silica precipitation activity of pR5 is comparable to the activity of silaffin-1A. However, in the acidic pH range pR5 no longer has any silica precipitation activity, whereas silaffin-1A activity exhibits a maximum around pH 5 and persists down to pH 4 (Fig. 11.9). Because the SDV is an acidic compartment, this result demonstrates that the lysine modifications present in silaffin-1A are required for silica-precipitating activity under physiological pH conditions. The oligo-*N*-methylpropylamine modification strongly increases the polycationic character of silaffin-1A, thus improving its flocculating property. More importantly, it has recently been shown that polyamines catalyze polycondensation of monosilicic acid at pH 8.5 [35]. The catalyzing effect may be due to the action of the amino groups, which can stabilize the transition state of the condensation reaction by aiding the proton transfer reactions. Figure 11.10 illustrates this model and shows how this catalysis could be accomplished by the oligo-*N*-methylpropylamine chain. A prerequesite for this model is the presence of both protonated and unprotonated amino groups within the oligo-*N*-methylpropylamine chain. This is probably the case even at acidic pH values, because charge repulsion and cooperative effects within the silaffin-1A molecule may lead to a decrease in pK_s values of amino groups. Indeed studies on polyethyleneimine (a synthetic polymer made of $-CH_2-CH_2-NH-$ and $-CH_2-CH_2-N(CH_2)_n-$ units) have shown that in the presence of 0.1 M electrolyte at pH 5.5 only 50–60% of the amino groups are protonated [36]. Therefore, even under acidic conditions the effective concentration of unprotonated amino groups would be sufficient to allow silaffin-mediated catalysis of silicic acid polycondensation.

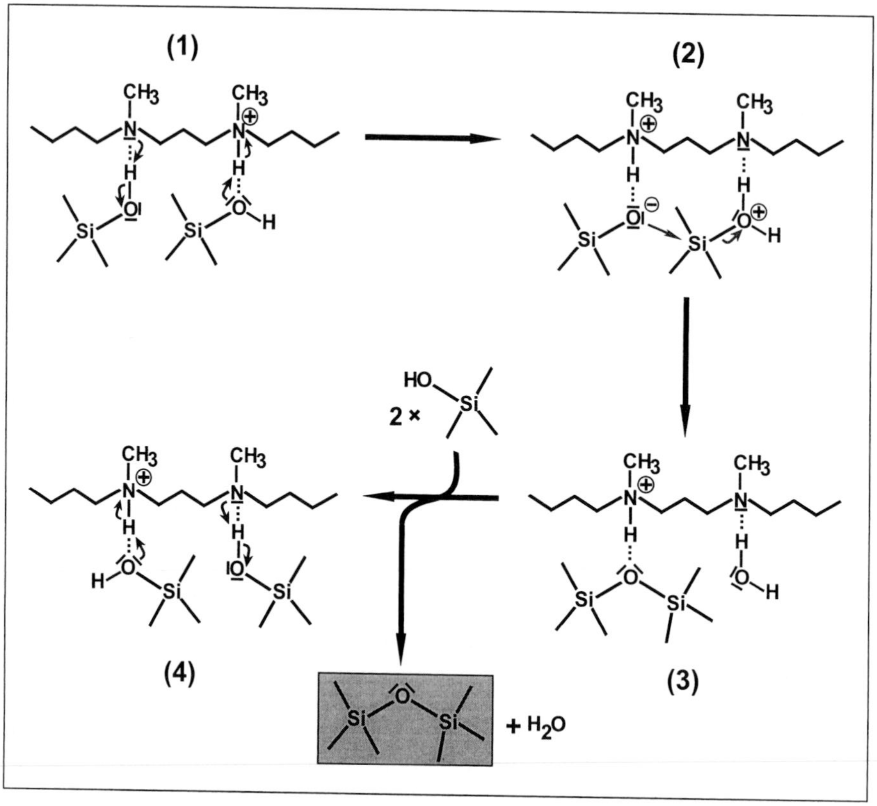

Figure 11.10 Proposed mechanism for siloxane bond formation catalyzed by the oligo-N-methyl-propylamine chain of silaffins. Two consecutive repeat units within the oligo-N-methyl-propylamine chain (polyamine) of a silaffin's modified lysine residue are depicted here. Protonated and unprotonated tertiary amino groups may alternate within the polyamine chain. (1) The amino group and the ammonium group each bind a silicic acid molecule by hydrogen bonding. The amino group becomes protonated by taking over a proton from the silanol group of the bound silicic acid molecule, whereas the ammonium group donates its proton to the silanol group of the other bound silicic acid molecule. (2) These proton exchange reactions result in formation of a reactive silanolate ion and transform the hydroxyl group of the neighboring silicic acid molecule into an oxonoium ion which is a good leaving group. Subsequently, in a nucleophilic reaction the silanolate group attacks the positively polarized silicon center of the neigboring protonated silicic acid molecule. (3) A siloxane bond forms between the two silicon centers and at the same time the oxonium group becomes split off as a water molecule. (4) The reaction products become displaced by two new silicic acid molecules and the catalytic cycle can start again. However, compared to the situation in (1), the orientation of protonated and unprotonated amino groups in (4) is reversed.

11.6 Conclusion and Future Prospects

Diatom biosilica is a complex composite material in terms of both morphology and composition. We propose that biogenesis of diatom biosilica is a hierarchical process that may be dissected at least into three subsequent steps. First, nanoscale silica particles with diameters of 10–50 nm are formed as basic building blocks. Second, these building blocks are arranged and fused in a way that creates the species-specific morphology of valve and girdle bands from which the diatom cell wall is constructed. Third, valve and girdle bands are assembled into a functional diatom cell wall. We suggest that each step in this process is guided by organic molecules. Silaffins are candidates for being involved in the first two steps: The polyamine modifications of silaffins may catalyze the polycondensation of silicic acid, thus forming nanoscale silica particles. Subsequently, the polypeptide moiety of silaffins may exert a guiding (nucleation?) function during supramolecular assembly and fusion of these basic building blocks. As an alternative to the pre-pattern model mentioned above (Section 11.3.3), silaffins together with an anionic partner could induce a phase separation within the SDV, creating a pattern of areas promoting or inhibiting polycondensation of silicic acid and flocculation of silica. Finally, HEPs appear to be involved in the assembly of girdle bands in the overlap region, forming an intact cell wall by combining an epitheca with a hypotheca.

Silaffins and HEPs are certainly not the only organic molecules involved in biogenesis of diatom cell walls, but currently they are the only components for which a function can be proposed based on experimental data. Further analysis of diatom cell wall (and SDV) components will reveal additional organic molecules that are involved in distinct steps of biosilica morphogenesis.

Only then will we be able to solve the most fascinating question about diatom cell wall formation: does biosilica morphogenesis rely on the templating function of an organic pre-pattern or are the silica patterns the result of dissipative structures that arise by non-linear processes acting far away from thermodynamic equilibrium?

Acknowledgements

This work was supported by the Deutsche Forschungsgemeinschaft (SFB 521).

References

[1] W. Smetacek, *Nature* 1999, *397*, 475–476.
[2] D. Werner, in *Biology of Diatoms, Botanical Monographs Vol. 13* (Ed. D. Werner), Blackwell Scientific Publications, Oxford, 1977, pp. 1–17.

[3] S. Mann, *Nature* 1993, *365*, 499–505.
[4] J. Pickett-Heaps, A. M. M. Schmid, L. A. Edgar, in *Progress in Phycological Research, Vol. 7* (Eds F. E. Round and D. J. Chapman), Biopress, Bristol, 1990, pp. 1–169.
[5] F. Azam, B. E. Volcani, in *Silicon and Siliceous Structures in Biological Systems* (Eds T. L. Simpson and B. E. Volcani), Springer-Verlag, New York, 1981, pp. 43–67.
[6] T. Nakajima, B. E. Volcani, *Science* 1969, *164*, 1400–1406; *Biochem. Biophys. Res. Commun.* 1970, *39*, 28–33.
[7] R. Hecky, K. Mopper, P. Kilham, T. Degens, *Mar. Biol.* 1973, *19*, 323–331.
[8] D. Swift, A. Wheeler, *J. Phycol.* 1992, *28*, 202–209.
[9] R. W. Drum, H. S. Pankratz, *J. Ultrastruct. Res.* 1964, *10*, 217–223.
[10] T. L. Simpson, B. E. Volcani (Eds) *Silicon and Siliceous Structures in Biological Systems*, Springer-Verlag, New York, 1981, pp. 3–12.
[11] C. W. Li, S. Chu, M. Lee, *Protoplasma* 1989, *151*, 158–163.
[12] E. G. Vrieling, W. W. C. Gieskes, T. P. M. Beelen, *J. Phycol.* 1999, *35*, 548–559.
[13] C. W. Li, B. E. Volcani, *Phil. Trans. R. Soc. Lond. B* 1984, *304*, 519–528.
[14] J. C. Lewin, in *The Physiology and Biochemistry of Algae* (Ed. R. A. Lewin), Academic Press, New York, 1962, pp. 445–455.
[15] C. C. Perry, in *Biomineralization: Chemical and Biochemical Perspectives* (Eds S. Mann, J. Webb and R. J. P. Williams), VCH, Weinheim, 1989, pp. 223–256.
[16] P. Tréguer, D. M. Nelson, A. J. van Bennekom, D. J. DeMaster, A. Leynaert, B. Quéguiner, *Science* 1995, *268*, 375–379.
[17] P. Bhattacharyya, B. E. Volcani, *Proc. Natl Acad. Sci. USA* 1980, *77*, 6386–6390.
[18] M. Hildebrand, K. Dahlin, B. E. Volcani, *Mol. Gen. Genet.* 1998, *260*, 480–486.
[19] M. Hildebrand, B. E. Volcani, W. Gassmann, J. I. Schroeder, *Nature* 1997, *385*, 688–689.
[20] M. A. Brzezinski, D. J. Conley, *J. Phycol.* 1994, *30*, 45–55.
[21] T. A. Villareal, L. Joseph, M. A. Brzezinski, R. F. Shipe, F. Lipschultz, M. A. Altabet, *J. Phycol.* 1999, *35*, 896–902.
[22] K. Shimizu, J. Cha, G. Stucky, D. E. Morse, *Proc. Natl Acad. Sci. USA* 1998, *95*, 6234–6238.
[23] D. H. Robinson, C. W. Sullivan, *TIBS* 1987, *12*, 151–154.
[24] B. E. Volcani, in *Silicon and Siliceous Structures in Biological Systems* (Eds T. L. Simpson and B. E. Volcani), Springer-Verlag, New York, 1981, pp. 157–200.
[25] R. Gordon, R. W. Drum, *Int. Rev. Cytol.* 1994, *150*, 243–372.
[26] N. Kröger, M. Sumper, *Protist* 1998, *149*, 213–219.
[27] N. Kröger, C. Bergsdorf, M. Sumper, *EMBO J.* 1994, *13*, 4676–4683.
[28] N. Kröger, C. Bergsdorf, M. Sumper, *Eur. J. Biochem.* 1996, *239*, 259–264.
[29] N. Kröger, G. Lehmann, R. Rachel, M. Sumper, *Eur. J. Biochem.* 1997, *250*, 99–105.
[30] H. Fischer, I. Robl, M. Sumper, N. Kröger, *J. Phycol.* 1999, *35*, 113–120.
[31] N. Kröger, R. Deutzmann, M. Sumper, *Science* 1999, *286*, 1129–1132.
[32] N. Kröger and R. Wetherbee, *PROTIST*, 2000, in press.
[33] R. Iler, in *The Chemistry of Silica*, John Wiley & Sons, New York, 1979, pp. 172–461.
[34] J. Pickett-Heaps, D. Tippit, J. Andreozzi, *Biol. Cell.* 1979, *35*, 199–206.
[35] T. Mizutani, H. Nagase, N. Fujiwara, H. Ogoshi, *Bull. Chem. Soc. Jpn* 1998, *71*, 2017–2022.
[36] G. M. Lindquist, R. A. Stratton, *J. Colloid. Interface Sci.* 1976, *55*, 45–59.

12 Silicic Acid Transport and Its Control During Cell Wall Silicification in Diatoms

Mark Hildebrand

12.1 Introduction

Perhaps the most outstanding examples of micro- and nanoscale structured materials in nature are the intricate and ornate silicified cell walls of diatoms. Although diatoms have been admired by microscopists for over 200 years, our understanding of the processes involved in making their cell walls is still very limited. With the application of electron microscopic and biochemical techniques, and more recently those of molecular biology, certain diatom species are being developed as model systems to study cellular silicon metabolism and silicified cell wall formation. Recent advances hold the promise of beginning to unravel the molecular details controlling these processes [1–7].

Our own research has focused on applying molecular biological approaches to understand diatom cellular silicon metabolism. We isolated the first cDNA clones derived from silicon-responsive genes [8]. From these clones, we isolated, functionally identified, and are continuing to characterize genes and proteins responsible for the transport of silicic acid into the diatom cell [5, 6]. The silicic acid transporters (SITs) were the first biological components shown to directly and specifically interact with silicon [5], and we hope to develop them as a model to understand how proteins in general may interact with silicon. More recently we have investigated the relationship of intracellular silicon pools with transport and cell wall silica incorporation [9]. We and others have demonstrated a tight coupling between uptake and incorporation in some diatom species, and in these species shown that silicic acid uptake is actually regulated by incorporation [9–11]. Thus, the mineralization process and transport are intimately linked. Because of the chemistry of silicic acid [12] and the need to deliver correct amounts into the silica deposition vesicle (SDV) where silicification occurs, transport of silicon into the cell, through the cytoplasm, and into the SDV are integral parts of the polymerization process. Therefore, a complete understanding of silicification cannot be gained without understanding silicic acid transport and its control. The purpose of this chapter is to develop this concept, and presents a review of our current understanding of silicic acid transport in the diatom and its relation to silicification of the cell wall, and identifies areas for future research.

12.2 Overall Considerations for Silicic Acid Transport During Diatom Cell Wall Synthesis

The diatom cell wall, which is also called a frustule, can contain a substantial amount of silica. Measurements range from 7 to 1100 fmole per cell, depending on the species and environmental conditions [9, 13], comprising up to 50% of the dry weight of the cell [14]. The silicified wall is formed after cytokinesis but prior to cell separation [15], and so the daughter cells are still attached during cell wall synthesis (Fig. 12.1). Silicification occurs within the silica deposition vesicle [16], which is enclosed by its own membrane (the silicalemma) and contained entirely within the plasma membrane of the cell (Fig. 12.1). After deposition is complete, exocytosis of the SDV occurs [15] to become part of the new cell wall. Because the new walls form on adjacent faces within the two attached daughter cells, they are shielded from the external environment (Fig. 12.1). This arrangement necessitates uptake of silicic acid from other parts of the cell which are in contact with the environment. Silicon must then be transported through the cytoplasm and into the SDV (Fig. 12.1). The requirement for cytoplasmic transport has important consequences in the overall scheme of cellular silicon metabolism, because the cell must deal directly with the large amounts of silicic acid required for the cell wall. In many diatom species the wall is made in an hour or two, and in some much less [13]. The substantial amounts of silicon required within these short time periods results in a high concentration in the cytoplasm, where silicon must be prevented from autopolymerizing [12] before reaching the SDV. Thus, not only must cellular uptake be rigorously controlled to allow enough silicon for the wall yet preventing excess, but intracellular transport must be adapted to maintain silicon in a soluble form. At the same time, the intracellular transport mechanism must deliver and release silicic acid into the SDV in a carefully controlled manner.

12.3 The Solution Chemistry of Silicon

The solution chemistry of silicon is complex, but has been thoroughly described in the monograph by Iler [12]. Because of its complexity and the fact that different silicon compounds may be present at different steps in cellular transport and deposition, some description may be useful. The term "silicon" not only refers to the element, but is used as a generic term when the specific form of a silicon compound is unknown [17]. The predominant form of silicon in aqueous solution at low concentrations is silicic acid, $Si(OH)_4$. This is a weak acid with a pK_a of 9.8 for the formation of $SiO(OH)_3^-$. By increasing the concentration in solutions of pH < 9, or decreasing the pH of a saturated solution, silicic acid will autopolymerize to form amorphous silica [12]. Polymerization occurs through stabilized intermediates,

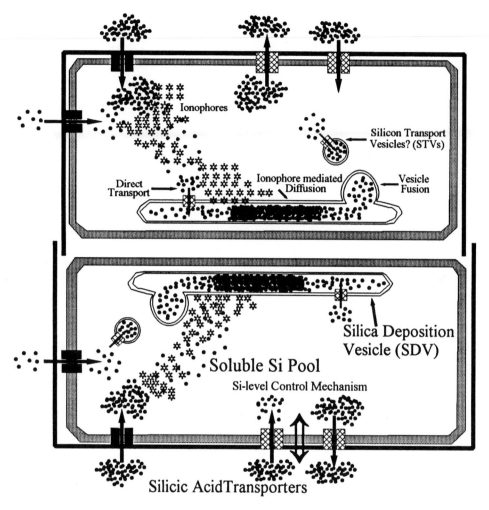

Figure 12.1 Features of silicic acid uptake, intracellular transport and deposition in a dividing diatom cell, depicting two daughter cells after cytokinesis still within the silicified walls of the mother cell during cell wall synthesis. Dark outer lines represent the silicified walls of the mother cell, large rectangles just inside with shaded borders represent the plasma membrane of each daughter cell. Other components are labeled. Dark dots represent silicic acid. The lower daughter cell illustrates three major components of transport and silicification: the silicic acid transporters, the soluble silicon pool, and the silica deposition vesicle. Different transporters may have differing affinity or capacity for transport (compare extreme left with lower left). As part of the cell's silicon level control mechanism the transporters can both take up and efflux silicic acid (bottom right of center). The upper daughter cell illustrates three possible means of intracellular transport: direct transport by intracellularly localized silicic acid transport proteins, ionophore-mediated transport by silicate ionophoretic activities, and transport via silicon transport vesicles, which can fuse with the SDV.

forming polysilicic acids, then colloidal silica particles of discrete sizes, and finally, depending on the pH, either a three-dimensional silica gel network or a sol [12, 18]. The solubility of silica is 1.7–2.5 mM over a broad range of pH values below 10 [12]. Although silicon is the second most abundant element in the Earth's crust, and is therefore plentiful, its availability for biological use is limited by its solubility in water. In fact, the growth of marine diatoms can so deplete surface waters of silicon that their further proliferation is prevented [19]. Prior to the evolution of siliceous plankton, oceanic silicic acid concentrations were near saturation [20], but now the global ocean average is 70 µM, and usually much less in surface waters [21]. $Si(OH)_4$ comprises about 97% of the dissolved silicon in seawater, and $SiO(OH)_3^-$ most of the remaining 3% [22]. Silicon in the aqueous environment is referred to as dissolved silicon (DSi), or in relation to the cell, extracellular or external silicon. Silicon in the diatom cell wall is amorphous silica, and silicon in intracellular pools is called intracellular soluble silicon.

12.4 Characterization of Diatom Silicic Acid Uptake

The first demonstrations of silicon uptake by diatoms were by Lewin [23, 24]. The first kinetic measurements of uptake, demonstrating active silicon transport, were by Paasche [25] and Azam *et al.* [26]. Numerous studies [25–31] have shown that the specific rate of uptake follows Michaelis–Menten or Monod [32] saturation functions, indicating that transport is carrier-mediated. Kinetic parameters for silicon uptake have been measured in many diatom species, and K_s values range from 0.2 to 7.7 µM, with V_{max} values of 1.2–950 fmol Si cell^{-1} h^{-1} [33].

Studies using the diatom *Phaeodactylum tricornutum* suggested that the ionized form of silicic acid [$SiO(OH)_3^-$] was the chemical form of silicon transported [34], but recent work [35] on three different diatom species has indicated that undissociated silicic acid [$Si(OH)_4$] was the transported form. Controls using *P. tricornutum* showed variable results, sometimes implicating silicic acid and sometimes the anion as the form transported [35]. The unique response of *P. tricornutum* may be related to its habitat, which can be in conditions of high pH where the anion is more prevalent [35]. Silicic acid transport is coupled to sodium in marine diatoms [31], and apparently coupled to sodium and perhaps potassium in freshwater species [29], although at much lower ionic strength. Thus transport in marine species has the characteristics of a sodium/silicic acid symporter [5, 31]. Indirect evidence suggests that transport is electrogenic, and that the $Si(OH)_4$:Na^+ ratio is 1:1 [31]. Metabolic energy is required for uptake [26], and silicic acid transport is inhibited by sulfhydryl blocking reagents [21, 29]. Germanium (germanic acid) and silicic acid are competitive inhibitors for uptake of each other [26, 36], which has led to the use of ^{68}Ge as a radiotracer analog of Si [36].

Three different modes of silicic acid uptake have been defined in diatoms: (i) surge uptake, (ii) internally controlled uptake, and (iii) externally controlled uptake [10]. Surge uptake occurs upon silicon replenishment of silicate-starved cells, when intracellular silicon pools are depleted, and the concentration gradient into the cell is maximum. Uptake rates are at a maximum during surge uptake. In internally controlled uptake, rates are controlled by the rate of utilization of silicon for cell wall deposition [9, 10]. When extracellular concentrations drop to very low levels, externally controlled uptake occurs, where rates are a function of decreasing substrate concentration [10].

Sullivan [30] showed that silicic acid uptake in diatoms did not occur continuously throughout the cell cycle, but in synchronized cultures of *Navicula pelliculosa* was induced just prior to cell wall synthesis. Sullivan [30] also showed that uptake was highly induced during silicate starvation, then decreased 2 h after silicate replenishment and was completely repressed after 4 h. He showed that cells took up enough silicon sometime between 1 and 2 h to allow complete cell wall synthesis and division [30]. Induction of transport was dependent upon protein synthesis [30]. Inhibition of protein synthesis paralleled the natural decrease in activity, and the timing of this suggested that the protein involved was turned over rapidly [30]. V_{max} and K_s values for uptake changed over the course of the cell cycle, indicating changes in the number and affinity of transport sites [30]. This was the first clear indication that silicic acid transport was controlled during cell wall formation and during the cell cycle, and that this occurred by the synthesis and degradation of a specific transporter protein.

12.5 Molecular Characterization of the Silicic Acid Transport System

The cDNA libraries that we generated from *Cylindrotheca fusiformis* [8] were derived from genes turned on or off between conditions of silicate starvation and 1 h after silicate replenishment in synchronized cultures. These conditions were similar to those in Sullivan's study [30], when silicic acid transport activity was induced and repressed. Therefore, it seemed reasonable that the libraries might contain a copy of a silicic acid transporter gene. We screened clones for expression of their corresponding mRNAs during the cell cycle and in response to silicate starvation and replenishment, and identified several that responded in accord with Sullivan's data. We initially isolated six of these clones and found that they had identical sequences (but differed in length), encoding proteins predicted to be rich in hydrophobic amino acids. After obtaining a full-length copy and determining the complete sequence, an open reading frame of 548 amino acids was identified, with what were predicted to be 12 hydrophobic stretches of amino acids indicative of membrane spanning segments [5]. The encoded protein also had a long hydrophilic carboxy-terminal region [5]. Multi-membrane spanning segments are characteristic of transport proteins [37, 38]. Because of this feature, and the expression response of the gene, we thought it possible that the clone might encode a silicic acid transporter.

The function of the protein encoded by the cDNA was tested directly by microinjecting RNA transcribed *in vitro* from the clone into *Xenopus laevis* oocytes and monitoring uptake. We showed that injected oocytes gained the ability to take up ^{68}Ge from the medium [5], and that uptake was dependent on factors that would control the expression levels of the protein (Hildebrand, unpublished). Competition experiments using unlabeled silicic acid demonstrated that the cloned protein was a specific silicic acid/germanic acid transporter [5]. Transport in oocytes was sodium-dependent, and sensitive to the sulfhydryl blocker N-ethyl malemide [5, and unpublished] as observed in diatoms [29, 31]. These results [5] directly demonstrated that the clone encoded a silicic acid transporter, which we called *SIT1* (*Si*licic acid *T*ransporter 1).

Using the *SIT1* cDNA as a probe, we identified multiple hybridizing bands in digests of *C. fusiformis* genomic DNA [6], suggesting the presence of a gene family. A total of five different types of *SIT* gene were isolated from *C. fusiformis* and sequenced, and each type was correlated to a hybridizing band [6]. The predicted amino acid sequences of the five SITs are presented in Fig. 12.2. The SITs have no identified homologs and thus represent a new class of transporter, although they contain most of a signature sequence (AX_3LX_3GR) for sodium symporters [38, 39] at residues 216–225 (Fig. 12.2). Although originally thought to have 12 membrane-spanning segments [5], two new predictive programs [40] suggest that only 10 are present (Figs. 12.2, 12.3), and agree on the location of each. A revised topological model of the SITs is presented in Fig. 12.3. The amino and carboxy termini are predicted to be located in the cytoplasm (Fig. 12.3), which has been shown directly in other even-numbered multi-transmembrane segment transporters [41]. The transmembrane domain must be where silicic acid passes through the membrane, and is highly conserved (87–99 % amino acid identity) comparing the five SITs (Fig. 12.2a, b). The C-terminal domain is less conserved (39–67 %), and has a high probability of forming a coiled-coil structure [6], suggesting that this portion of the transporters interacts with other proteins. In other transporters, the C-terminus can be involved in regulating the activity of transport, in controlling conformational changes in the protein, and in some cases in differential intracellular targeting [42–48]. By analogy, the C-terminus of the SITs is likely to be important in controlling their activity or intracellular location, and the lower degree of conservation in this region suggests that these parameters will vary in the different SITs.

SIT gene mRNA expression was induced four-fold just prior to cell wall synthesis in *C. fusiformis* [6] and the overall pattern of expression correlated with silicic acid transport activity during cell wall synthesis in *N. pelliculosa* [30], quickly reaching high levels prior to maximum silica deposition, and then decreasing rapidly. An increase in silicic acid uptake lagged between 20 and 40 min after induction of the genes [Hildebrand, unpublished], which was probably due to the time required for *SIT* mRNA to be translated and the protein to reach the plasma membrane. Although the pattern of mRNA expression of four of the genes (*SIT2–5*) was almost identical, levels differed by 24-fold. The pattern differed for *SIT1*. Thus, the transporters were required in different amounts during cell wall synthesis, suggesting that they played specific roles in the overall process of transport. These results indicated that transcriptional control was a major component in the regulation of silicic acid transport, but the coiled-coil structures at the C-termini suggested that transport was also controlled via protein–protein interactions.

A.

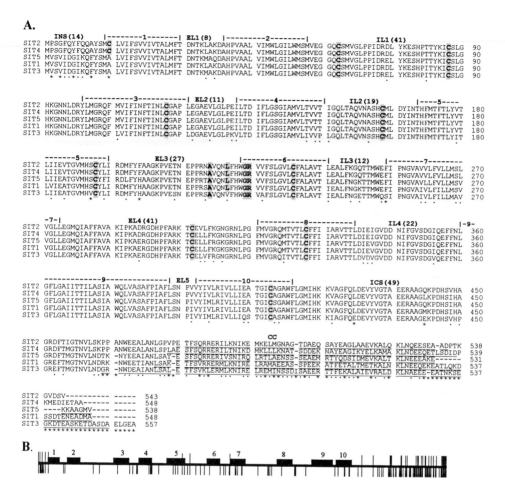

B.

Figure 12.2 Comparison of silicic acid transporter sequences from *C. fusiformis*. (a) Predicted amino acid sequence from the five *SIT* genes were aligned using the MAP program [49]. Sequences are identified on the left, and the amino acid residue number is given on the right. Gaps in the sequences are placed every 15 residues to aid in determining residue number, dashes (–) facilitate maximal alignment. Below the sequences, an asterisk (*) represents a non-conservative replacement in at least one of the sequences, a dot (.) is a conservative replacement, and no symbol indicates identity in all of the sequences. The locations of potential membrane-spanning segments [40] are numbered (1–10) and delineated by brackets above the sequences. Between these segments are intra- and extracellular loops (IL and EL, respectively) numbered according to order, and in parentheses are the number of amino acids in each loop. INS is an intracellular amino segment, and ICS is an intracellular carboxy segment, with parentheses indicating the number of residues in each. Underlined residues in the carboxy-terminal regions have a high probability of forming coiled coils (CC), as determined by the COILS program [50] using a window of 28 residues. Cysteine residues are highlighted, as well as conserved residues in a portion of a signature sequence for sodium coupled symporters [37, 38] at residues 216–225 (AX_3LX_3GR). (b) Schematic diagram of the amino acid conservation pattern in the five *C. fusiformis* SIT sequences. The thick horizontal line represents the coding region. The dark blocks above the thick line represent the location of transmembrane segments, numbered consecutively. Thinner vertical lines below the thick line locate a conservative amino acid replacement in at least one of the sequences, and vertical lines spanning the thick line locate non-conserved residues in at least one of the sequences.

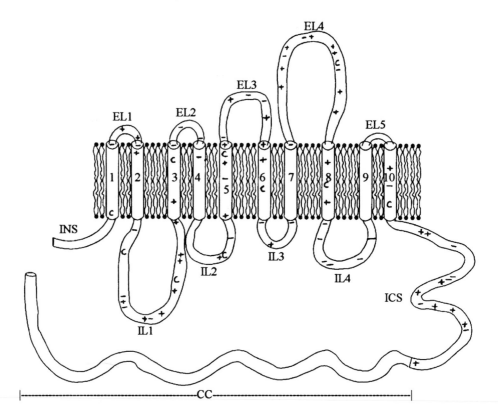

Figure 12.3 Topological model of *C. fusiformis* silicic acid transporters, depicting an extended model of the SITs, based on SIT4, residing in a lipid bilayer membrane. Bottom would be intracellular. Membrane-spanning segments are shown as tubes (1–10), which are connected by intracellular (IL) and extracellular (EL) loops. The intracellular amino segment (INS) and intracellular carboxy segment (ICS), as well as the coiled coil region (CC) are shown. Pluses (+) and minuses (−) locate positively and negatively-charged amino acids, and C locates cysteine residues. The location of charged amino acids is not presented in the coiled coil region because of substantial variation within the five SITs.

From the sequence and expression analysis data, and based on the precedence of other transporters [43, 44, 46, 47] and the documented changes in K_s and V_{max} during diatom cell wall synthesis [30], we proposed [6] that the different forms of SITs have differing affinity and capacity for transport (K_s and V_{max} values), or different intracellular localization (Fig. 12.1). We believe that the regulated expression and possibly localization of the SITs, as well as control at the protein level, are key factors in the overall control of silicic acid uptake in diatoms. Hybridization experiments showed that multiple *SIT* gene copies were present in all diatom species tested [6], suggesting that similar mechanisms may operate.

12.6 Intracellular Silicon Pools

Intracellular pools of soluble silicon in diatoms were first identified by Werner [51]. Studies since then have shown that pool silicon can account for a sizable fraction (up to 50%) of the total cellular silicon in some species under certain conditions [9, 11, 33]. The actual concentration of intracellular soluble silicon depends upon measurement of intracellular water volume, which has resulted in some discrepancies. Pool concentrations (assuming monosilicic acid) of 438–680 mM in *N. pelliculosa* were reported by Sullivan [52]. However, cell water (by weight) was estimated as 20%, and recent measurements suggested a value of 85% for this species [9]. New direct measurements of cell water volume and intracellular silicic acid [9] were in accord with an appropriate adjustment of Sullivan's data, indicating that intracellular concentrations ranged between 58 and 162 mM in this species. Comparison of several diatom species using direct determination of intracellular water volume indicated that pool concentrations (assuming monosilicic acid dispersed equally throughout the cell) ranged from 19 to 340 mM [9]. All results [9, 11, 52, 53] have indicated that diatoms can maintain extremely high intracellular concentrations of soluble silicon, well above saturation for silica solubility [12].

An important question is how can such high intracellular concentrations be maintained? An extremely high Si concentration gradient from inside to outside the cell has been identified [9, 52], which is consistent with intracellular silicon having a different chemical form than extracellular, or being bound or sequestered by some means. Azam *et al.* [26] and Sullivan [52] suggested that partially polymerized or colloidal silica, or organosilicon compounds, could account for high pool levels. However, intracellular concentrations are so high in most species [9] that maintenance of lower molecular weight polysilicic acids should not be possible without some mechanism of stabilization. Substantial amounts of colloidal silica are also not likely because this would be detrimental to the integrity of cellular membranes [12]. Measurement of molybdate-reactive silicon from sonicated *Thalassiosira weissflogii* cells suggested that the predominant form was mono- or disilicic acid [53]. The results of Blank *et al.* [54] also indicated that most intracellular Si was molybdate-reactive in the first hour after silicate addition to starved *Navicula saprophilia*, but from 1–6 h the fraction of molybdate-reactive Si decreased. If the measured form of Si was truly monomeric silicic acid in the cell, then because of the high concentration [9] it would have to be maintained by interaction with other cellular components. However, one must consider that in the molybdate assay intracellular contents are substantially diluted and in some cases heated to 100 °C [9, 53, 54], which may dissolve polymerized forms of silicic acid into molybdate-reactive monomers [12]. More substantial evidence that suggested an association of silicon with organic material or proteins was obtained by Werner [51], and by Azam *et al.* [26], who found that 80% of the pool of soluble silicon in *Nitzschia alba* was precipitable with trichloroacetic acid. Thus, pool silicon could be maintained in soluble form at high concentrations by organic silicon-binding components in the cell. An alternative mechanism proposed that silicon was sequestered in specialized vesicles with conditions in the lumen conducive to maintaining solubility [55, 56].

However, there was no evidence that these intracellular vesicles contained silicon, and even if they did, they would also have to have organosilicon complexes or extremely high pHs to maintain it in a soluble form. In summary, it has not been unambiguously shown what form of silicic acid is present in intracellular pools, which may in part be due to the complexity of silicic acid solution chemistry [12]. However, it is probably safe to say that pool silicon consists of mono- or lower molecular weight polysilicic acids, maintained in soluble form by complexation with organic material [26, 51, 56], or perhaps sequestration [55, 56].

Pool sizes differ in different diatom species, part of which may be due to the relative timing of silicic acid uptake and silica incorporation into the frustule. In many species, uptake and deposition are almost simultaneous, resulting in a relatively small soluble pool of Si, but in others uptake and deposition are temporally uncoupled, allowing the accumulation of sizable pools [11]. *T. weissflogii* can accumulate intracellular soluble silicon sufficient for the synthesis of an entire cell wall [53, 57]. Even in species where pools remain small, levels are above saturation for silica solubility [9]. Pool sizes vary within a species [9, 30, 53]; in *C. fusiformis* pool levels changed in a regular manner during the course of cell wall synthesis [9].

12.7 The Relationship of Intracellular Pools and Incorporation to Uptake

In diatom species where uptake and incorporation were coupled temporally, transport was regulated by the intracellular utilization of silicon, which presumably was the rate of silica incorporation into the wall [10]. We recently showed directly that uptake was controlled by incorporation in *C. fusiformis* [9]. Germanium is a specific inhibitor of diatom cell wall silica incorporation [58], and a competitive inhibitor of silicic acid uptake [26]. In *C. fusiformis* cultures growing in medium with a Ge:Si ratio of 0.1:1, cell wall incorporation was inhibited to 22–25% of the control [9, 59], but according to the K_i value determined in another diatom species, silicic acid transport would not have been appreciably directly inhibited [26, 60]. We showed, however, that uptake was inhibited by blocking incorporation with germanium, and the level of inhibition was identical for both processes, indicating that they were coupled [9]. Thus, in some, and perhaps most [11] diatom species, not only is there a close coupling between the timing of silica incorporation and silicic acid uptake, but incorporation actually exerts control over uptake [9, 10]. In other words, rather than silicic acid being pumped into the cell and driving the silicification process, silicic acid is drawn into the cell upon demand. This would allow the cell to take up enough silicic acid at a given time, but not an excess, keeping intracellular pools to a minimum.

The control of uptake by incorporation was proposed to occur by a mechanism involving intracellular soluble pools, whereby pools were assumed to increase to maximum levels and then feedback to control transport [10]. However, in *C. fusi-*

formis, soluble pools did not have to increase to maximum levels before uptake was controlled [9]. In fact, soluble pool levels changed only gradually over long time periods, and did not transiently expand to accommodate uptake (except under certain conditions of surge uptake). This suggested that other cellular factors were involved in the mechanism controlling transport. We proposed that soluble pool levels were determined by the capacity of intracellular silicon-binding components [9]. We do not know what these components are, and use the term "binding" only to indicate some means of maintaining Si in a soluble form. However, the proposed mechanism should work regardless of how this is accomplished. More or less of the silicon-binding components could be present, explaining the observed range of pool levels over long time periods, but at a given time, soluble pool sizes would be determined by the amount of binding component. We proposed that transport was controlled by the relative amounts of bound and unbound silicon [9]. When unbound silicon-binding component was in excess, uptake would be favored, and with excess unbound Si, uptake would be inhibited or efflux-induced (Fig. 12.4). Incorporation could control uptake by drawing silicic acid from the intracellular pool, altering the ratio of bound to unbound silicon (Fig. 12.4). The driving force for efflux could be the unbound silicic acid concentration gradient from inside to outside the cell.

This mechanism is consistent with data measuring silicic acid efflux from diatom cells [26, 29]. Efflux did not occur in the absence of external silicate [29], indicating that the cell has some means of preventing this, perhaps by binding or sequestering all intracellular silicic acid. Consistent with this, intracellular pools were not depleted even upon prolonged silicate starvation [9, 53]. Increasing amounts of external silicate up to a saturating level actually increased efflux [29]. An explanation for this is that surge uptake should be higher with increasing external Si (due to the higher concentration gradient into the cell), and the greater intracellular excess at higher

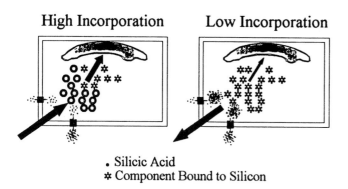

. Silicic Acid
* Component Bound to Silicon

Figure 12.4 Model for control of silicic acid uptake by cell wall silica incorporation through the intermediary of intracellular silicon-binding components. Large rectangles represent diatom cells. Arrows spanning the left side of cells represent net uptake or efflux (left cell and right cell, respectively) of silicic acid under conditions of high and low silica incorporation into the cell wall. The structures at inside top of cells represent silica deposition vesicles; arrows pointing at these denote the relative rate of cell wall incorporation.

external Si should result in more efflux. These data [29] are consistent with our hypothesis [9], in that they suggest that intracellular levels are maintained by silicon-binding components and that efflux results from a transient imbalance when the level of transport exceeds the capacity of these components.

Because of the requirement to control the silicic acid concentration in the cytoplasm to prevent autopolymerization, we propose that silicon-binding components are part of an intracellular silicon-level control mechanism operating in the diatom (Fig. 12.4). This mechanism senses the amount of silicon required for deposition, and through the silicon-binding components links this to the uptake process (Fig. 12.4). Because the carboxy-terminal portions of the SITs are very likely to interact with other proteins [6], and the C-termini of other transporters can regulate activity [43, 61], it may be that the control mechanism operates at this step through proteins that bind to and affect the activity of the SITs (Fig. 12.1).

12.8 Intracellular Transport of Silicon

The mechanism of intracellular silicon transport, and transport into the SDV (Section 12.9) are very poorly understood. Silicon is found in all major cellular organelles in the diatom [62], and surprisingly in many organelles in rat tissue cells [63]. A direct comparison of levels in all organelles was not possible, but silicon was particularly abundant in the mitochondria and chloroplasts of the diatom (the SDV could not be isolated), and nucleus, microsomes and mitochondria in rat liver cells [62, 63]. Lipid bilayer membranes are almost impermeable to $Si(OH)_4$ [12], so the enrichment of Si in organelles must be due to a transport process. If intracellular transport involved silicic acid transport proteins, then a specific form of SIT would have to be located in each major organelle not only in the diatom but in rat tissue cells as well. Unless each organelle has a specific use for silicon, this is unlikely. The available data [62, 63] suggest a general intracellular transport mechanism without specific targeting to a particular location.

Silicate ionophoretic activities, enabling the transport of silicic (and germanic) acid across lipid bilayers and bulk organic phases, have been isolated from *Nitzschia alba* [64]. The ionophoretic activities were induced six-fold in silicate-starved cultures, suggesting that they were directly linked to cellular silicon metabolism [64]. Bhattacharyya and Volcani [64] proposed that intracellular transport could occur by ionophore-mediated diffusion, which would be consistent with silicon's presence in all organelles or at least within organellar membranes [62]. They proposed that, through binding, these ionophoretic activities could also be involved in maintaining high intracellular levels of unpolymerized silicic acid [64]. These activities differed from classical ionophores in that they apparently bound two different ions, silicon and sodium, and required both to translocate across an organic layer [64]. The ionophores were isolated by organic extraction, therefore they should preferentially partition into cellular lipid membranes. It is possible that in the cell they translocate silicic acid by binding and releasing on either side of a membrane. What then would prevent them from transporting silicic acid through the plasma membrane out of the cell, which would be favorable due to the high concentration gradient in this direction? Perhaps because the Na^+ concentration gradient is opposite, and the ionophore binds Na^+, this may favor an inward facing orientation, preventing silicic acid efflux. In intracellular organelles or vesicles, binding and release of silicic acid by the ionophore may be facilitated by conditions (Na^+, pH or other factors) on one or other side of the membrane. The structures of these ionophoretic activities were not determined, although it was suggested that they contained vicinal hydroxyl groups [64]. If one ionophore per Si were required, their synthesis may be energetically expensive to the cell. However, the fact that silicon is found in many organelles in rat tissue cells suggests that the activities responsible are derived from compounds normally present in all cells. For example, catechols readily bind and dissolve silica [12], and catecholamines such as epinephrine and derivatives are components of normal cellular metabolism. Perhaps in the diatom these types of compounds have been adapted or modified, including amplifying their levels, for use in intracellular silicon transport. Desired properties of these compounds would be specific binding, but with not too high a binding constant, because silicic acid would have to be eventually released for polymerization.

An alternative hypothesis regarding intracellular transport is that it could occur by the movement of vesicles in which silicic acid has been sequestered [55, 65]. Small vesicles, called silicon transport vesicles (STVs) by Schmid and Schulz [55], have been identified near actively silicifying portions of the SDV in several electron micrographic studies [55, 66, 67]. These vesicles have been seen to fuse with the SDV, and the size of the vesicles matched reasonably well with the size of silica particles adding to the growing frustule [55]. Presumably, these vesicles would be transported through the cell along microtubule networks. There are some considerations in regard to this hypothesis. Most important is that these vesicles have not been directly shown to contain silicon, either in soluble or polymerized form. This should be possible by X-ray microprobe analysis [63]. Also, the identification of specific silicic acid transport proteins [5, 6] precludes the uptake of silicic acid by binding at the plasma membrane and vesicle invagination, so silicic acid pumped into the cytoplasm by the SITs would then have to be transported into the vesicles. This could be feasible, because both plasma membrane and vesicular localization of transporters occur [47]. To maintain silicic acid solubility in the lumen of the vesicles, they would either have to contain silicic acid binding components and/or maintain a very high pH. However a low pH has been found in the SDV [7], which should not be the case if appreciable amounts of high pH vesicular contents were being continuously released into it. If silicon-binding components in the vesicles were released into the SDV, these might also interfere with polymerization. An alternative hypothesis regarding the STVs is that they do not specifically deliver silicon but rather provide a way of adding large amounts of membranous material (along with proteins and other components) during rapid expansion of the SDV [66], and that the silica particles identified in the SDV result from the colloidal growth [4, 18] of silica. Thus, it is unclear at present what the potential contribution of STVs to the maintenance of soluble pools and intracellular transport is.

There may be other as yet unidentified cellular components involved in maintenance of pools and intracellular transport. Proteins that could bind multiple silicic acids would be energy efficient for the cell to produce, and there is ample evidence for the involvement of proteins in binding and sequestration of other ions such as Ca^{2+} [68].

Most of the evidence to date is consistent with an ionophoretic or organic silicon-binding component being responsible for maintenance of intracellular soluble pools and intracellular transport. However, because these compounds have been insufficiently characterized, and not enough work has been done on the alternatives, we are still ignorant of the actual mechanisms.

12.9 Transport into the Silica Deposition Vesicle

The possibility that silica deposition and transport are coupled in some diatom species has been discussed above. In addition, there is a mass transport effect on the extent of silicification, which is determined by the extracellular silicon concentration. It has been shown that diatom frustules are more heavily silicified with higher extracellular silicon [9, 25, 69], and under conditions of silicon limitation, frustules are thinner and siliceous spines (if normally present) are smaller or absent [13, 70].

The molecular details of the formation of silicified structures in the SDV are not well known. How silicon is transported into the SDV is also unclear, and understanding this depends upon understanding the intracellular transport mechanism, about which we have little hard evidence. Thus, we can currently only present speculative possibilities, which hopefully will guide future research in these areas.

As far as the possibility of direct transport of silicic acid into the SDV by the SITs is concerned, it should be possible to test this by determining the intracellular location of the SITs using specific antibody probes. If the SITs are involved in intracellular transport, they would have to pump in the reverse direction compared with plasma membrane localized forms, which could occur because transporters can work in both directions, although it is not presently clear what the driving force for intracellular transport would be. Although vesicle fusion with the SDV has been directly demonstrated [55, 66, 67], as discussed, the presence of silicon in these vesicles needs to be substantiated, and there are other considerations as described. There are several favorable observations regarding the ionophore-mediated transport hypothesis. The cationic, lipophilic dye Rhodamine 123 (R123) accumulates in electronegative intracellular compartments such as the mitochondria [71]. It also accumulates in the SDV in diatoms, where it is incorporated into actively polymerizing silica [72] and can actually be used to quantitatively monitor silica incorporation [57]. R123 transport into mitochondria is independent of ΔpH, but dependent on the transmembrane potential [73]. The compound electrophoretically moves across the mitochondrial membrane, and the concentration ratio inside:out is 4000:1 [73]. Thus, intracompartmental electronegativity can provide a highly efficient mechanism for concentrating compounds. Perhaps such a mechanism works for silicate ionophores in the SDV and mitochondria. This would imply that the ionophores were positively-charged and lipophilic, the latter of which we know is true and the former of which has not been tested [64]. In this hypothesis, if electro-negativity concentrated silicic acid in an organelle, what then would prevent silica polymerization in certain organelles and promote polymerization in the SDV? Perhaps the SDV has specific intralumenal chemical conditions and/or specific proteins or carbohydrates not found in other organelles that promote the release of Si from organic transporting components and polymerization. It could be that the acidic environment identified in the SDV [7] favors the dissociation of organosilicon complexes, whereas in the slightly basic pH of the mitochondria [74] this may not be favored. By itself, this may not explain why silicic acid is not released in other acidic organelles, where at high enough concentrations the pH would promote polymerization [12]. Perhaps proteins are also involved. Models proposing a templating or polymerization-enhancing role of SDV proteins in diatoms have been described [75]. The identification of proteins occluded within the silica of a sponge spicule that have templating and polymerization-enhancing activity *in vitro* [76, 77, and Chapter 14], indicates that polymerization is mediated by these proteins *in vivo*. Interestingly, not only was silicic acid a substrate for the sponge proteins, but also silicon-organic (silsesquioxanes) derivatives normally requiring extremes of pH for polymerization [77]. It is tempting to see this as an analogy to a silicic acid–ionophore conjugate in the SDV, in that SDV proteins may catalyze the release of silicic acid from the ionophores and polymerization. Alternatively, removal of silicic acid from the intralumenal SDV solution by the polymerization process may drive the release of silicic acid from an organic complex. Proteins occluded within diatom silica have been identified [1, 3, 4, 78], and the recently characterized poly-cationic polypeptides known as silaffins [4, and Chapter 11] promoted condensation of uncomplexed silicic acid under acidic conditions, which would be consistent with the latter hypothesis for silicic acid release.

12.10 Summary

In this chapter we have developed the concept that cell wall silicification in diatoms is an integrated process. Not only silica polymerization, but silicic acid uptake, maintenance of intracellular silicon pools, intracellular transport, and release into the SDV are essential for the formation of the cell wall. Indeed, in the model we have developed, intracellular silicon-binding components may play a pivotal role in the process because they would not only maintain Si in soluble form, but would be part of the mechanism of controlling intracellular levels, be responsible for the coupling between deposition and uptake, and deliver silicic acid to and into the SDV for polymerization.

Although the model is consistent with the available data, studies on many aspects of silicic acid transport and deposition are clearly limited and more investigation is required. We are now poised to pursue an in-depth study of the mechanism and control of silicic acid uptake by the SITs. Having cloned these genes enables us to manipulate them *in vitro*, and analyze them after reintroduction into the diatom [79], or into a heterologous host such as yeast. By comparing SIT gene sequences in different diatom species [6], we can identify regions containing amino acids that are essential for transport. Using site-directed mutagenesis, we can directly test the role of conserved amino acids in transporter function and the control of transport activity by the carboxy-terminus. Because we believe that proteins interacting with the C-terminus of the SITs are part of the intracellular sensing and control mechanism, isolating and characterizing these proteins may bring us one step further towards understanding how this mechanism works.

It is not clear what maintains Si in soluble form at high concentrations in the cytoplasm, nor what the mechanism of intracellular transport is. The silicate ionophoretic activities [64] may be the most promising candidates for being involved in these processes, and should be reisolated and characterized in detail. To evaluate the validity of the silicon transport vesicle hypothesis [55], the contents of these vesicles need to be identified, and their movements in the cell tracked. Approaches aimed at identifying other silicon-binding components may also be useful.

These investigations are not only important in understanding the cellular and molecular biology of silicification in the diatom, but may have applications in *in vitro* chemical or materials syntheses. The efficiency of biological processing of silicon undoubtedly stems from the specific molecular interaction between cellular components and silicon. Understanding the nature of these interactions at the molecular level could provide insight into how to control or optimize chemical reactions. Controlling the transport of silicic acid across lipid bilayers or bulk organic layers may have direct application in chemical or materials syntheses. By developing purification schemes for the SITs, and understanding how their activity is controlled, perhaps these could be used to deliver specific amounts of silicic acid into or out of lipid bilayer systems. Ionophoretic activities could be used for transport across both lipid bilayers and bulk organic layers [64]. Biology has had billions of years to develop and refine sophisticated mechanisms for controlling chemistry, including spatial and temporal control of reactions, and an understanding of the biological control of silicon chemistry is likely to be extremely beneficial.

Acknowledgements

We are grateful for support over the years from Ethyl Corp. and Whitby Research, the Kellogg Family of La Jolla, CA, Dow Corning Corp., and the US Army Research Office.

References

[1] N. Kröger, C. Bergsdorf, M. Sumper, *EMBO J.* 1994, *13*, 4676–4683.
[2] N. Kröger, C. Bergsdorf, M. Sumper, *Eur. J. Biochem.* 1996, *239*, 259–264.
[3] N. Kröger, G. Lehmann, R. Rachel, M. Sumper, *Eur. J. Biochem.* 1997, *250*, 99–105.

[4] N. Kröger, R. Duetzmann, M. Sumper, *Science* 1999, *286*, 1129–1132.

[5] M. Hildebrand, B. E. Volcani, W. Gassmann, J. I. Schroeder, *Nature* 1997, *385*, 688–689.

[6] M. Hildebrand, K. Dahlin, B. E. Volcani, *Mol. Gen. Genet.* 1998, *260*, 480–486.

[7] E. G. Vrieling, W. W. C. Gieskes, T. P. M. Beelen, *J. Phycol.* 1999, *35*, 548–559.

[8] M. Hildebrand, D. R. Higgins, K. Busser, B. E. Volcani, *Gene* 1993, *132*, 213–218.

[9] M. Hildebrand, in preparation.

[10] (a) H. L. Conway, P. J. Harrison, C. O. Davis, *Mar. Biol.* 1976, *35*, 187–199.
(b) H. L. Conway, P. J. Harrison, *Mar. Biol.* 1977, *43*, 33–43.

[11] S. W. Chisholm, F. Azam, R. W. Eppley, *Limnol. Oceanogr.* 1978, *23*, 518–529.

[12] R. K. Iler, The Chemistry of Silica: Solubility, Polymerization, Colloid and Surface Properties, and Biochemistry, John Wiley & Sons, New York, 1979.

[13] M. A. Brzezinski, R. J. Olson, S. W. Chisholm, *Mar. Ecol. Prog. Ser.* 1990, *67*, 83–96.

[14] J. C. Lewin, R. R. L. Guillard, *Annu. Rev. Microbiol.* 1963, *17*, 373–414.

[15] J. Pickett-Heaps, A.-M. M. Schmid, L. A. Edgar, in *Progress in Phycological Research, Vol 7.* (Eds F. E. Round and D. J. Chapman), Biopress, Bristol, 1990, pp. 2–168.

[16] R. W. Drum, H. S. Pankratz, *J. Ultrastruct. Res.* 1964, *10*, 217–223.

[17] T. L. Simpson, B. E. Volcani (Eds) *Silicon and Siliceous Structures in Biological Systems*, Springer-Verlag, New York, 1981, p. 6.

[18] R. G. Greer, in Scanning Electron Microscopy, Part I, Proceedings of the 4th Annual Scanning Electron Microscopy Symposium. IIT Research Institute, Chicago, 1971, pp. 153–160.

[19] (a) J. J. Goering, D. M. Nelson, J. A. Carter, *Deep-Sea Res.* 1973, *20*, 777–789.
(b) F. Azam, S. W. Chisholm, *Limnol. Oceanogr.* 1976, *21*, 427–435.
(c) M. A. Brzezinski, D. M. Nelson, *Deep-Sea Res.* 1989, *36*, 1009–1030.

[20] H. D. Holland, *The Chemical Evolution of the Oceans*, Princeton University Press, Princeton, New Jersey, 1984.

[21] T. Tréguer, D. M. Nelson, A. J. Van Bennekom, A. Leynaert, B. Queguiner, *Science* 1995, *268*, 375–379.

[22] (a) N. Ingri, in *Biochemistry of Silicon and Related Problems.* (Eds G. Bendz and I. Lindqvist), Plenum Press, New York, 1978, pp. 3–51.
(b) W. Stumm, J. J. Morgan, *Aquatic Chemistry*, Wiley Interscience, New York, 1970.

[23] J. C. Lewin, *J. Gen. Physiol.* 1954, *37*, 589–599.

[24] J. C. Lewin, *J. Gen. Physiol.* 1955, *39*, 1–10.

[25] E. Paasche, *Mar. Biol.* 1973, *19*, 262–269.

[26] F. Azam, B. B. Hemmingsen, B. E. Volcani, *Arch. Microbiol.* 1974, *97*, 103–114.

[27] D. M. Nelson, J. J. Goering, S. S. Kilham, R. R. L. Guillard, *J. Phycol.* 1976, *12*, 246–252.

[28] S. S. Kilham, C. L. Kott, D. Tilman, *J. Great Lakes Res.* 1977, *3*, 93–99.

[29] C. W. Sullivan, *J. Phycol.* 1976, *12*, 390–396.

[30] C. W. Sullivan, *J. Phycol.* 1977, *13*, 86–91.

[31] P. Bhattacharyya, B. E. Volcani, *Proc. Natl Acad. Sci. USA* 1980, *77*, 6386–6390.

[32] J. Monod, *Annu. Rev. Microbiol.* 1942, *3*, 3–71.

[33] V. Martin-Jezequel, M. Hildebrand, M. Brzezinski, *J. Phycol.*, in press.

[34] G. F. Riedel, D. M. Nelson, *J. Phycol.* 1985, *21*, 168–171.

[35] Y. Del Amo, M. A. Brzezinski, *J. Phycol.* 1999, *35*, 1162–1170.

[36] (a) F. Azam, *Planta* 1974, *121*, 205–212.
(b) F. Azam, B. E. Volcani, *Arch. Microbiol.* 1974, *101*, 1–8.

[37] M. A. Hediger, *J. Exp. Biol.* 1994, *196*, 15–49.

[38] J. Reizer, A. Reizer, M. J. Saier Jr, *Biochim. Biophys. Acta* 1994, *1197*, 133–166.

[39] Y. Deguchi, I. Yamoto, Y. Anraku, *J. Biol. Chem.* 1990, *265*, 21704–21708.

[40] (a) T. Hirokawa, S. Boon-Chieng, S. Mitaku, *Bioinformatics* 1998, *14*, 378–379.
(b) TMpred, Baylor College of Medicine Protein Secondary Structure Prediction, Web address: http://dot.imgen.bcm.tmc.edu:9331/seq-search/struc-predict.html

[41] (a) A. Davies, K. Meeran, M. T. Cairns, S. T. Baldwin, *J. Biol. Chem.* 1987, *262*, 9347–9352.
(b) R. Seckler, J. K. Wright, P. Overath, *J. Biol. Chem.* 1983, *258*, 10817–10820.

[42] T. Caspari, R. Stadler, N. Sauer, W. Tanner, *J. Biol. Chem.* 1994, *269*, 3498–3502.

[43] A. D. Due, Q. Zhinchao, J. M. Thomas, A. Buchs, A. C. Powers, J. M. May, *Biochemistry* 1995, *34*, 5462–5471.

[44] G. D. Holman, S. W. Cushman, *BioEssays* 1994, *16*, 753–759.
[45] H. Katagiri, T. Asano, H. Ishihara, K. Tsukuda, J.-L. Lin, K. Inukai, M. Kikuchi, Y. Yazaki, Y. Oka, *J. Biol. Chem.* 1992, *267*, 22550–22555.
[46] Y. Oka, T. Asano, Y. Shibasaki, J.-L. Lin, K. Tsukuda, H. Katagiri, Y. Akanuma, F. Takaku, *Nature* 1990, *345*, 550–553.
[47] K. J. Verhey, S. F. Hausforff, M. J. Birnbaum, *J. Cell Biol.* 1993, *123*, 137–147.
[48] K. J. Verhey, J. I. Yeh, M. J. Birnbaum, *J. Cell Biol.* 1995, *130*, 1071–1079.
[49] K. C. Worley, R. Smith, B. Weise, P. Culpepper, *Genome Res.* 1996, *6*, 454–462.
[50] A. Lupas, M. Van Dyke, J. Stock, *Science* 1991, *252*, 1162–1164.
[51] D. Werner, *Arch. Mikrobiol.* 1966, *55*, 278–308.
[52] C. W. Sullivan, *J. Phycol.* 1979, *15*, 210–216.
[53] B. J. Binder, S. W. Chisholm, *Mar. Biol. Lett.* 1980, *1*, 205–212.
[54] G. S. Blank, D. H. Robinson, C. W. Sullivan, *J. Phycol.* 1986, *22*, 382–389.
[55] A.-M. Schmid, D. Schulz, *Protoplasma* 1979, *100*, 267–288.
[56] C. W. Sullivan, in *Silicon Biochemistry*, Wiley, Chichester, 1986, pp. 59–89.
[57] M. A. Brzezinski, D. J. Conley, *J. Phycol.* 1994, *30*, 45–55.
[58] J. Lewin, *Phycologia* 1966, *6*, 1–12.
[59] W. M. Darley, B. E. Volcani, *Exp. Cell Res.* 1969, *58*, 334–342.
[60] F. Azam, B. E. Volcani, in *Silicon and Siliceous Structures in Biological Systems* (Eds T. L. Simpson and B. E. Volcani), Springer-Verlag, New York, 1981, pp. 43–67.
[61] E. Carafoli, *FASEB J.* 1994, *8*, 993–1002.
[62] C. W. Mehard, C. W. Sullivan, F. Azam, B. E. Volcani, *J. Physiol.* 1974, *30*, 265–272.
[63] C. W. Mehard, B. E. Volcani, *Cell Tiss. Res.* 1976, *166*, 255–263.
[64] P. Bhattacharyya, B. E. Volcani, *Biochem. Biophys. Res. Commun.* 1983, *114*, 365–372.
[65] M. Lee, C.-W. Li, *Bot. Bull. Acad. Sin.* 1992, *33*, 317–325.
[66] C.-W. Li, B. E. Volcani, *Phil. Trans. R. Soc. Lond. B* 1984, *304*, 519–528.
[67] (a) C.-W. Li, B. E. Volcani, *Protoplasma* 1985, *124*, 10–29.
      (b) D. Schulz, G. Drebes, H. Lehmann, R. Jank-Ladwig, *Eur. J. Cell Biol.* 1984, *33*, 43–51.
[68] (a) R. Spencer, M. Charman, P. Wilson, D. E. M. Lawson, *Nature* 1976, *263*, 161–163.
      (b) S. S. Jande, S. Tolnai, D. E. M. Lawson, *Nature* 1981, *294*, 765–767.
      (c) J. H. Henson, D. A. Begg, S. M. Beaulieu, D. J. Fishkind, E. M. Bonder, M. Terasaki, D. Lebeche, B. Kaminer, *J. Cell Biol.* 1989, *109*, 149–161.
[69] (a) D. Tilman, S. Soltau-Kilham, *J. Phycol.* 1976, *12*, 375–383.
      (b) P. J. Harrison, H. L. Conway, R. W. Holmes, C. O. Davis, *Mar. Biol.* 1977, *43*, 19–31.
[70] (a) E. Paasche, *J. Exp. Mar. Biol. Ecol.* 1975, *19*, 117–126.
      (b) C. O. Davis, *J. Phycol.* 1976, *12*, 291–300.
      (c) P. J. Harrison, H. L. Conway, R. C. Dugdale, *Mar. Biol.* 1976, *35*, 177–186.
[71] L. V. Johnson, M. L. Walsh, L. B. Chen, *Proc. Natl Acad. Sci. USA* 1980, *77*, 990–994.
[72] C.-W. Li, S. Chu, M. Lee, *Protoplasma* 1989, *151*, 158–163.
[73] R. K. Emaus, R. Grunwald, J. J. Lemasters, *Biochim. Biophys. Acta* 1986, *850*, 436–448.
[74] J. Llopis, J. M. McCaffery, A. Miyawaki, M. G. Farquhar, Y. Tsien, *Proc. Natl Acad. Sci. U.S.A.* 1998, *95*, 6803–6808.
[75] (a) R. E. Hecky, K. Mopper, P. Kilham, E. T. Degens, *Mar. Biol.* 1973, *19*, 323–331.
      (b) K. D. Lobel, J. K. West, L. L. Hench, *Mar. Biol.* 1996, *126*, 353–360.
[76] K. Shimizu, J. Cha, G. D. Stucky, D. E. Morse, *Proc. Natl Acad. Sci. USA* 1998, *95*, 6234–6238.
[77] J. N. Cha, K. Shimizu, Y. Zhou, S. C. Christiansen, B. F. Chmelka, G. D. Stucky, D. E. Morse, *Proc. Natl Acad. Sci. USA* 1999, *96*, 361–365.
[78] D. M. Swift, A. P. Wheeler, *J. Phycol.* 1992, *28*, 202–209.
[79] (a) T. G. Dunahay, E. E. Jarvis, P. G. Roessler, *J. Phycol.* 1995, *31*, 1004–1012.
      (b) K. E. Apt, P. G. Kroth-Pancic, A. R. Grossman, *Mol. Gen. Genet.* 1996, *252*, 572–579.
      (c) H. Fischer, I. Robl, M. Sumper, N. Kröger, *J. Phycol.* 1999, *35*, 113–120.

13 The Nanostructure and Development of Diatom Biosilica

Richard Wetherbee, Simon Crawford, Paul Mulvaney

13.1 Introduction

The degree of complexity and hierarchical structure displayed by biomineralized composites has never been matched in artificial materials, as a diverse range of biological organisms possess mechanisms for the nanofabrication of ornately sculptured silicates under ambient conditions and at near-neutral pH [1, 2]. The success of biological systems in processing silica must result from specific interactions between the silica and the associated biomolecules produced by the cells. Templating molecules exist in nanogram or smaller concentrations within the complex organic–inorganic matrix of a range of biologically mineralized composites [3–6], and there is now evidence that a similar template may exist during the development of diatom biosilica [7]. However, the identity and location of the organic–inorganic interface that chemically and spatially directs the polymerization of silica in diatoms is largely unknown. In addition, there remain enormous ambiguities and mysteries about the mechanisms used by diatoms to absorb silicon, and then to process and template siliceous structures with such speed and precision. In this chapter we summarize the state of our knowledge on the development and nanostructure of diatom biosilica.

13.2 General Features of the Diatom "Glass House"

The diatoms from the algal class Bacillariophyceae (Heterokontophyta) are microscopic, unicellular protists that are major primary producers in most marine and freshwater habitats. Diatoms are immediately recognizable due to the presence of a highly ornate, silicified cell wall. Called a frustule, the morphology of the "glass house" is species-specific and comprised of both silicified components and a range of organic layers and casings. Recent reviews cover the ultrastructure and cell biology of diatom wall formation in great detail [8–10], and we only summarize some of these features below. In addition, two other chapters in this book are devoted to various aspects of diatom wall biogenesis, and many of the issues involved with diatom wall formation and silicification are discussed here.

The diatom frustule consists of two halves that typically overlap much like a petri

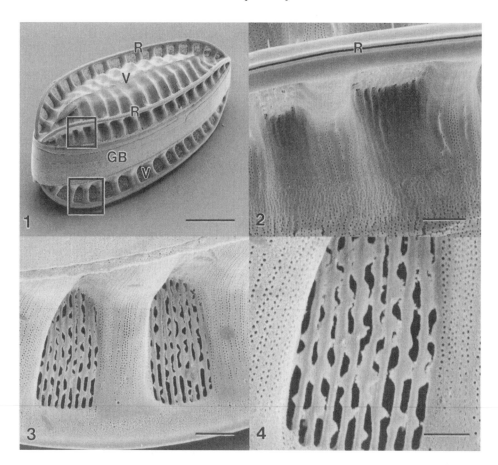

Figures 13.1–13.6 Field emission scanning electron microscope (FESEM) images illustrating the ornate morphology of the silicified frustule of *Surirella sp.* Figure 13.1 depicts a whole cell, illustrating basic diatom structure with two valves (V) joined by girdle bands (GB). In this species each valve has two raphes (R). Scale bar = 20 μm. Figures 13.2–13.4 show higher magnification images of the boxed regions of the cell shown in Fig. 13.1. The major structural features of the valves are a result of macromorphogenesis, such as the raphe fissure shown in Fig. 13.2. The tiny pores and some of the additional detail seen in Figs. 13.4–13.6 probably results from micromorphogenesis. Scale bars = 2 μm for Figs. 13.2, 13.3, 13.5; 1 μm for Fig. 13.4; 500 nm for Fig. 13.6.

dish, a larger epitheca and a marginally smaller hypotheca (Fig. 13.1). Each theca consists of a highly patterned valve plus one or more girdle bands that run around the circumference of the cell. The silicified valves and accompanying girdle bands are precisely attached to one another by organic layers (or adhesives) to form each theca. In addition, the silicified components are also associated with a range of organic casings that function in a variety of ways, including protection from desilicification [8–10]. As living cells must constantly interact with their environment, the valves and, to a lesser degree, girdle bands are formed with a myriad of openings (pores, slits, etc.) that facilitate such exchanges (Figs. 13.1–13.10) [8, 9]. These

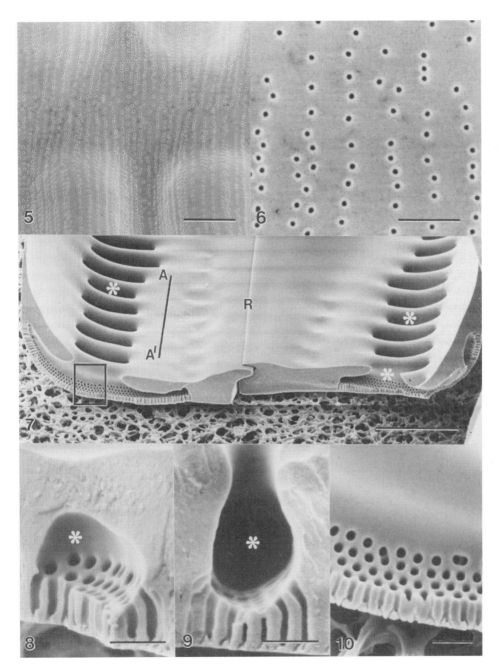

Figures 13.7–13.10 FESEM images of *Pinnularia sp.* Figure 13.7 shows a fractured valve in cross-section and viewed on the cytoplasmic surface, while the enclosed chambers seen in Figs. 13.8 and 13.9 (asterisk) would result from a fracture along the AA′ axis seen in Fig. 13.7. Figure 13.10 is a high magnification image of the boxed region in Fig. 13.7. Note the results of moulded macro-morphogenesis including the raphe (R) in Fig. 13.7 and the larger chambers (asterisks) in Figs. 13.7–13.9. In Figs. 13.8–13.10, the tiny pores organized in precise arrays result from micro-morphogenesis. Scale bars = 5 μm for Fig. 13.7; 500 nm for Figs. 13.8–13.10.

openings are not randomly oriented but, together with other sculptured features of the valve, typically display a complex pattern and symmetry that is the basis of their taxonomy. Two major groups of diatoms are discerned, the radially symmetric "centric" diatoms and the bilaterally symmetric "pennate" diatoms.

The biogenesis and silicification of individual valves and girdle bands occurs within a single silica deposition vesicle (SDV) which is surrounded by a distinct membrane, the silicalemma [9]. The complex environment of the SDV determines the nanostructure and nanofabrication of diatom biosilica, and is discussed in more detail below.

13.3 The Chemistry of Biosilica Formation

Biomineralization studies seek to explore the mechanisms by which soft molecular structures such as surfactants and organic polymers (e.g. proteins, polysaccharides, glycoproteins) are able to template and direct the crystal growth or precipitation of "harder" inorganic minerals such as $CaCO_3$ and silica. The major problem with silica is the fact that it is comparatively inert chemically. Silicon forms few bonds with elements other than oxygen, and direct Si–C bonds are unusual in the natural world. Hence a method of sequestering silicon has remained largely undetermined. There are several points of chemical interest in the sequence of events required for the formation of diatom biosilica within a cytoplasmic SDV.

Firstly, the silica walls appear to be assembled from colloidal silica [11–13], although evidence suggests that diatoms do not directly transport colloidal silica, but use monomeric silica as a substrate [14–16] and then "gel" it within the SDV.

The exact nature and location of the organic–inorganic interface within the SDV is unknown, although a "matrix" material has been hypothesized to be involved in both the morphogenesis and silicification of the sculptured walls [7, 17].

It is not known how well the pH, salt and metal ion concentrations, can be regulated within the SDV, as a means of controlling the silica nucleation and growth processes.

13.3.1 Parameters Affecting Silicon and Silicification

Sillén and coworkers [18] carried out potentiometric titrations of silicate ion in the early 1960s as part of their research into global weathering and pH control through geochemical processes. Much of their data was assimilated and analysed by Baes and Mesmer in their classic text, *Hydrolysis of Cations* [19]. More recently, Sjoberg *et al.* [20, 21] have remeasured the hydrolysis constants of silicon(IV) in water at high NaCl concentrations to mimic seawater. It is important to recognize that the speciation of silicon, like other species in solution, is determined firstly by resolving whether the solution is saturated or not with respect to the element. This is determined by the equilibrium:

$$SiO_2(s) + 2H_2O \Leftrightarrow Si(OH)_4(aq) \qquad K_{s10} \qquad (1)$$

For quartz, $\log K_{s10} = -4.00$ (6 ppm), whereas for amorphous silica $\log K_{s10} = -3.0$ to -2.7, about 10 times higher. Note that the silicic acid molecule has not been formally isolated; its existence has been inferred on the basis of NMR, titration data and molybdate complexation. In view of the prolific abundance of sand, one can assume that marine waters close to land are saturated with respect to silicate ion. Hence at all times the $Si(OH)_4$ concentration is about 6 ppm. In laboratories the silicate ion concentration may be artificially reduced by keeping the diatoms in polythene bottles rather than glass vessels. Sufficient silicate will leach out of most commercial glasses within several hours when in contact with water at $pH > 9$.

Under these saturated conditions, the concentration of $Si(OH)_4$ is always given by Equation (1). Depending on pH, electrolyte and chelate concentrations, the silicic acid will dissociate into a number of ions and oligomers. All the earlier researchers seem to concur that at low [Si], the hydrolysis of silicon is simple and involves only monomeric silicon species. The two important ionization equilibria are:

$$Si(OH)_4 + H_2O \Leftrightarrow SiO(OH)_3^- + H^+ \qquad \log K_{11} = -9.86 \qquad (2)$$

At 0.5 M $NaClO_4$, this is lowered slightly to $\log Q_{11} = -9.46$, where the change to Q indicates the use of a supporting electrolyte to maintain constant activity coefficients. The second deprotonation has the following equilibrium constant:

$$Si(OH)_4 \Leftrightarrow SiO_2(OH)_2^{2-} + 2H^+ \qquad \log K_{12} = -22.92 \qquad (3)$$

This species is only significant above pH 12, and can be ignored in studies of biosilica uptake. The total silicon dissolved in solution, Si_T is hence the sum of the individual concentrations:

$$Si_T = [Si(OH)_4] + [SiO(OH)_3^-] = [Si(OH)_4]^*\{1 + K_{11}/[H^+]\} \qquad (4)$$

The speciation vs pH diagram for 10 μM Si(IV) is shown in Fig. 13.11a. Note that at pH 8, $Si(OH)_4$ accounts for over 90% of the silicon in solution. At higher silicon concentrations or under conditions of supersaturation, polymers form. The stability constants for the two most well-established polymers have been measured to be

$$4Si(OH)_4 \Leftrightarrow Si_4O_6(OH)_6^{2-} + 2H^+ + 4H_2O \qquad \log K_{12} = -13.44 \qquad (5)$$

$$4Si(OH)_4 \Leftrightarrow Si_4O_8(OH)_4^{4-} + 4H^+ + 4H_2O \qquad \log K_{12} = -35.80 \qquad (6)$$

In natural waters, the silicon levels are low enough to preclude significant amounts of polynuclear species, and at pH 6–9, such species are only metastable anyway, and will condense to form colloidal silica. The speciation diagram for

Si(IV) at 0.1 M is shown in Fig. 13.11b. Note that below pH 10, the calculated concentrations are non-equilibrium values. Silica is thermodynamically favoured below pH 10 at 0.1 M Si(IV).

13.3.2 Hypothetical Effects of Chelating Agents on Silica Deposition

The above calculations fail to take into account the role of complexation and chelation by ligands, biomolecules and templating molecules that might exist in organisms, such as within the cytoplasm and SDV of diatoms. Thus, normal chemistry fails to predict the nucleation conditions for silica precisely when it is needed most. Without tabulated values of the binding constants of amino acids, proteins, proteoglycans, sugars and membrane surfactants to silicate ions under various pH regimes, it is impossible to calculate the actual silicate concentration within the diatom, nor can we be sure about the conditions needed for supersaturation. Ligands that form soluble silicates will increase the amount of total dissolved silica and make precipitation more difficult. Conversely, cations that can form very insoluble complexes, e.g. Ca or Mg, can render silica less soluble in water and assist deposition in biological systems.

The difficulties of interpretation can be highlighted by considering the binding of a silicate substrate to a hypothetical biomolecule or transporter. We will assume that the silicate is bound through an amine functional group on the biomolecule, which we will denote by NH_2R. Firstly, the biomolecule could be aiding deposition by formation of *soluble* amino-silicates which are used to help build up a high, local concentration of silicate ions within the SDV. By simple modulation of the pH, it could release the silicate ion, which would then rapidly precipitate within the confines of the SDV.

Scheme 1:

$$Si(OH)_4 \text{ (low, CP)} + NH_2\text{--R (transporter)}$$
$$\Leftrightarrow Si(OH)_3NH\text{--R (high, SDV)} + H_2O \tag{7}$$

This equation denotes silicates being coerced into SDVs by complexation with transporter amines. CP refers to the cytoplasm and low and high refer to relative silicate concentrations.

$$Si(OH)_3NH\text{--R} + H_2O + H^+ \Leftrightarrow Si(OH)_4 \text{ (high, SDV)} + NH_3\text{--R}^+ \tag{8}$$

(pH lowered in SDV from 9 to 7; the amine protonates and releases silicic acid)

$$Si(OH)_4 \text{ (high, SDV)} \Leftrightarrow SiO_2 + H_2O \tag{9}$$

(supersaturated silica precipitates)

Scheme 2:

Alternatively, the biomolecule could form a very *insoluble* complex with silicate ions, lowering its solubility within the SDV. This could be used to spatially direct nucleation. In this case, a transporter brings silicate ions to the SDV. As the concentration increases, nucleation on the silicalemma occurs:

$$Si(OH)_4 \text{ (high, SDV)} + NH_2-R \text{ (template director molecule)}$$
$$\Leftrightarrow Si(OH)_3NH-R \text{ (insoluble, SDV wall)} + H_2O \tag{10}$$

Spatial localization of a polyamine on the inner surface of the silicalemma and facing the SDV lumen, for example, could allow it to direct silicate nucleation. In this case silicate ion is simply transported at high flux into the SDV. Precipitation of the least soluble silicate is thermodynamically favored. If the solubility of the biosilicate complex is less than that of pure, colloidal biosilica, it will precipitate preferentially. This keeps the total silicate ion concentration below the threshold for silica nucleation, and ensures homogeneous growth of the templated nuclei. Thereafter, the SDV silicate ions will condense onto the nascent silica surfaces. Hence complexation of silicate by biomolecules can be utilized in several ways to serve biomineralization processes.

13.3.3 Silica Chemistry in Seawater

The uptake of silica as part of diatom valve/girdle band formation occurs at pH values in the range 6–9. At these pH values, the dominant monomers are $Si(OH)_4$ and $SiO(OH)_3^-$.

It is important to note that only monomers are significant even in alkaline solution, and that these two species successfully account for silicon speciation from pH 7 to 10.5. Hence under conditions where the sodium silicate concentration is kept below 100 µM, no insoluble silica exists, and biogenic uptake must involve sequestration of one of these monomers.

The high sodium ion concentrations in seawater could, in principle, complicate the equilibria by formation of sodium silicate complexes. However, Sjoberg *et al.* [20, 21] find these can be neglected at low [Si] and at up to 0.5 M NaCl. Their values for Equation (11) suggest that sodium binding is not important, although sodium levels may be critical in gel and colloid formation.

$$Na^+ + SiO(OH)_3^- \Leftrightarrow NaSiO(OH)_3 \qquad K = 0.37 \text{ M}^{-1} \tag{11}$$

The solubility of Si(IV) increases from $K_{s10} = 10^{-2.74}$ to $10^{-2.64}$ as the temperature is varied from 25 °C up to 35 °C, and hence in normal marine environments there is no strong effect of temperature on the bioavailability of silicon.

13.4 Silica Uptake by Diatoms

Recent work has shown that the silica walls of diatoms consist of colloidal aggregates, gelled to form porous, non-crystalline layers [13]. This suggests that diatoms may in fact take silica up in monomeric form, or even directly as a colloid building block. At least three factors dictate against this latter mechanism. Firstly, the larger size makes intercellular and intracellular transport difficult. Secondly, mechanisms for biologically selecting colloidal particles and binding them would be difficult. Thirdly, the concentration of "active" particles capable of being taken up is capricious, being controlled by solution phase kinetic factors, ionic strength, complexation, ligand concentrations, etc., and pH fluctuations [22]. Finally, silica growth is observed even when no insoluble silica is present in solution!

The rates of silica valve growth are so fast that normal diffusion is not considered adequate, and "active" transport of silica in some form is necessary [14, 15] to provide the flux of silicon to the SDV for deposition. For the reasons outlined above, it seems likely that active transport is used to transfer silica monomers into the diatom and to the SDV. Colloidal silica could then be generated by controlled polymerization of silica monomer within the confines of the SDV, using pH, chelators and alkali metal ion concentrations to modulate gelation and colloid growth processes.

The rate of growth of diatoms appears to be incredibly fast, and hence many authors refer to the miracle of "active" transport. If convection is assumed to be low, the maximum rate at which silicon monomers could be taken up from the solution by the growing diatom is fixed by diffusion. We can estimate this "diffusion-limited" rate of growth as follows. The diffusive flux to a single diatom is given by

$$J_{\mathrm{diff}} = \frac{4\pi a D}{10^3} [\mathrm{Si}]_\infty \ \mathrm{mol \ s}^{-1} \tag{12}$$

where a is the diatom radius, D the silicate ion diffusion coefficient and $[\mathrm{Si}]_\infty$ is the bulk solution silicate concentration. Taking the diatom to be a 5 μm radius sphere with a silica shell thickness of 2 μm and taking $D \sim 10^{-5} \ \mathrm{cm}^2 \ \mathrm{s}^{-1}$ for the $\mathrm{Si(OH)}_4$ diffusion coefficient, yields a value of about $J_{\mathrm{diff}} \sim 5 \times 10^{-16} \ \mathrm{mol \ s}^{-1}$. We have assumed a silica monomer concentration of just 10 μM. The total volume of silica required for the shell described is $4 \times 10^{-10} \ \mathrm{cm}^3$. Hydrated silica has a molar volume of about $V_{\mathrm{m}} = 35 \ \mathrm{cm}^3 \ \mathrm{mol}^{-1}$, so the number of moles of silica needed to be absorbed by the diatom is roughly 1×10^{-11} moles. The time to accrue this silica under optimal conditions is hence:

$$t_{\mathrm{growth}} = \text{moles required/diffusive flux} = 1 \times 10^{-11}/5 \times 10^{-16}$$
$$= 2 \times 10^4 \ \mathrm{s} \tag{13}$$

If the silica concentration were 100 times higher at 1 mM, this would reduce to just 200 s! Based on these figures it is easy to see that provided diatoms are able to effectively sequester silicon by surface complexation and to actively transport it into the cytoplasm, there is little difficulty in explaining the growth of diatoms on a time scale of minutes to hours. In fact, Sullivan has estimated that total soluble silica concentrations inside the diatom lie in the range 450–700 nM [23, 24], which is very high. This suggests that uptake of Si by the diatom is very efficient and the rate of growth could be pushed towards the diffusion-limited rate.

13.5 Nanostructure of Diatom Biosilica

Early structural evidence suggested that the silica valves/girdle bands had a nano-structure resembling globular particles, and they appeared to be constructed of colloidal silica [11, 12]. Our own AFM data fully support this result, i.e. that the walls show globular silica in cross-section (Figs. 13.12–13.14, 13.16, 13.17). However, it is interesting that the surface of many frustule components can be perfectly smooth (e.g. Fig. 13.15) while other surface images reveal the globular silica. This in turn raises the question whether diatoms are able to manipulate colloidal silica, and whether they take up colloidal silica directly. Two factors dictate against the direct uptake of colloidal material. The first is simply that diatoms can grow even under conditions of strongly undersaturated silicon, e.g. at micromolar concentrations, albeit it more slowly. As highlighted above, under these conditions, only mono-meric silicon species exist in solution, as determined from Fig. 13.11. The second fact is that oligomeric silica exists in a range of sizes from 0.5 to 4 nm in diameter. In this size regime, it redissolves rapidly, undergoes Ostwald ripening, and the par-ticle size is strongly affected by both pH and electrolyte levels. It seems unlikely that active uptake via a membrane transporter species would be reliable if the speciation is so prone to size fluctuations. Growth would technically only be possible in satu-rated silica solutions, which would be a colossal disadvantage to the diatom. It seems more likely that the colloidal silica that is observed in cell walls is a product of internal morphogenic processes.

A number of key papers have been published in this area. By monitoring the uptake kinetics via radiolabeling, it has recently been shown that the most likely species to be the active substrate for transport is $Si(OH)_4$. Again growth was ob-served at silicon levels down to 15 μM, well below the level where any colloidal silica could be present. This implies that at some point during internal transport, or directly within SDVs, the silicon is then assembled into colloidal particles for in-corporation into the growing valve. Hildebrand *et al.* [14] have now identified a protein which they claim is the active silicon transporter that brings $Si(OH)_4$ [16] into the cell cytoplasm. It is not clear whether the same transporter or a relative is then responsible for transfer of silicon to the SDV and "developing" wall. This cytoplasmic transport issue has not been directly addressed as yet, and is discussed in detail in another chapter in this book.

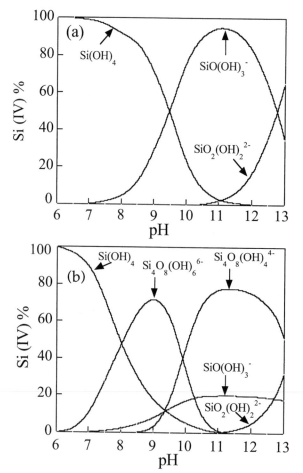

Figure 13.11 (b) Distribution of silicate species at an ionic strength of I = 3m and 25 °C, for solutions containing 0.1 M Si(IV). The silicate species below pH 10 are supersaturated with respect to silica formation. (a) Distribution of silicate species at an ionic strength of I = 3m and 25 °C, for solutions containing 10 μM Si(IV).

13.6 Development of Diatom Biosilica within a Confined Space – the Silica Deposition Vesicle (SDV)

The intricate and highly patterned valves of diatoms are species-specific, and therefore genetically determined, although alterations to the chemical composition and thickness of valves can be induced by the cell's environment [9]. All of this information must impact on the SDVs, which seemingly regulate all aspects of valve/girdle band nanofabrication. Two primary hypotheses have been proposed to explain the processes of diatom wall morphogenesis within SDVs. The first is that patterning and silicification are nucleated and controlled simply by the physico-chemical environment and constraints provided by the SDV and the silicalemma

[25, 26]. According to this hypothesis, cytoplasmic components imprint on the outer surface of the silicalemma to mould and shape the wall, followed by autopoly-condensation of silica within the SDV [26]. These cytoplasmic moulding processes are referred to as macromorphogenesis [9, 27], and have been observed during the expansion of the SDV in a range of diatoms. A highly patterned arrangement of organelles (e.g. mitochondria, endoplasmic reticulum, "spacer vesicles") as well as cytoskeletal components (e.g. microtubules and microtubule-associated molecules and complexes) become closely associated with the silicalemma, restricting SDV expansion to form structural features such as pores, slits and chambers of various size and shape. The larger compartments seen in Figs. 13.1–13.3 and 13.7–13.10 are examples of macromorphogenesis, including the slit-like raphe (Fig. 13.2) that is responsible for diatom adhesion and motility. The exact way in which moulding occurs is not known, and it is also possible that silica deposition is selectively controlled within the SDV by cytoplasmic components interacting through the silicalemma. Therefore, the macromorphology of the enclosed, developing valve may be affected in a number of ways.

There is extensive evidence in support of SDV moulding during valve formation, but far less information about events that occur within the silicalemma. It has been postulated that silicification and fine patterning may be induced by the presence of an organic matrix or template located within the SDV [7, 9, 17]. Presumably, this template could be attached to the inner surface of the silicalemma, within the SDV lumen, or both. If organic matrix material exists within the SDV lumen, it may exist embedded or encased within the globular silica particles, as surface strands or coatings surrounding the colloidal silica, or both. The processes isolated to the inner silicalemma have been referred to as "membrane-mediated morphogenesis" [9, 27], while we use the term micromorphogenesis to refer to all processes that occur within the SDV lumen, including those generated from the silicalemma. The tiny pores that often dominate the valves, and occasionally the girdle bands, of several diatoms illustrate a structure that must result from a form of micromorphogenesis (Figs. 13.3–13.6, 13.8–13.10). In *Pinnularia*, Pickett-Heaps *et al.* [28] showed that the highly patterned, tiny pores of the valve (Figs. 13.7–13.10) were not surrounded by the silica membrane during their formation, but arose within the lumen during valve development. We assume here that the pore openings must have resulted from a template on the inner surface of the expanding silicalemma, blocking silicification at those points, or perhaps by the precise deposition of an organic component that prevents mineralization. Alternatively, it is possible that a cytoplasmic component acts through the silicalemma via a transmembrane connector (i.e. macromorphogenesis). Another example of micromorphogenesis might be the perfectly flat surfaces of some silica components (Figs. 13.12, 13.15), which may result from a specific molecule (or template) that only exists on the inner surface of the SDV at the completion of valve formation. If micromorphogenesis is the mechanism, it must be possible to identify and characterize the responsible molecules.

It should be noted that no organic component has thus far been localized within the developing SDVs of diatoms, although there are many hypotheses about their existence [9, 27]. A protective coating of organic material is known to encase the mature silicified components of the frustule, although there is no evidence at present

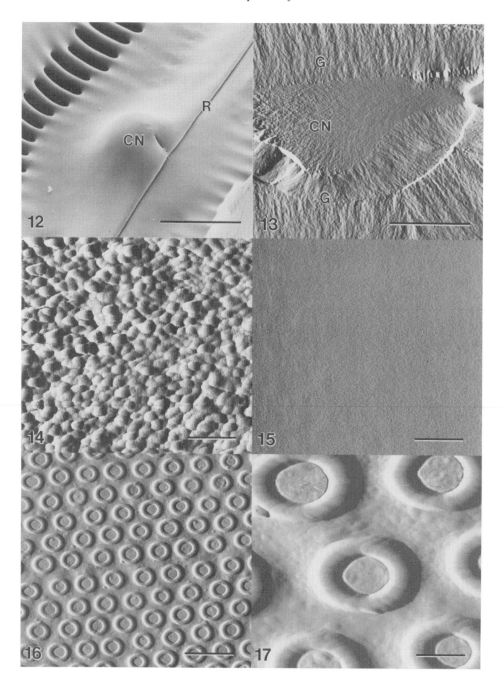

that these coatings are synthesized within the SDV or that they are in any way involved in morphogenesis or silicification. If present, embedded matrix molecules might be isolated from mature walls. Although this approach has resulted in the characterization of a number of wall-associated proteins (i.e. frustulins and HEP proteins) [30–32], these proteins have been localized to the surface of the silicified wall components and not within developing SDVs. It appears that the frustulins and HEP proteins are unlikely to be involved in mineralization or morphogenesis. However, in the closely related chrysophycean alga *Mallomonas*, we have shown a protein and glycoprotein to be localized to the mature silicified scales and bristles and within their developing SDVs [33], suggesting that similar organic molecules will be found within diatom SDVs as well. The recent discovery of silaffins isolated from diatom biosilica [7] indicates that an organic template may have a role in diatom wall silicification. However, the silaffins have yet to be formally localized to the SDV, although they have very interesting properties, as demonstrated by their ability to generate silica spheres of uniform morphology when added to a solution of silicic acid.

In many organisms, interactions between an organic matrix and mineral results in the controlled nucleation, growth and final morphology of the biomineralized structure [34], and such mechanisms have been shown to operate during biosilicification in molluscs and sponges [3, 34, 35]. Biochemical and molecular analyses of materials isolated from demineralized diatom frustules provides the only evidence to date for an organic matrix or template [7, 17, 24, 35–37 (note, concentration units are incorrect in 24, p. 151)], and these studies indicate the presence of proteins and/or polysaccharides associated with the frustule. In most cases the organic coverings could not have been totally removed from the silicified components with the extraction conditions used, and the resulting "matrix" material was undoubtedly contaminated with these surface materials. Regardless, the published data provide no basis from which to infer mechanisms of morphogenesis and silicification in diatoms.

Figure 13.12 FESEM image of the internal valve surface of *Pinnularia sp.* showing the raphe fissure (R) and central nodule (CN). Scale bar = 5 µm.

Figures 13.13–13.17 Images taken with the atomic force microscope (AFM). Figure 13.13 is an AFM image of the *Pinnularia* valve in cross-section, mounted in epoxy glue (G) and fractured through the region of the central nodule (CN) by a technique described by Egerton-Warburton *et al.* [29]. Figure 13.14 shows a high resolution AFM image of the *Pinnularia* valve in cross-section revealing the globular particle structure. Figure 13.15 is an AFM image of the internal surface of the valve seen in Fig. 13.12, and reveals a perfectly smooth surface where the particulate structure is no longer observed. Figures 13.16 and 13.17 are AFM images of the outer valve surface of the centric diatom *Coscinodiscus* sp. showing larger pores that result from macromorphogenesis. The surface of this species is not smooth as seen in Fig. 13.15, but Fig. 13.17 reveals that the globular particulate structure can be seen in high magnification. Scale bars = 5 µm for Fig. 13.13; 2 µm for Fig. 13.16; 500 nm for Figs. 13.14, 13.15, 13.17.

13.7 Transport of Silica to the SDV

Reimann [11] and later Dawson [12] both suggested that the silica is transferred into the SDV by fusion of smaller vesicles containing either colloidal or polymeric silica. Once within the SDV that directs silica deposition, the major biomineralization events take place. Some of the critical observations are as follows:

1. Silica deposition stops after the silicalemma is lost – hence the particular environment offered by the SDV is critical. After the new valve is secreted onto the cell surface, silica dissolution becomes possible. However, this is offset by the presence of an organic coating that inhibits chemical dissolution. This layer is apparently 8–10 nm thick and of unknown origin [9].
2. The SDV expands as the silica wall grows, so the growing silica surface is always in the same environment with respect to any organic component. This may be essential to ensuring consistent deposition.
3. Various studies have reported silica spheres to be present in the nanostructure of diatom biosilica [25, 37, 38]. The spheres are usually 10–50 nm in size, but are mostly 30–50 nm. Many appear to be aggregates of smaller spheres, and occasionally some are up to 200 nm.
4. These spheres have been found also in early walls with smaller sizes of 3–20 nm and 40–50 nm. The implication is that the smaller particles are primary particles [12]. In some cases, the spheres appear to be organized into regular arrays.
5. According to Schmid and Schulz [25], all silica grows by accretion of silica spheres, which become compacted with age. Schmid [38] had earlier found that the younger growing zone consisted of a loose assemblage of small silica spheres, and the older zone was compacted. They claimed that there was a clear younger/older zone visible in the walls. Subsequently they also presented evidence for the silica spheres originating in 30–40 nm cytoplasmic vesicles [25]. Leaching with SDS or acid was used to support the presence of silica in these extraneous vesicles. Finally, they also suggested that a second type of silica was involved in filling in the spaces [25].

Once again, this data comes from microscopic images of diatom thin sections, and information on the chemistry and the chemical environment of the SDV is sparse. If silicate ion is accumulated against the natural concentration in solution, this requires energy. In principle, deposition of silica is only thermodynamically feasible once it reaches saturation levels of 6 ppm within the SDV. However, precipitation is possible at lower concentrations if it occurs via precipitation of insoluble metal silicates [39], or as an organic chelated complex (e.g. an amino acid–silicate). The evidence from elemental composition is that the walls are predominantly amorphous silica, so the precipitation through reaction with Mg (or Ca, Fe, Al, Mn) within the SDV via

$$Ca^{2+} + SiO(OH)_3^- \Rightarrow CaSiO_3 + H_2O + H^+ \qquad (14)$$

does not seem plausible at this time. If the valve is more or less silica then, at the time of nucleation and growth, the local silicate concentration must exceed its solubility at that intracellular pH value. Silicate deposition can be accelerated by decreasing the overall solubility of silicate ions by acidifying the SDV in the region of valve growth.

$$SiO(OH)_3^- + H^+ \Rightarrow SiO_2 + 2H_2O \tag{15}$$

From Fig. 13.11a, b, we see that below pH 6 the only species in solution is silicic acid with a solubility of 6 ppm. The major question is then how to rationalize the fact that the diatom walls are clearly made up of globular particles in cross-section (Figs. 13.12–13.14), and are not molecularly smooth as would be predicted if they were assembled from $Si(OH)_4$ by a molecular accretion process.

The monodispersity of the globular particles suggests a mechanism that is independent of the vagaries of nucleation and is controlled. This would be more easily explained by concentration of the silica in a spatially and chemically controlled microenvironment such as SDVs. The prerequisites appear to be:

1. The cell needs to prevent random precipitation of silica within the cytoplasm.
2. The Si(IV) must at some point exceed 3–4 mM if it is to precipitate.
3. The pH must be kept reasonably low to ensure rapid polymerization and nucleation/growth.
4. It needs to be moved to the walls without redissolving in low [Si] environments.

The actual morphology of the silica is influenced by pH and pNa. At high pH, surface charge causes the silicate ions to condense as colloidal spheres, whereas at intermediate pH, where the surface charge is lower, cross-linking of silicate polymers to form networks and gels occurs via siloxane formation:

$$SiOH + SiO^- \Rightarrow Si-O-Si + OH^- \tag{16}$$

The necessity for anionic surface sites, which facilitate nucleophilic attack of the hydroxylated silicon by the negative charge on the ionized silicon requires $4 < pH < 9$. Above pH 9, the silicate solubility increases due to formation of anionic complexes, and the reaction in Equation (16) tends to run to the left, i.e. there are increasing rates of silica dissolution, and uncontrolled reprecipitation. If the Na or K level is high, gelation occurs. Consequently, the ideal conditions for silica deposition as globular particles or colloidal silica are low pH and low alkali metal concentrations. Vrieling *et al.* [40] have recently reported that the SDV is acidic, which they established via fluorescence experiments with a pH-dependent fluorophore.

As valve formation is completed, it appears they may undergo some further cross-linking and ripening, which leads in some species to incredibly smooth silica surfaces (Fig. 13.15). This could be rationalized by simple control of pNa levels and by Ostwald ripening. Once the initial silica globules are deposited, addition of Na or acid to the SDV will drive silica gelation, and lead naturally to the pores being filled. In addition, transition metals will react with silicate within the valves to form insoluble silicates.

13.8 Micromorphogenesis and an Organic Matrix?

The question remains as to whether the diatom valve is assembled from colloidal silica, which is itself constructed from monomers within vesicles, or does the globular particle structure come about as a result of rapid deposition of monomers by a template-directing organic matrix? The final answer here is likely to depend on the nature of the biomineralizing matrix [e.g. proteins?]. Two templating candidates have recently appeared for two different types of silicified structure. Shimizu *et al.* [4] have isolated a protease from sponge spicules (silicatein) where the active site contains a serine. It is proposed that this hydroxyl-containing amino acid "participates in the organization of silicic acid precursors" [4]. This supports a computer model based on hydroxyl chelation [41].

Perhaps a more chemically satisfactory candidate has now been isolated from a diatom [7]. This polycationic peptide directs silica formation via pendant polyamines that are grafted biochemically onto a protein backbone. This molecule drastically alters the rates of silicate precipitation. Importantly, these results correlate strongly with the known catalytic effects of polyamines on sodium silicate nucleation [42–44]. Such amines would lower the solubility of silicate ion, provide a template for nucleation and would control the silica colloid size within the SDV. In fact we may argue that the globular particles of silica observed by AFM, TEM and SEM reflect the chain lengths of the polyamines used to direct silica deposition. However, once again, at this stage the silaffins have yet to be formally localized to the SDV and growing walls.

13.9 Conclusions

Ultimately, it seems obvious that if active transport is capable of sequestering monomeric silica in one of its pH-dependent forms, and transferring it through the plasma membrane and into the SDV, then such molecules or enzymes may also be able to directly cement each monomer into the growing valve. The active transport molecule/enzyme can be guided into position via membrane proteins. However, ultimately the cross-linking of the silica monomer to the nascent wall is governed by silica chemistry. Thus strong limits on pH and salt are still necessary. Conversely the deposition may simply be due to the excess monomer concentration within the SDV due to the active transport enzyme causing accumulation of silicic acid that then deposits into the silicalemma-lined mould. In either case, identification of the active moieties responsible for silica complexation is vital for further elucidation of the mechanisms of silicification by diatoms.

Acknowledgements

R. W. thanks the Office for Naval Research (USA) and all three authors thank the Australian Research Council for financial support.

References

[1] S. Mann, *Nature* 1993, *365*, 499–505.
[2] S. Oliver, A. Kupermann, N. Coombs, A. Lough. G. A. Ozin, *Nature* 1995, *378*, 47–50.
[3] J. N. Cha, K. Shimizu, Y. Zhou, S. C. Christiansen, B. F. Chmelka, G. D. Stucky, D. E. Morse, *Proc. Natl Acad. Sci.* USA 1999, *96*, 361–365.
[4] K. Shimizu, J. N. Cha, G. D. Stucky, D. E. Morse, *Proc. Natl Acad. Sci. USA* 1998, *95*, 6234–6238.
[5] A. M. Belcher, X. H. Wu, R. J. Christensen, P. K. Hansma, G. D. Stucky, D. E. Morse, *Nature (London)* 1996, *381*, 56–58.
[6] H. A. Lowenstam, S. Weiner, *On Biomineralization*, New York, Oxford University Press, 1989.
[7] N. Kröger, R. Deutzmann, M. Sumper, *Science* 1999, *286*, 1129–1132.
[8] F. Round, R. Crawford, D. Mann, *The Diatoms. Biology and Morphology of the Genera*, Cambridge University Press, Cambridge, 1990. pp. 1–129.
[9] J. D. Pickett-Heaps, A.-M. M. Schmid, L. A. Edgar, in *Progress in Phycological Research, Vol 7* (Eds F. E. Round and D. J. Chapman), Biopress, Bristol, 1990, pp. 2–168.
[10] N. Kröger, M. Sumper, *Protist* 1998, *149*, 213–219.
[11] B. E. F. Reimann, J. C. Lewin, B. E. Volcani, *J. Phycol.* 1966, *2*, 74–84.
[12] P. Dawson, *J. Phycol.* 1973, *9*, 353–365.
[13] E. G. Vrieling, T. P. M. Beelen, R. A. van Santen, W. W. C. Gieskes, *J. Phycol.* 2000, *36*, 146–159.
[14] M. Hildebrand, B. E. Volcani, W. Gassmann, J. I. Schroeder, *Nature* 1997, *385*, 688–689.
[15] M. Hildebrand, K. Dahlin, B. E. Volcani, *Mol. Gen. Genet.* 1998, *260*, 480–486.
[16] Y. Del Amo, M. A. Brzezinski, *J. Phycol.* 1999, *35*, 1162–1170.
[17] D. M. Swift, A. P. Wheeler, *J. Phycol.* 1992, *28*, 202–209.
[18] L. G. Sillén, in *Oceanography* (Ed. M. Sears), American Association for Advancement of Science, Washington DC, 1961.
[19] C. F. Baes, R. E. Mesmer, *The Hydrolysis of Cations*, Robert Krieger Publishing Company, Malabar, Florida, 1986.
[20] S. Sjöberg, Y. Hägglund, A. Nordin, N. Ingri, *Mar. Chem.* 1983, *13*, 35–44.
[21] S. Sjöberg, A. Nordin, N. Ingri, *Mar. Chem.* 1981, *10*, 521–532.
[22] R. K. Iler, The Chemistry of Silica: Solubility, Polymerization, Colloid and Surface Properties, and Biochemistry, John Wiley & Sons, New York, 1979.
[23] C. W. Sullivan, *J. Phycol.* 1976, *12*, 390–396.
[24] D. H. Robinson, C. W. Sullivan, *Trends Biochem. Sci.* 1987, *12*, 151–154.
[25] A.-M. Schmid, D. Schulz, *Protoplasma* 1979, *100*, 267–288.
[26] S. Mann, in *Biomineralization in Lower Plants and Animals* (Eds B. S. C. Leadbeater and R. Riding), Oxford, Clarendon, 1986, pp 39–54.
[27] A.-M. M. Schmid, *Protoplasma* 1994, *181*, 43–60.
[28] J. D. Pickett-Heaps, D. H. Tippit, F. A. Andreozzi, *Biologie Cellulaire*, 1979, *35*, 199–206.
[29] L. M. Egerton-Warburton, S. T. Huntington, P. Mulvaney, B. J. Griffin, R. Wetherbee, *Protoplasma* 1998, *204*, 34–37.
[30] N. Kröger, C. Bergsdorf, M. Sumper, *EMBO J.* 1994, *13*, 4676–4683.

[31] N. Kröger, C. Bergsdorf, M. Sumper, *Eur. J. Biochem.* 1996, *239*, 259–264.
[32] N. Kröger, G. Lehmann, R. Rachel, M. Sumper, *Eur. J. Biochem.* 1997, *250*, 99–105.
[33] (a) M. Ludwig, J. L. Lind, E. A. Miller, R. Wetherbee, *Planta* 1996, *199*, 219–228.
     (b) T. F. Schultz, L. M. Egerton-Warburton, S. Crawford, R. Wetherbee, 2000, submitted.
[34] H. A. Lowenstam, *Science* 1981, *211*, 1126–1131.
[35] T. L. Simpson, B. E. Volcani (Eds) *Silicon and Siliceous Structures in Biological Systems*, Springer-Verlag, New York, 1981, p. 6.
[36] R. E. Hecky, K. Mopper, P. Kilham, E. T. Degens, *Mar. Biol.* 1973, *19*, 323–331.
[37] B. E.Volcani, in *Silicon and Siliceous Structures in Biological Systems* (Eds T. L. Simpson and B. E. Volcani), Springer-Verlag, New York, 1981, pp. 157–200.
[38] A. M. Schmid, *Nova Hedwigia* 1976, *28*, 309–351.
[39] W. Stumm, J. J. Morgan, *Aquatic Chemistry*, Wiley-Interscience, New York, 1970.
[40] E. G. Vrieling, W. W. C. Gieskes, T. P. Beelen, *J. Phycol.* 1999, *35*, 548–559.
[41] K. D. Lobel, J. K. West, L. L. Hench, *Mar. Biol.* 1996, *126*, 353–360.
[42] G. M. Lindquist, R. A. Stratton, *J. Coll. Interface Sci.* 1976, *55*, 45–59.
[43] T. Mizutani, H. Nagase, N. Fujiwara, H. Ogoshi, *Bull. Chem. Soc. Jpn* 1998, *71*, 2017–2022.
[44] T. Mizutani, H. Nagase, H. Ogoshi, *Chem. Lett.* 1998, 133–134.

14 The Biological and Biomimetic Synthesis of Silica and Other Polysiloxanes

Katsuhiko Shimizu, Daniel E. Morse

14.1 Introduction

Silicon dioxide, the most abundant material in the Earth's crust, is used in a wide range of industrial products. Since ancient times, glasses and ceramics have been used in our daily life, and the more recent development of silicon dioxide-based catalysts, molecular sieves, insulators and semiconductors has transformed many industrial process and products used today. Furthermore, the silicones, organically modified siloxane polymers, appear in hundreds of products ranging from hydraulic fluids, plastics, resins, sealants and adhesives to personal care products. Geological silicon dioxide is produced at high temperatures and pressures, and conventional methods of silicon dioxide and silicone manufacturing require high temperatures and extremes of pH. Although recently developed sol–gel methods reduce the requirements for heat [1], acid- or base-catalysis of polymerization is still required. Recent advances in organosilicon chemistry offer the prospect of new high-performance silicon-based polymeric materials displaying properties such as photoluminescence, electroluminescence and electrical conductivity [2–4]. Synthesis under still milder conditions will be required to incorporate complex and still-functional biochemical macromolecules such as nucleic acids and proteins.

A wide range of living organisms, including diatoms, protozoans, sponges, molluscs and higher plants are also capable of producing hydrated amorphous silica, often in exquisitely controlled micro- and nanoscale architectures [5]. The precision of structural control of biogenic silica produced by such species as diatoms (single-celled microalgae) is a remarkable manifestation of the genetic control of this material synthesis, pointing to the important role of the proteins (primary products of gene expression) in this process [6, 7]. Globally, the scale of biological silica production (e.g. as Diatomite and Spongillite) is measured in Gigatons per year. This synthesis, which exceeds by several orders of magnitude the scale of anthropogenic production, is all the more impressive when one considers that it occurs under ambient conditions of pressure and temperature (often near 0 °C, as in the polar seas, where the production of diatoms is greatest), and in the absence of caustic chemicals, strong acids or bases. The source of the silicon used for this biological synthesis is silicic acid, present in low concentrations in all the world's oceans, fresh water and ground water.

Elucidation of the mechanisms of biological silica synthesis and structural control thus offers the exciting prospect that it might lead to new, low-temperature and environmentally benign routes to the manufacture of high-performance silicon-based materials. Through analysis of the proteins and genes controlling these processes, the underlying molecular mechanisms governing silica synthesis in two different living systems have recently been revealed, and the mechanisms adapted for new routes to polysiloxane synthesis *in vitro*. We have called this approach and its applications "silicon biotechnology" [6].

14.2 Sponges

Marine and freshwater sponges produce large quantities of silica in species-specific forms ranging from simple macroscopic or microscopic rods, fibers and struts to microscopic needles (called spicules), hooks, harpoons and stellate forms. The smaller and more elaborate structures are generally found in the class of organisms known as demosponges, while the macroscopic fibers and rods are found in the hexactinides. This latter group includes the Venus' basket sponge, which produces a beautifully ornamental assembly of fiber and the *Monorhaphis* sponges whose silica rods are 8 mm in diameter and a remarkable 3 m in length [8].

Sponges synthesize silica in specifically differentiated cells called sclerocytes; in the demosponges, single sclerocytes are believed to synthesize each spicule [9, 10]. Electron micrographic observations reveal that a lipid bilayer membrane called the silicalemma surrounds each spicule while it is being formed, suggesting that the production of silica in sponges occurs in a specific membrane-enclosed organelle equivalent to the silica deposition vesicles seen in diatoms (see below). This observation indicates that a specific and precisely controlled environment is required for biological silica synthesis. The process of extrusion of the silica spicule from the silicalemma has not been observed and remains poorly understood. Furthermore, the possibilities that synthesis of spicules may also occur in some species either in the extracellular environment or under the collaborative control of more than one cell cannot be excluded. Each sponge spicule always contains a central organic core or axial filament of protein [11]. Minute stellate silica "microscleres" made by sponges also apparently contain axial filaments running from the center toward each tip [5]. Because these protein filaments are wholly occluded within the bio-silica, their involvement in the formation of the mineral had long been suspected.

14.2.1 Silicateins

The marine sponge *Tethya aurantia* (Fig. 14.1), an abundant species in the shallow waters off the Californian Coast, produces abundant needle-like silica spicules (megascleres) and a small proportion of stellate silica microscleres. The spicules (2 mm in length × 30 μm in diameter) constitute 75% of the dry weight of this

Figure 14.1 The marine sponge, *Tethya aurantia*. This organism was the rich source of silica spicules that led to the characterization and genetic engineering of the silicatein proteins and their DNAs. (Photograph reprinted with permission from D. W. Gotschall and L. L. Laurent, *Pacific Coast Subtidal Marine Invertebrates*, Sea Challengers, Los Osos, California, 1979.)

sponge, making this species uniquely tractable for analysis of occluded proteins and their mechanisms of action [12–16]. The axial filaments (2 mm in length × ~2 μm in diameter) are readily purified following dissolution of the silica spicules with buffered HF. These protein filaments can then be dissociated to yield their three constituent protein subunits, which we have called silicatein (for silica protein) α, β and γ (12). The molar stoichiometry of these three constituents in the filaments is $\alpha:\beta:\gamma = 12:6:1$. Molecular biological analysis of the cloned cDNAs coding for the two most abundant subunits revealed the primary structures (amino acid sequences) of silicatein α and β, which were shown to be very similar, containing 218 and 217 amino acid residues, respectively [12, 16]. The amino acid composition and molec-

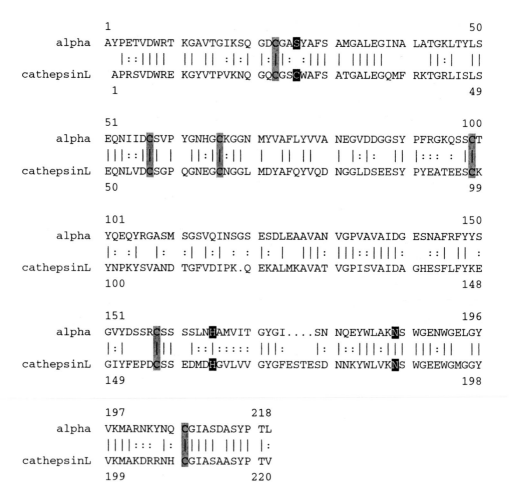

Figure 14.2 Alignment of amino acid sequences deduced from cDNA sequences for silicatein α and human cathepsin L. Identical amino acids are indicated by vertical bars; similar residues are indicated by colons. Cysteine residues involved in disulfide bonds in cathepsin L are shaded. Catalytic triad amino acids of the active site of human cathepsin L and corresponding amino acids in silicatein α are highlighted. Numbers above and below sequences represent residues from the amino terminus of silicatein α and cathepsin L, respectively (from [17, 18]).

ular weight of silicatein γ also are similar to those of α and β, indicating that the three silicatein subunits are all closely related, and that the axial filament is composed of proteins belonging to a single family. Analyses of the protein and DNA sequence databases revealed the surprising fact that the silicateins are members of a well-known family of hydrolytic enzymes – the cathepsin L family of the papain-like protease superfamily (Fig. 14.2). Identical amino acids are found at 50% of the corresponding positions in cathepsin L and silicateins α and β, while functional or genetic similarities increase the correspondence to 75%. Cathepsin L is an ubiquitous animal lysosomal protease with six cysteine residues forming three pairs of covalent disulfide bridges that constrain the three-dimensional structure of the protein [17]. The positions of these six cysteine residues are fully conserved in silicatein α and β, suggesting that the three-dimensional structures of the silicateins is very similar to that of the well-known protease. Using local energy minimization algorithms to refine the structures based on the silicatein amino acid sequences, we have constructed three-dimensional models for silicatein α and β based on the three-dimensional structure of human cathepsin L, known from X-ray crystallography to the 2.3 Å level of resolution (Fig. 14.3). The remarkable degree of similarity between the silicateins and cathepsin L, and the fact that both the silicateins and cathepsin L are vectorially transported into membrane-enclosed subcellular organelles in which they act, make it a virtual certainty that the silicateins and cathepsin L evolved from a shared molecular ancestor [7, 12].

Small-angle X-ray diffraction reveals a regular, repeating structure within the intact silicatein filaments (periodicity = 17.2 nm), as would be predicted if a simple repeating subassembly of the protein subunits underlies the macroscopic filament [12]. This finding is consistent with the electron micrographic evidence for para-crystallinity of the protein filaments from silica spicules of other sponges [18, 19]. Calculation of the hydropathy [20] of the silicateins reveals a unique hydrophobic domain on the molecular surface of silicatein α, and a distinct hydrophobic domain on the surface of silicatein β; these are not found on human cathepsin L, which functions as a monomer. These observations suggest that the macroscopic silicatein filaments may form as a result of interactions between the hydrophobic surface contacts of the subunits.

14.2.2 Activity of Silicateins

The discovery that the silicateins are highly homologous to members of a family of hydrolytic enzymes raised the possibility that they might function catalytically during silicification, in addition to any templating-like role they might have in organizing the structure of the resulting silica product. This suggestion was strengthened by the finding that two of the three amino acids of the catalytic active site of cathepsin L are exactly conserved in silicateins α and β, with the third residue (a nucleophilic cysteine, bearing a sulfhydryl sidechain) conservatively replaced in the silicateins with a nucleophilic serine (hydroxyl sidechain) [12]. Consistent with this suggestion, the silicatein filaments and subunits were found to be capable of catalyzing the hydrolysis and subsequent condensation of silicon alkoxides to form silica at neutral pH and low temperatures [13]. When silicatein filaments were mixed with tetraethoxysilane (TEOS) in buffered solution at pH 7, the surfaces of the filaments became covered with the silica that was produced. The filaments exhibit both a catalytic activity (catalyzing polymerization at neutral pH and low temperature, in a reaction that otherwise requires acid or base catalysis) and a "scaffolding" or template-like activity, in which they direct the nanoscale assembly of the polymerized silica to follow the molecular counters of the underlying protein filament. The silicatein filaments also exhibit these activities with organically substituted sili-

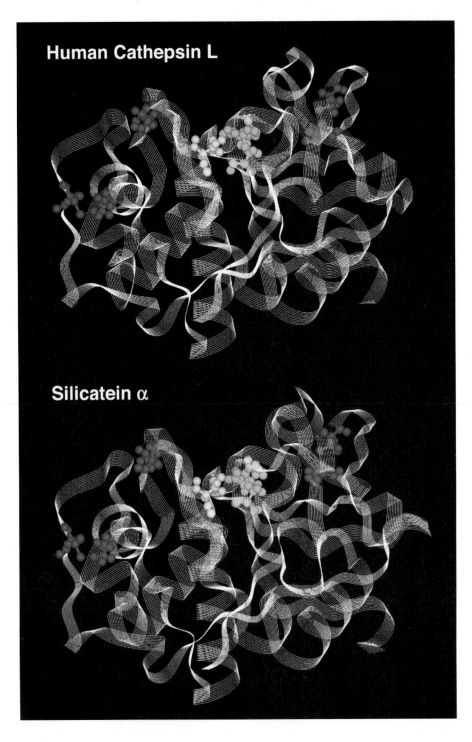

con alkoxides, catalyzing and spatially directing the polymerization of the corresponding phenyl- and methyl-silsesquioxanes (silicone polymer networks). Identification of the silica and silsesquioxanes produced *in vitro* was unequivocally established by solid-state MAS–NMR. The catalytic activity of the silicateins was abolished by thermal denaturation, proving that it depends on the intact three-dimensional conformation of the protein. Other biological filaments, including silk and cellulose, proved to be inactive with TEOS, further indicating the specificity of the reaction. The silicatein subunits obtained by disaggregation of the macroscopic silicatein filaments also exhibited this catalytic activity, which was inactivated by prior heat denaturation of the proteins.

Genetic engineering was used to produce the silicatein α subunit from a recombinant DNA template cloned in bacteria [13, 14]. The purified and reconstituted recombinant silicatein α exhibited a catalytic activity comparable to that of the silicatein in the native protein filaments, thus proving that the single subunit alone is capable of catalyzing the hydrolysis and polycondensation of TEOS to yield silica at neutral pH. This activity was also destroyed by prior heat denaturation of the protein. Site-directed mutagenesis (another genetic engineering technique) of the cloned silicatein α DNA was then used to probe the requirements for catalytic activity, by producing variants of the protein with substitutions of amino acids at the suspected catalytic active site [14]. Replacement of either the serine (with a hydroxyl side chain at position 26 in the protein chain) or histidine (with an imidazole side chain at position 165) at the suspected active site resulted in the loss of catalytic activity, thus substantiating the role of these two side chains in the catalytic activity and supporting the mechanism proposed for catalysis [7, 13, 14]. This mechanism, similar to that of the related proteolytic and hydrolytic enzymes, suggests that the identified active site hydroxyl and imidazole nitrogen form a hydrogen bond across the active site cleft in the protein, thus enhancing the nucleophilicity of the hydroxyl group and facilitating its attack on the silicon of a silicon alkoxide to form a transitory protein–silicon covalent intermediate. Hydrolysis of this intermediate by water then generates the reactive silanol species that initiates the condensation reaction with another silicon alkoxide, while regenerating the hydrogen-bonded pair at the protein's active site. Kinetic analyses confirm that silicatein α reacts with TEOS with typical, enzyme-like (Michaelis–Menten) substrate saturability, although the velocity of the reaction catalyzed is quite low [15].

A major uncertainty remaining in these and all other studies of biosilicification concerns the identity of the proximate precursor for silica formation. Studies in diatoms (see below) suggest that there may be some carrier or conjugate of silicic acid

---

Figure 14.3 Models for three-dimensional structures of cathepsin L and silicatein α. The models were constructed based on the three-dimensional structure of human cathepsin L determined by X-ray crystallography with resolution at 2.3 Å [17]. The side chains of the catalytic cysteine-25, histidine-163 and asparagine-187, and the cysteine residues forming the disulfide crosslinks of the cathepsin L, are presented with yellow, green, blue and red, respectively. Corresponding residues in silicatein α, in which the catalytic serine replaces cysteine, are shown with the same colors as those in cathepsin L.

formed intracellularly, and efforts are under way to identify such species in a variety of biosilicification systems. The occurrence of catechol–silicate complexes (analogous to the alkoxides) has been implicated in biosilicification in higher plants [21] and Perry and Lu [22] have produced silica from such a complex *in vitro*. While we do not yet know the identity of the substrate for biosilicification in any living system, and thus cannot yet know whether the silicateins function catalytically *in vivo*, the results described above illustrate one way in which nature can control the synthesis and nanofabrication of silica under physiological conditions, and have pointed the way to the design of synthetic catalysts that can direct the synthesis of structure-controlled silicon-based materials under environmentally benign conditions (see below).

14.2.3 Biomimetic Synthesis

Based on the structures revealed in the analyses of the natural and genetically engineered silicateins, "biomimetic" diblock copolypeptides have been synthesized that are capable of catalyzing the synthesis of shape-controlled polysiloxanes from the alkoxides at neutral pH and low temperatures [23]. These self-assembling diblock copolypeptides include both nucleophilic and amino functionalities analogous (but not identical) to those demonstrated by site-directed mutagenesis to be required for the efficient catalysis of TEOS hydrolysis and polycondensation by silicatein α These peptides act simultaneously as catalysts of silica polymerization at neutral pH (from the silicon alkoxide precursors) and as powerful structure-directing agents (like the silicatein on which they are based.) Control of the three-dimensional structure of the self-assembled catalytically active peptides makes it possible to control the structure of the resulting silica–polymer composites, with structures ranging from fibrillar bundles to glassy, transparent mesoporous spheres.

14.3 Diatoms

Long viewed as a paradigm for the precise biological control of nanofabrication of silica structures [24], diatoms have proved to be especially useful model organisms for studies of the mechanisms of biosilicification because they are unicellular and grow rapidly in the laboratory [5], and because new approaches are making them tractable for genetic manipulation [25–27]. New microscopic techniques have recently revealed remarkable details of the silica structures produced by diatoms [28–30], further stimulating the search to uncover the secrets of their nanostructural control.

14.3.1 Transporters

Pioneering studies by Hildebrand, Volcani and their colleagues resulted in the first cloning and characterization of a silicon transporter from any living system [25].

Dramatic proof that this molecule mediates the sodium-dependent influx of silicic acid across the cell membrane was obtained by enzymatically transcribing the cloned DNA *in vitro*, and injecting the resulting messenger RNA molecules into living frog eggs. The egg cells, which previously were incapable of taking up silicic acid from the medium, suddenly acquired this ability once the injected molecules were translated to make the transporter protein. Deduction of the transporter protein's sequence from that of the cloned DNA allowed this team to model the three-dimensional structure of the transporter molecule. Analogous to other ion transporters, its 12 helical transmembrane domains are thought to form a cylindrical barrel spanning the lipid bilayer of the cell membrane, with movement of silicic acid from the external environment into the cell, possibly regulated by intracellular concentrations and by the demand for silicification. Recent studies have revealed a family of these transporter molecules and their genes [26], raising the possibility that some might function in the outer cell membrane, while others might regulate transport between intracellular compartments, including the silica deposition vesicle (see below).

14.3.2 Frustulins and Other Outer Proteins

Kröger and his colleagues have made great progress in using molecular biological approaches to resolve and characterize a range of proteins from the silicified walls of diatoms, including those they have named frustulins, HEPs and silaffins [31–35]. The frustulins (named for their location on the outer surfaces of the frustules, or silica walls) are a highly conserved family of closely related glycoproteins with amino acid compositions similar to those of the diatom cell wall proteins previously analyzed by Hecky *et al.* [36], who proposed that the hydroxyl side chains of these proteins might serve to orient silicic acid moieties during polymerization. Computational modeling of the energetics of this process supported this suggestion [37]. As a result, for many years the only role of the proteins was thought to be a "templating-like" function, in which the coordination of silicic acid monomers in favorable proximity facilitated their spontaneous condensation. (It was in this context that the finding of catalytic activity in the sponge silicateins [13] was at first so surprising.) Although the precise roles of the diatom frustulins is yet to be elucidated, recent immunohistochemical studies suggest that they are not in intimate contact with the silica surface, but instead constitute one of several layers of an outer, varnish-like casing that may serve to protect the silica from dissolution [38]. Recent immunohistochemical studies by Wetherbee and his colleagues have also revealed the presence of other classes of glycoproteins in the diatom cell walls [39, 40].

14.3.3 HEPs and Silaffins

Analogous to the silicateins occluded in the biosilica of the sponge discussed above [6, 7, 12–16], it is the proteins and peptides occluded in the silica of the diatoms that have thus far yielded the greatest insights into mechanisms controlling biosilicification in these organisms. Kröger and his colleagues first characterized this class of proteins (named HEPs, for hydrofluoric acid-extractable proteins) and the DNAs that code for their synthesis [31, 32], showing for the first time that one of these silica-occluded proteins is uniquely associated with only certain specific domains within the silica architecture. In dramatic recent studies, this team has now demonstrated that a family of occluded cationic peptides (named silaffins, for their high affinity for silica) contain unusual post-translational modifications, making them uniquely capable of directing the formation of silica nanoparticles from silicic acid [35]. These peptides contain multiple repeats of a modified lysine–lysine pair never seen before in biology; the first lysine side chain is modified as a polyamine containing multiple N-propylamine repeats, and the second is modified as N,N-dimethyllysine. Because these amino acids are not incorporated by the genetic code-dependent protein synthesizing machinery of the ribosome and tRNAs, it is clear that the unique modifications are produced enzymatically after the peptides have been synthesized. Most significantly, the peptides bearing these modifications rapidly direct the formation of nanoparticulate silica from silicic acid *in vitro*, in much the same way that other polyamines had been observed to display this activity [41]. Replacement of the polyamine modifications abolished this activity.

There are interesting differences and similarities between the diatom silaffin and sponge silicatein systems. Both require amines for their activities, yet at this point their activities have been characterized *in vitro* with different substrates (silicic acid or the silicon alkoxides, respectively). The extent to which the observations with these two different systems may reflect underlying similarities or fundamental differences between independently evolved mechanisms of biosilicification remains to be elucidated. Further evaluation of the significance of the two systems and mechanisms will also require identification of the proximate precursor for silicification in these and other organisms, as discussed above.

14.3.4 Silica Deposition Vesicles

Biosilicification in sponges and diatoms occurs within a membrane-enclosed subcellular organelle known as the silicalemma (in sponges) or the silica deposition vesicle (SDV, in diatoms). Theoretical considerations suggested that for spontaneous condensation of silicic acid, the intravesicular pH within the SDV would be acidic [42, 43]. Microscopic investigations of the diatom SDV with chromogenic pH indicators were recently presented by Vrieling and his colleagues in support of this suggestion [44], although influences of the hydrophobic confinement on the ionization of the indicating chromophore may introduce some uncertainty to these measurements. It is interesting to note that the intravesicular pH of the lysosome, the vesicle in which cathepsin L functions, is also acidic [45]. Since the intracellular concentration of unpolymerized silicic acid in the diatom cell has been found to exceed the threshold for polymerization, it has been suggested that some carrier or conjugate (e.g. an alkoxide or catecholate) might stabilize the silicic acid. If this is the case, it remains to be determined whether the precursor transported into the SDV is the complexed form, or free silicic acid.

In addition to providing the appropriate microenvironment, pH and concentration of reactants and proteins necessary to initiate silicification, the SDV may also function as a mold to help determine the final structure of the polymerized silica. Thus, biological silicification may be controlled by catalysis, templating and molding, all acting in concert [7].

14.4 Silica in Higher Living Systems

Certain tissues of higher plants such as grasses (including bamboos), horsetails, rice and other grains, and even the leaves of oak trees, are rich in amorphous granular deposits of silica called phytoliths (literally, "plant rocks"). In fact, rice hulls are a commercial precursor to the valuable material, silicon carbide. Carole Perry and her colleagues have shown that the polysaccharides isolated from developing plant tissues at the stages in which silicification begins can dramatically influence the formation of amorphous silica *in vitro*, thus suggesting a possible role for these biopolymers in the biological process [46]. The higher plant systems of biosilicification differ considerably from those in the diatoms and sponges discussed above, in that the phytoliths display relatively little species-specific structure and none of the intricate nanostructural control characteristic of the diatoms and sponges.

Depletion of dietary silica results in defects in bone formation in developing chickens and mice, indicating a requirement for trace amounts of silica in vertebrate animals [47–49]. Silica-containing materials are used as artificial bone implants in human surgery [50–52]; following implantation, these materials are gradually resorbed and replaced by natural hydroxyapatite in the body. These facts suggest a potential for silicon metabolism in humans and other vertebrates, although interpretations have remained controversial and far from resolved.

14.5 Summary and Future Prospects

A wide variety of living organisms have evolved unique capabilities for building structures of silica, an amorphous network of silicon and oxygen. These capabilities are most impressive in certain marine organisms, in which the precision of nanoscale architectural control of the resulting silica structures often exceeds the present capabilities of human engineering. Equally remarkable is the fact that the conditions of biological silicification are those of the living cells, typically involving low temperature, ambient pressure and near-neutral pH, in marked contrast to the conditions for anthropogenic manufacturing with the same material. Recently, the proteins, genes and molecular mechanisms underlying biological silicification in two different groups of marine organisms – sponges and diatoms – have been partially elucidated. While there are interesting differences between these two, some similarities are also seen. These findings, and recent progress in harnessing these mechanisms to develop new and environmentally benign routes to the synthesis and structural control of silica and silicones (organically substituted polysiloxanes) are summarized in this chapter.

Powerful new methods of genetic engineering, including combinatorial mutagenesis and directed protein evolution, can help further define the determinants of polysiloxane synthesis and structural control of the silicateins, silaffins, and other yet-to-be discovered controllers of silica synthesis and nanofabrication in biological systems, and optimize these activities for a wide range of substrates [7, 53]. The results of these efforts should make it possible to harness the potential of biomolecular recognition and nanostructural control for the synthesis of new high-performance silicon-based composites suggested by the remarkable biosilica structures found in nature [6, 24, 28]. New and environmentally benign routes to the synthesis of silica and silicones have already been developed through biomimetic adaptation of the biological mechanisms [6, 7, 23]. Use of the mild conditions that are central to the biochemical and biomimetic catalysis and control of polysiloxane synthesis offers new prospects for the incorporation of functional biomolecules and cells with minimal loss of function, for new levels of efficiency and miniaturization in sensors and catalytic reactors. This path may also facilitate the direct coupling between biological macromolecules (such as DNA and proteins) and the electronic and optoelectronic properties of silicon-based semiconductor materials.

Acknowledgements

We thank our colleagues C. Lawrence, J. N. Cha, B. F. Chmelka, G. D. Stucky and Y. Zhou for allowing us to report the results of our collaborative research, and M. Brzezinski, U. Ciesla, Y. Del Amo and J. Gaul for helpful discussions. This work was supported by grants from the US Army Research Office Multidisciplinary University Research Initiative (DAAH04-96-1-0443), the US Office of Naval Research (N00014-93-1-0584), the Materials Research Division of the National Science Foundation (MCB-9202775), the NOAA National Sea Grant College Program, US Department of Commerce, under grant NA36RG0537, Project E/G-2, through the California Sea Grant College System, the MRSEC Program of the National Science Foundation under award No. DMR-96-32716 to the UCSB Materials Research Laboratory, and a generous donation from the Dow Corning Corporation. The US Government is authorized to reproduce and distribute copies for governmental purposes.

References

[1] C. J. Brinker, G. W. Scherer, *Sol–Gel Science: The Physics and Chemistry of Sol–Gel Processing*, Academic Press, San Diego, 1990.

[2] N. Auner, J. Weis (Eds) *Organosilicon Chemistry III: From Molecules to Materials*, Wiley-VCH, Weinheim, 1998.
[3] N. Auner, J. Weis (Eds) *Organosilicon Chemistry I: From Molecules to Materials*, Wiley-VCH, Weinheim, 1999.
[4] W. Ando, *Bull. Chem. Soc.* 1996, *69*, 1–16.
[5] T. L. Simpson, B. E. Volcani, *Silicon and Silicious Structures in Biological Systems*, Springer-Verlag, New York, 1981.
[6] D. E. Morse, *Trend. Biotechnol.* 1999, *17*, 230–232.
[7] D. E. Morse, in *Organosilicon Chemistry IV: From Molecules to Materials* (Eds N. Auner and J. Weis), Wiley-VCH, Weinheim, 1999.
[8] C. Levi, J. L. Barton, C. Guillemet, E. Le Bras, P. Lehuede, *J. Mat. Sci. Lett.* 1989, *8*, 337–339.
[9] T. L. Simpson, *The Cell Biology of Sponges*, Springer, New York, 1984.
[10] G. Imsiecke, R. Steffen, M. Custodio, R. Borojevic, W. E. G. Müller, *In Vitro Cell Dev. Biol.* 1995, *31*, 528–535.
[11] R. E. Shore, *Biol. Bull.* 1972, *143*, 689–698.
[12] K. Shimizu, J. Cha, G. D. Stucky, D. E. Morse, *Proc. Natl Acad. Sci. USA* 1998, *95*, 6234–6238.
[13] J. N. Cha, K. Shimizu, Y. Zhou, S. Christiansen, B. F. Chmelka, G. D. Stucky, D. E. Morse, *Proc. Natl Acad. Sci. USA* 1999, *96*, 361–365.
[14] Y. Zhou, K. Shimizu, J. N. Cha, G. D. Stucky, D. E. Morse, *Angew. Chem. Int. Ed. Engl.* 1999, *38*, 779–782.
[15] Y. Zhou, K. Shimizu, J. N. Cha, G. D. Stucky, D. E. Morse, in preparation.
[16] K. Shimizu, C. Lawrence. D. E. Morse, in preparation.
[17] R. Coulombe, P. Grochulski, J. Sivaraman, R. Ménard, J. S. Mort, M. Cygler, *EMBO J.* 1996, *15*, 5492–5503.
[18] D. F. Travis, C. F. Francois, L. C. Bonar, M. J. Glimcher, *J. Ultrastruct. Res.* 1967, *18*, 519–550.
[19] R. Garrone, *J. Microscop.* 1969, *8*, 581–598.
[20] J. Kyte, R. F. Doolittle, *J. Mol. Biol.* 1982, *157*, 105–132.
[21] A. Weiss, A. Herzog, in *Biochemistry of Silicon and Related Problems* (Eds G. Bendz and I. Lindqvist), Plenum Press, New York, 1978, pp. 109–125.
[22] C. C. Perry, Y. Lu, *J. Chem. Soc. Faraday Trans.* 1992, *88*, 2915–2921.
[23] J. N. Cha, G. D. Stucky, D. E. Morse, T. J. Deming, *Nature* 1999, in press.
[24] J. Parkinson, R. Gordon, *Trend. Biotechnol.* 1999, *17*, 190–196.
[25] M. Hildebrand, B. E. Volcani, W. Gassmann, J. I. Schroeder, *Nature* 1997, *385*, 688–689.
[26] M. Hildebrand, K. Dahlin, B. E. Volcani, *Mol. Gen. Genet.* 1998, *260*, 480–486.
[27] H. Fischer, I. Robl, M. Sumper, N. Kröger, *J. Phycol.* 1999, *35*, 113–120.
[28] E. G. Vrieling, T. P. M. Beelen, R. A. van Santen, W. W. C. Gieskes, *J. Biotechnol.* 1999, *70*, 39–51.
[29] E. G. Vrieling, T. P. M. Beelen, R. A. van Santen, W. W. C. Gieskes, *J. Phycol.*, in press.
[30] L. M. Egerton-Warburton, S. T. Huntington, P. Mulvaney, B. J. Griffin, R. Wetherbee, *Protoplasma* 1998, *204*, 34–37.
[31] N. Kröger, M. Sumper, *Protist* 1998, *149*, 213–219.
[32] N. Kröger, G. Lehmann, R. Rachel, M. Sumper, *Eur. J. Biochem.* 1997, *250*, 99–105.
[33] N. Kröger, C. Bergsdorf, M. Sumper, *Eur. J. Biochem.* 1996, *239*, 259–264.
[34] N. Kröger, C. Bergsdorf, M. Sumper, *EMBO J.* 1994, *13*, 4676–4683.
[35] N. Kröger, R. Deutzmann, M. Sumper, *Science*, in press.
[36] R. Hecky, K. Mopper, P. Kilham, T. Degens, *Mar. Biol.* 1973, *19*, 323–331.
[37] K. Lobel, J. West, L. Hench, *Mar. Biol.* 1996, *126*, 353–360.
[38] W. H. van de Poll, E. G. Vrieling, W. W. C. Gieskes, *J. Phycol.* 1999, *35*, 1044–1053.
[39] M. Ludwig, J. L. Lind, E. A. Miller, R. Wetherbee, *Planta* 1996, *199*, 219–228.
[40] R. Wetherbee, J. L. Lind, J. O. Burke, R. S. Quatrano, *J. Phycol.* 1998, *34*, 9–14.
[41] T. Mizutani, H. Nagase, N. Fujiwara, H. Ogoshi, *Bull. Chem. Soc. Jpn* 1998, *71*, 2017–2022.
[42] R. K. Iler, *The Chemistry of Silica*, Wiley, New York, 1979.
[43] R. Gordon, R. W. Drum, *Int. Rev. Cytol.* 1994, *150*, 243–372.

[44] E. G. Vrieling, W. W. C. Gieskes, T. P. M. Beelen, *J. Phycol.* 1999, *35*, 548–559.
[45] H. Kirschke, J. Langner, B. Wiederanders, S. Ansorge, P. Bohley, *Eur. J. Biochem.* 1977, *74*, 293–301.
[46] C. C. Perry, M. A. Fraser, N. P. Hughes, in *Surface Reactive Peptides and Polymers* (Eds C. S. M. Sikes and A. P. Wheeler), American Chemical Society, Washington DC, 1991, pp. 316–339.
[47] E. M. Carlisle, *Fed. Proc.* 1974, *33*, 1758–1766.
[48] K. Schwarz, *Fed. Proc.* 1974, *33*, 1748–1757.
[49] M. G. Voronkov, G. I. Zelchan, E. J. Lukevits, *Silicon and Life*, 2nd edn, Zinatne Publishing, Riga, 1977.
[50] L. L. Hench, *J. Am. Ceram. Soc.* 1991, *74*, 1487–1510.
[51] T. Kokubo, *Biomaterials* 1991, *12*, 155–163.
[52] B. Ratner, A. S. Hoffman, F. J. Schoen, J. E. Lemons, *Biomaterials Science*, Academic Press, San Diego, 1996.
[53] L. Giver, A. Gershenson, P.-O. Freskgard, F. H. Arnold, *Proc. Natl Acad. Sci. USA* 1998, *95*, 12809–12813.

15 Protein Components and Inorganic Structure in Shell Nacre

Angela M. Belcher, Erin E. Gooch

15.1 Introduction

Living organisms produce a wide variety of minerals with exquisite shape and form that exhibit exceptional strength and regularity. The study of the dynamic process of biomineralization could elucidate new routes to the synthesis of materials with increased hardness, toughness and fracture resistance. There are more than 40 different minerals produced by organisms, the most common being salts of calcium carbonate and phosphate [1] like the minerals forming the basis of bones, teeth and shells. Other common mineral salts are barium, strontium, iron and silicon. These mineral salts are formed in conjunction with organic biopolymers in the form of protein, glycoproteins and carbohydrates. These biopolymers aid in the formation of minerals with uniform crystal size and nanostructural regularity. These polymers also help stabilize specific crystallographic phases, morphologies and orientations. Understanding the process by which biominerals are made could aid in the design and synthesis of materials with increased toughness and biological and environmental compatibility [2–6].

Haliotis rufescens is a marine gastropod that produces an adult shell consisting of two different polymorphs of calcium carbonate, calcite and aragonite. The nacreous, mother of pearl, layer of the abalone shell has been a candidate for such studies because of its highly regular aragonitic crystalline repeating domains. The shell is a microlaminate composite material composed of 98% by mass calcium carbonate and 2% by mass organic material. The minerals are predominately formed with prismatic calcite on the outside of the shell and stacked tablets of aragonite, nacre, on the inside of the shell. The highly organized and hierarchical structure of the organic–inorganic composition of the nacre helps it achieve a fracture resistance of 3000 times that of the pure mineral of aragonite alone [7, 8].

Calcium carbonate exists in three different polymorphs – calcite, aragonite and vaterite. The triangular nature of the carbonate ion dominates the structure of these minerals. The carbonate salts can exist with a trigonal or orthorhombic symmetry, as in calcite and aragonite, respectively. The fact that the abalone can make both polymorphs of calcium carbonate under the same temperature and pressure conditions is an excellent example of biological control over mineralization. Remarkably, this final biomineralized product contains a small weight percent of the biopolymer phase. The abalone can secrete inorganic ions and biopolymers to control crystal shape and produce different polymorphs in close proximity. This polymorph selection could be controlled by changes in pH or addition of additives. According to the Ostwald step rule in polymorph precipitation, the less stable phase often occurs first, and is then subsequently transformed into a more stable phase [9]. The kinetics of this transformation can be hindered or even suppressed by the addition of additives such that the less thermodynamic phase is stabilized.

The influence of organic biopolymers has been shown to be important in the regulation of crystal growth [10–15], nucleation, particle aggregation and polymorph selection [16, 17]. The activation energy for nucleation in biological systems can be reduced by lowering the interfacial energy or increasing the supersaturation [1]. Supersaturation can be controlled by extracellular compartmentalization and ion-specific pumping. Interfacial energies can be lowered by having organic molecules on the surface at nucleation sites. These biopolymers involved in mineral growth must have areas of high local charge where electrostatic, dipolar and hydrogen bonding can take place during nucleation. If molecular recognition is achieved then the structure and the size of the mineral can be controlled.

The aragonite crystals that make up the nacre are polygonal or pseudohexagonal in shape. These tablets are 10–20 μm across (a–b plane) and uniformly 300–500 nm thick (c direction). These tablets have been reported to be approximately single crystals [18–21]. The nacre tablets are truncated in the c-axis and elongated in the a–b plane. This crystal habit and orientation is different from that of aragonite formed in seawater [22] or geologically, which is typically elongated along the c-axis.

In nacre, these tablets are stacked on top of each other along the c-axis. These tablets are vertically coherent and crystalline from one layer to the next over distances of at least ten tablets, as shown by electron diffraction and atomic force microscopy (AFM) [23–26]. Each tablet is surrounded by interlamellar organic sheets (Fig. 15.1). The tablets are offset along the a–b plane such that adjacent stacks interdigitate (Fig. 15.2). The combination of the "brick wall" like interdigitation and the dampening effect of the organic interlamellar sheets gives the nacre its greatly enhanced toughness and mechanical strength. The uniform thickness of the stacked tablet, which is of the order of the wavelength of visible light, gives the nacre its iridescent "mother of pearl" appearance.

15.2 Flat Pearls: An *in Vitro* Study of Abalone Shell Growth

The abalone shell is a highly organized, multiphase organic–inorganic composite material. For the organism to produce such an organized composite it must have a mechanism or mechanisms to control the spatial and temporal regulation of mineral deposition, mineral polymorph, crystal size, shape and orientation. In order to reproduce such structures or design new materials based on these structures, Fritz and co-workers developed an *in vivo* model system to elucidate the spatial and temporal sequence of both the mineral deposition and the corresponding organic biomolecule secretion. This model is called the flat pearl system [27, 28].

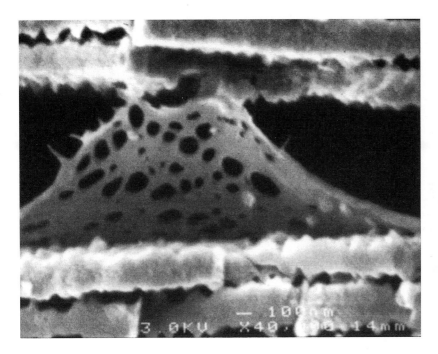

Figure 15.1 Scanning electron micrograph of a vertical cleavage of nacre from adult shell coated with Au/Pd, showing two of the aragonite tablets that form the "brick wall structure" of nacre with the organic matrix in between. The c-axis of the aragonite lies in the plane of the paper. SEM images were taken using a JEOL-6300F instrument equipped with a cold cathode field-emission source.

Typically, the sequential progression of nacre has been studied by removing a marginal portion of the shell of a live mollusc and allowing it to be regenerated [29, 30]. This process, referred to as shell regeneration, is studied by sacrificing animals at intervals and analyzing the regenerated shell region. The flat pearl system facilitates spatial and temporal studies of the shell during growth *in vivo* without sacrificing the animal.

The growth front of the shell is regulated by the epithelial cells of the mantle, which secrete both the organic macromolecules and inorganic ions into the extra-pallial space between the mantle and the growing shell. The mantle is retractable and can be gently lifted away from the shell without harming the animal. An abiotic inorganic substrate can then be inserted between the mantle in the extrapallial space of the shell, and the new shell is biofabricated on the inorganic substrate. In these experiments, the substrate was removed at various times in the shell growth process, and the flat pearl was analyzed and characterized by physical and biochemical techniques. The biomineralization process is thus analyzed in a native dynamic state, and the various organic–organic, organic–inorganic and inorganic–inorganic phase transitions are characterized as they occur.

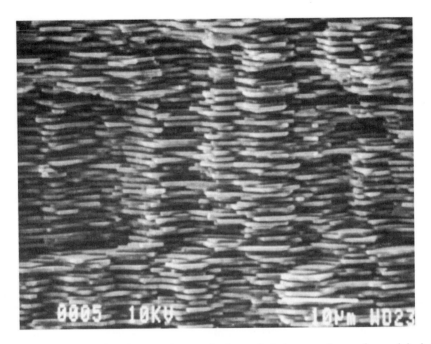

Figure 15.2 Scanning electron micrograph of a vertical cleavage of nacre from adult shell, showing the inter-digitation of aragonite tablets that form the "brick wall structure" of nacre. The c-axis of the aragonite lies in the plane of the paper.

15.2.1 Spatial Resolution

Spatial resolution of mineral deposition on an 18 mm glass coverslip extracted after 14 days revealed extensive mineralization across the surface facing the mantle tissue. Three distinct regions could be observed (Fig. 15.3): a red outer band corresponding to the outer calcite region of the shell, an iridescent central zone, and a green organic inner layer. The red outer band was 1–3 mm in width and was the smallest zone. The central iridescent region was pigmented and resembled the nacreous region of the natural shell. After 14 days this central region was often 6–10 mm in width and 300 µm in thickness; after 30 days it could cover the entire coverslip. The green inner proximal region, composed primarily of water-insoluble organic material, was 8–10 mm in width and was only partially mineralized after 14 days. This green organic material visually resembled an interstitial organic sheet observed in natural shell. After 30 days on a fully mineralized coverslip, the entire surface resembled the nacreous structure of the shell and the green organic layer could not be observed from the top view. The green layer was still visible by looking underneath the

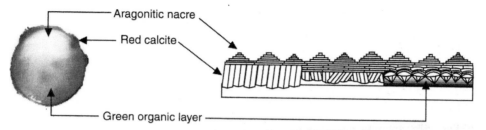

Figure 15.3 A mature flat pearl. The c-axis of the nacre is perpendicular to the paper. The three different regions of the flat pearl are labeled red calcite, aragonitic nacre and green organic layer. The schematic diagram on the right shows the spatial organization of the regions.

coverslip. X-ray diffraction studies using a power diffractometer revealed the polymorph of each zone described above (Fig. 15.4).

The crystalline mineral component of the red layer was determined to be calcium carbonate in the polymorph of calcite. The reflections corresponding to the polymorph calcite (2θ and (hkl) values) are 23.09° (102), 29.414° (104), 36.005° (110), 39.483° (113), 42.23° (202), 47.535° (108), 48.554° (116). No aragonite was found in this region. Because of the curvature of this red region the mineral was powdered, and so the diffraction pattern represented a random pattern and no orientation data were obtained for this region by this method. This region was similar in both color and crystallographic mineralogy to the outer prismatic region of natural shell. If the flat pearl is removed at a stage where the red prismatic calcite is still flat (< 12 days), orientation data can be collected. Pole figure X-ray diffraction data of the flat prismatic calcite surface reveals a preferred (001) orientation. The calcite (006) reflection is centered about the flat pearl surface normal, with a broad full width at half maximum (FWHM) of 24.6° indicating out of plane c-axis orientation. A pole figure of (hkl) = (104) gave a very spotty pattern, indicative of a coarse grain size. If the assumption is made that each peak is associated with one polycrystalline prism, then the prism diameter can be estimated. When the X-ray beam illuminates 3 mm² of the sample, 40 peaks are observed in the pole figure. This would give a grain size of 160 μm. This dimension agrees with Mutvei's [31] measurements of 100–200 μm based on scanning electron microscopy (SEM) measurements.

The diffraction pattern for the iridescent region of the flat pearl gave the reflections consistent with the polymorph aragonite. The 2θ and the (hkl) values are 31.12° (002) and 33.13° (012) (Fig. 15.4). The presence of only two reflections indicates that the crystals in this region are preferentially oriented approximately along the [001] (c-axis) direction. This diffraction pattern correlated precisely with that of nacre from adult shell, which also only reveals two reflections (002) and (012). Rietveld [27, 32] refinement of the diffraction data from the aragonite iridescent layer of the flat pearl gave an index of orientation of $r = 0.065$.

The green region was determined to be lightly mineralized with calcite. Only the most intense (104) reflection ($2\theta = 29.41°$) was observed at this time point under these conditions.

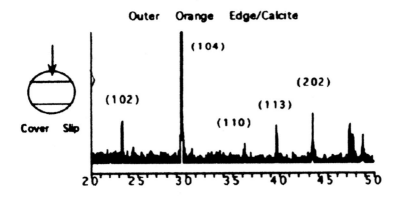

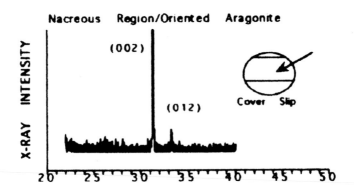

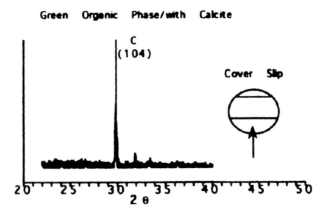

Figure 15.4 X-ray diffraction of the spatial deposition of flat pearl mineral phases. The top diffraction pattern is from the outer calcite region, the middle diffraction pattern is from the nacreous region, and the lower diffraction pattern is from the green organic sheet. Powder diffraction X-ray analysis was performed using a Scintag PADX equipped with a liquid nitrogen-cooled germanium solid-state detector using Cu Kα radiation.

15.2.2 Temporal Development

Temporal studies of mineral deposition showed that the first component to be secreted on the substrate was an organic layer. On this organic layer, the first mineral to be detected by X-ray diffraction using a powder diffractometer was calcite (Fig. 15.5). This calcite showed some orientation about the {104} crystallographic plane (2θ 29.41°) with an index of orientation $r = 0.3$ based on Rietveld refinement, with all other planes being randomly oriented. Nucleation of aragonite crystals occurred at discrete sites on the calcite primer after about 7 days. X-ray diffraction data

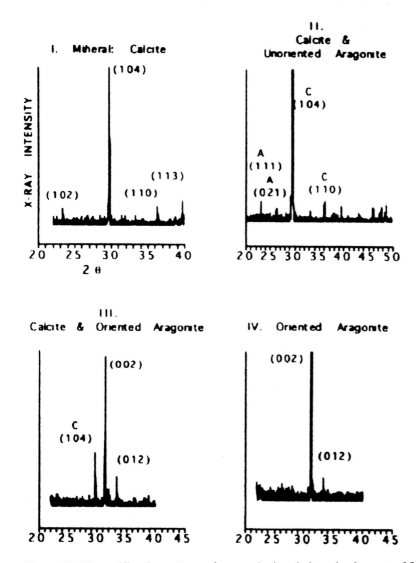

Figure 15.5 X-ray diffraction patterns of temporal mineral phase development of flat pearl.

revealed a non-oriented layer of aragonite with 2θ and (hkl) values of 26.25° (111), 27.31° (021), 33.13° (012), 37.93° (112), 38.51° (130), 38.69° (022). The oriented layer of calcite was still detected with a 2θ and (hkl) value of 29.41° (104). After 10 days mainly oriented aragonite [31.12° (002), 33.13° (012)] with some oriented calcite was still detected (29.41°, (104). After 14 days, only oriented aragonite was observed by X-ray diffraction on the non-green layer and non-red prismatic layer [31.12° (002), 33.13° (012)].

SEM images taken at this time show the characteristic stacks of coins of aragonitic nacre [28]. This iridescent zone showed a microlaminant structure of pseudo-hexagonal aragonite tablets with approximate dimensions of 10–20 μm wide and 0.35 μm thick, intercalated between organic polymer sheets. Over 200 layers of these aragonitic tablets were observed in samples removed from the animal after 14 days. Nacre deposition was observed after approximately 7 days, therefore the growth was approximately 200 nacreous layers per 7 days. This value is approximately 26 times faster than the average rate of nacreous growth of pearls in bivalve molluscs [27].

There are two types of organic sheets deposited on the abiotic substrates prior to the initial mineral deposition. The first was described above as the green sheet and the second is called the transparent sheet. These organic sheets differ visually in terms of color and texture and exhibit different Fourier transform infrared (FTIR) spectra. Both contain a protein component and a carbohydrate component. Interestingly, both of these organic layers preferably nucleate (104) calcite in flat pearls.

Temporally, the transparent organic sheet is deposited before the green organic sheet. This transparent sheet is deposited within the first 24 h after implantation of the abiotic substrate. The transparent sheet was shown to be a protein in composition by staining with Coomassie blue and a general protein stain, and by IR spectroscopy. This transparent sheet is generally found in the center of the substrate and can cover an area of 5–15 mm in diameter. The calcite is nucleated on the transparent organic sheet to a thickness of ~ 4 μm. These polycrystalline calcite crystals grow to form a confluent surface. Polarized optical microscopy revealed that polycrystalline domains of the order of 100 μm undergo uniform extinction suggesting in-plane orientation within these regions.

Pole figure X-ray analysis reveals that these calcite crystals have a very high degree of out-of-plane orientation. Measurements taken keeping the 2θ angle constant at 29° for the (hkl) of (104) and scanning about ψ and φ show (104) orientation of calcite on the transparent sheet of a 3 day flat pearl with a FWHM of only 3.1°. Therefore, there is only ~ 3° of misalignment (crystals not aligned with this orientation) with respect to the [104] direction. The calcite grows to a thickness of ~10–50 μm before a phase transition occurs and the first aragonite is deposited.

Pole figure analysis on 8 day flat pearls show a preferred orientation with the [104] direction centered about the flat pearl surface normal. However, there is misalignment about the [104] direction, as indicated by the broadness of the peak. The full width at half maximum is 21.2°. The degree of (104) alignment increases with flat pearl implantation time. These calcite spherulites nucleate off or in the green sheet and are spaced at distances of 25–100 μm apart. The nucleation density is approximately 150 spherulites per mm². The first nucleation sites appear to be embedded within the green sheet, as shown by the visual growth scars, analyzed by light microscopy, left after demineralization of the green sheet, and also in the transverse section of the natural shell where the tip of the mineral appears to be embedded a few microns below the surface.

The sperulites on the green sheet grow to confluence, forming a flat surface. The thickness of the calcite layer is 10–20 μm before the initial deposition of aragonite. The switch from calcite to aragonite is abrupt and is accompanied by a switch in soluble proteins. No evidence for a pre-organized sheet structure separating the two phases prior to switching has been observed using either SEM or a general fluorescent protein probe and fluorescent light microscopy. Although this evidence supports a mineral–mineral phase transition, organic molecules could be absorbed at the calcite–aragonite interface that were not detected.

The first aragonite to be deposited is either submicron size granular crystals that tend to stack in a conical fashion up to a thickness of a few microns, or a sharp transition to nacre tablets. The granular transition tends to occur in regions where the surface was not uniformly flat. This granular layer does not show the preferred (001) orientation of nacre as evident from the powder X-ray diffraction pattern. After the granular layer, normal nacre forms on the surface. The orientation of the nacre on the flat pearl is the same as that of natural shell, as seen by pole figure X-ray analysis. Pole figure analysis at (hkl) = (002) shows that the nacre has a high degree of out-of-plane orientation with a FWHM of 12°, showing that the c-axis is aligned relative to the surface normal. In-plane orientation (hkl) = (111) shows random orientation. This in-plane diffraction pattern is ~ 1 cm^2 for a large sample size and reflects the orientation of the entire sample. This, however, does not mean that the stacks of adjacent coins are not oriented relative to each other or twin related, as proposed by Sarikaya [33].

15.2.3 Protein Components

The results of the flat pearl studies indicate that biomineralization on inorganic substrates can be spatially and temporally organized such that the patterning of the composite closely resembles the growth front of the natural shell. Moreover, biofabrication of the flat pearl involves the secretion of the same proteins as those involved in natural shell growth and allows us to monitor the spatial and temporal secretion of these proteins from the mantle epithelial cells. At least nine distinct water-soluble proteins were isolated from the demineralized flat pearl. The mineral of the natural shell or flat pearl was crushed and demineralized with ethylene-diaminetetraacetic acid (EDTA), and the proteins were then separated by gel electrophoresis. These proteins are highly acidic [34–38] and optical detection after electrophoresis was achieved by staining with a cationic carbocyanin dye (stains all) [34]. The electrophoretic mobilities of the acidic proteins closely match proteins extracted from the intact natural shell. Differences in relative intensity were detected and reflect the fact that some of the proteins are unique to the calcite composites, while others are unique to the aragonitic nacre. The flat pearl samples analyzed contain different proportions of these mineralized composites than the thicker native shell, as verified by X-ray diffraction. Therefore the relative intensities of the mineral-specific proteins differ between a flat pearl and natural shell of similar mass under these conditions. Electrophoresis under denaturing conditions and Western blot analysis also revealed that proteins extracted from the flat pearl have relative molecular masses indistinguishable from those of proteins isolated from intact shell.

Mineral-specific proteins were found associated with both the aragonitic nacre fraction and the prismatic calcite fractions. Proteins were isolated by demineralization of an early flat pearl before nacre was deposited. The pattern and relative mobility of these proteins corresponds to that of the proteins isolated from the pure calcite fraction of the natural shell. However, demineralization of older flat pearls yields proteins that correspond to proteins in both of the pure mineral fractions, suggesting that the newly secreted aragonite proteins have a role in the mineral phase transition.

No evidence of withdrawal of the mantle edge was observed, suggesting that epithelial cells adjacent to areas of nacre deposition in the natural shell essentially remain spatially fixed in the presence of the coverslips. Therefore, the epithelial cells can control nacre production by down-regulation until the calcite–protein matrix is established on the abiotic inorganic surface. This change in mineral from aragonite to calcite and back to aragonite is in conjunction with a change in the secretion of the acidic proteins.

This mineral switching is consistent with information obtained on the process of shell regeneration in molluscs subjected to damage [29, 30]. Under shell regeneration conditions, epithelial cells engaged in nacre deposition adopt processes normally restricted to cells at the mantle edge. Therefore, the biological fidelity of the molluscan shell architecture is a consequence of the dynamic interaction at the cell–mineral interface, in which cells' recognition of the abiotic surface governs a genetic switch controlling the structure of the mineral that is formed. Although the nature of this interfacial recognition is not known, it is clear that severing the communication between the mantle and mineral results in a defaulting of cell activity and re-initiation of the mineralization sequence.

15.2.4 Green Organic Nucleation Sheet

The green organic sheet is also a protein–carbohydrate composite, as shown by amino acid analysis, carbohydrate analysis by gas chromatography, and infrared spectroscopy. The green sheet can be isolated by demineralization in acetic acid and physical removal from the rest of the components of the shell or from the abiotic substrate. This green organic composite is a tough material that can be easily manipulated without tearing or breaking. Comparison of total amino acid composition of the green organic sheets from both the flat pearl and natural shell revealed that the two are almost identical with regards to amino acid composition. Table 15.1 shows the amino acid composition. The amino acids with the most interesting percent mole fraction are tyrosine ($\sim 16\%$), proline ($\sim 19\%$) and glycine ($\sim 16\%$). These amino acid compositions are high compared with the average percentage occurrence in proteins estimated from the composition of 207 unrelated proteins [39],

Table 15.1 Amino acid composition of the green sheet from natural shell and flat pearl

Amino acid	Green sheet – shell (%)	Green sheet – flat pearl (%)
Asx	7.2	5.9
Glx	7.6	7.7
Ser	8.0	9.7
Gly	16.3	14.8
His	3.9	3.6
Arg	4.3	4.9
Thr	4.3	5.8
Ala	2.0	3.4
Pro	19.5	16.0
Tyr	16.0	13.2
Val	3.6	4.7
Met	0.7	1.1
Ile	0.9	1.8
Leu	2.2	4.0
Phe	0.9	1.7
Lys	2.6	2.2

which are tyrosine ($\sim 3.5\%$), proline ($\sim 4.6\%$) and glycine ($\sim 7.5\%$). It is also interesting to note that the other amino acids that have non-polar side chains (excluding proline) are relatively low in composition compared with the average percentage. These are listed as (% green sheet, % average): alanine ($\sim 2\%$, 9.0%), valine ($\sim 3.6\%$, 7%), isoleucine ($\sim 0.9\%$, 4.6%), leucine ($\sim 2.2\%$, 7.5%) and phenyalanine ($\sim 0.9\%$, 3.5%).

The green sheet is an insoluble matrix in seawater and deionized water, in chaotropic agents such as urea and guanadinium, and is relatively inactive to protease digestion. This suggests that there is a covalent crosslinking component to this organic sheet. The high percentage composition of the tyrosine suggests the existence of tyrosine–tyrosine crosslinking. This has been described in other systems [40]. This tyrosine crosslinking is also known to produce a dark color, similar to this organic sheet. IR studies on the green sheet also revealed a high percentage of tyrosine and tyrosine–tyrosine bonds (Table 15.1).

This organic sheet also contains a carbohydrate component. Isolation of the green sheet followed by desiccation, analytical weighing and hydrolysis with 6 M HCl and amino acid analysis revealed approximately 541 µg protein per 1000 µg of sheet. This non-protein component is assumed to be carbohydrate. Analysis of the green sheet by IR also reveals a high content of carbohydrate component. However, this method predicts a slightly lower carbohydrate component compared with the amino acid analysis. Hydrolysis of the green sheet with 4 M HCl for sugar analysis of neutral sugars [41] shows that the green sheet is composed of sugars that have standard retention factors of arabinose ($\sim 14\%$), xilose ($\sim 24\%$), mannose ($\sim 4\%$), galactose ($\sim 13\%$), glucose (13%), N-acetyl glucosamine (13%), other (19%).

15.2.5 Effects of Other Inorganic Substrates

To investigate charge, hydrophobicity and surface effects on green sheet deposition, different abiotic substrates were used for flat pearl generation. Mica was cleaved to leave a negatively-charged surface and inserted in the extrapallial space. SEM images of a mica flat pearl resemble the growth on glass. However, the X-ray data of the initial stages of mineral deposition show some orientation effects. The calcite is deposited as [104] oriented spherulites on glass. However, the initial stages of aragonite deposition show some orientation about the [012] at $2\theta = 33.1°$.

Comparing the initial stages of green sheet and mineral deposition on the hydrophobic substrate MoS_2 yields different results. The green sheet appears to be rippled on the surface of the MoS_2 and does not form a flat, even surface across the substrate. The calcite that is nucleated off this green sheet has a wide range of growth directions relative to the surface. This may arise from the altered orientation of the organic molecules that make up the green sheet. These organic components could self-assemble with their hydrophobic components in contact with the MoS_2 surface. This would be in contrast to the more polar surfaces of the organic components interacting with the surface of glass or mica.

X-ray diffraction patterns of a MoS_2 flat pearl after aragonite has been deposited revealed that the major calcite peak is at $2\theta = 36.1°$, corresponding to the (110) reflection. SEM images of the MoS_2 flat pearl show that the shell structure can eventually correct itself over a distance of a few microns, giving [001] oriented nacre. However, X-ray diffraction data of aragonite deposition prior to highly oriented nacre growth shows a high degree of orientation about [012] at $2\theta = 33.1°$. This is not observed in natural shell or flat pearls grown on glass coverslips. The fact that the orientation of both the green sheet and mineral differ on this hydrophobic substrate indicate that cooperative surface interactions are important in the formation of the flat pearl and the natural shell. Moreover, the animal adjusts to the different surface of the substrate by first depositing the correct mineral on top in a non-oriented arrangement until it has a surface on which to build the oriented mineral. Once that surface is achieved, a normal flat pearl can be biofabricated.

On all the above inorganic abiotic substrates studied, a green organic sheet was deposited prior to mineral deposition and nacre growth. To test if the green sheet deposition was in response to a foreign substrate (i.e. non-calcium carbonate) or just an abiotic substrate from different sources, polymorphs and orientations of calcium carbonate were implanted in the abalone.

When naturally terminated inorganic [001] c-axis oriented calcite was used as an abiotic substrate, a thick green sheet was deposited over the surface. Calcite spherulites, resembling those nucleated off from green sheet deposited on glass were then deposited. Pole figure X-ray diffraction, keeping 2θ constant about the [104], revealed that the calcite spherulites that nucleated off from the green sheet grown on the (001) calcite were oriented about [104] with an angular distribution of $\sim 25°$ about ψ. Because the calcite nucleated on the green sheet in all the experiments listed is oriented about the [104] direction, freshly cleaved inorganic [104] oriented calcite was implanted to see if the animal would by-pass the green sheet deposition and start depositing nacre. However, in this case a green sheet was also deposited, followed by [104] oriented calcite. The morphology of this calcite was different than the calcite from other substrates, with rhombohedron calcite highly oriented out of plane and with a high degree of in-plane orientation. Powder X-ray diffraction only revealed the (104) peak of calcite, and pole figure X-ray analysis was not able to distinguish the substrate calcite from the green sheet nucleated calcite. In summary for calcite substrates, for the two orientations of inorganic calcite studied, a green sheet is always deposited prior to mineralization, thus triggering shell regeneration. This suggests that the animal does not recognize just any calcite as "self".

When inorganic [001] calcite is implanted, [104] calcite is deposited on the green sheet suggesting that this calcite is nucleated off from the green sheet and not grown epitaxially from the inorganic calcite substrate. This nucleated calcite is spherulitic in morphology, resembling the spherulites of calcite from glass. Both the glass and the [001] calcite have negatively-charged surfaces, suggesting that the organic molecules of the green sheet are interacting at the surface of these two substrates in a similar manner.

When an inorganic calcite [104] substrate is implanted, the animal still undergoes shell regeneration, suggesting that not just any [104] calcite surface is sufficient to directly form nacre. However, this [104] substrate surface gives rise to highly oriented calcite crystals on the green sheet surface. This suggests that the green sheet does nucleate [104] calcite, and by providing a [104] surface the organic molecules are deposited on the green sheet and nucleate highly ordered calcite crystals. Thus the orientation information is passed through the green sheet by inorganic–organic–inorganic interactions. No evidence of epitaxial growth was detected using optical polarizing microscopy.

To test if inorganic aragonite could by-pass shell regeneration and continue making nacre, two different crystallographic orientations of aragonite were implanted in the flat pearl. [001] naturally terminated aragonite and aragonite cut parallel to [010] were tested. In both cases a green sheet was deposited prior to mineralization and calcite was deposited on top. On [001] aragonite two distinct morphologies were observed, either rhombohedral calcite [104] that showed no in-plane orientation or triangular-shaped calcite that showed some alignment in-plane. This triangular shape resembles [012] calcite synthetically grown by Berman [42] on thin films of polydiacetylene. [010] aragonite had very few crystals nucleated from the green sheet. These crystals had rod-shaped morphologies.

It is particularly interesting to note that [001] aragonite by itself is not recognized as a surface suitable to form nacre. However, if a flat pearl that has a nacre surface is removed from one animal and implanted in another animal the flat pearl is recognized as "self" and acts as a suitable surface for nacre deposition without the interruption of green sheet deposition. It is also interesting that calcite, not aragonite, is the polymorph deposited when aragonite is the substrate. According to Lippmann [43], aragonite should always form as the preferred polymorph when the seed crystal is aragonite in solution conditions that favor either calcite or aragonite. This supports our hypothesis that soluble proteins are the key component in polymorph selection and recognition.

In all the substrates above, with the exception of the homologous biotic substrate of a pre-formed flat pearl, the green sheet was deposited prior to mineralization. An exception to this can be seen with the implantation of a heterogeneous biotic substrate of chitin derived from a squid pen. When this substrate is inserted, highly ordered c-axis aragonite needles are formed within the layers of the chitin. Then a green sheet is deposited on this aragonite followed by nucleation of calcite.

15.3 Soluble Shell Proteins and *in Vitro* Crystal Growth

Biomineralization is thought to be directed, in part, by preformed organic arrays. An equally important structure-directing factor is the role of cooperative interactions between water-soluble protein molecules and the inorganic phase during crystal nucleation and growth. These interactions can efficiently control phase, morphology and growth dynamics on a time domain basis, allowing the organism to rapidly introduce structural changes in the growing biomineralized composite over spatial scales ranging from angstroms to microns. Members of two families of proteins from the mineralized microlaminate composites of the abalone shell and flat pearl have been purified and characterized. Their roles in controlling biomineralization have been resolved. These include a nucleating protein sheet, responsible for the selective nucleation of oriented calcite to form a "primer" in the first step of biomineralization, and members of two distinct populations of soluble polyanionic proteins that define the crystal phase, morphology and growth dynamics of the growing crystals. These proteins alone can be used, during *in vitro* crystal growth, to abruptly switch crystallographic phase from calcite to aragonite and vice versa, thus controlling sequential switching between phases. This process has been used to grow multiphase composites *in vitro* with micron scale phase domains. It is proposed that this cooperative assembly of calcium carbonate and differentially secreted specific proteins is responsible for the abrupt phase transitions from calcite to aragonite observed during biofabrication of abalone flat pearls and shell.

X-ray diffraction and electron microscopy revealed that the nucleating protein sheet is responsible *in vivo* for the highly oriented [104] nucleation of calcite in the first step of biomineralization of the flat pearl. However, this protein sheet does not determine the phase of the crystals that are nucleated. Instead, crystal phase and morphology during growth on the nucleating protein sheet *in vitro* are controlled by the addition of mineral-specific polyanionic proteins purified from the shell.

Two independent research groups [17, 24], have shown the production of the aragonitic-specific polyanionic proteins (one occluded in the growing crystals and one or two bound to the external crystal faces [17]) can explain the genetic switch from the production of the calcite to the subsequent production of aragonite observed both in the formation of the flat pearl and in the synthesis of the native shell. Sequential switching back and forth between the calcite and aragonite phases, independent of the sheet protein substrate, is readily achieved by changing the concentrations of these mineral-specific soluble proteins. The addition of soluble shell proteins is sufficient to promote formation of the thin plate morphology and "stack of coins" growth of aragonite crystals resembling that in abalone nacre.

Soluble polyanionic proteins were extracted with EDTA, purified and characterized from the calcite and aragonite regions of the shell. These mineral–organic composites were physically separated and finely ground, and the purity of the two minerals verified by powder X-ray diffraction. Gel electrophoresis under denaturing conditions demonstrated that the calcite and aragonite composites contained different populations of crystal-associated, EDTA-solubilized proteins. The aragonite composite yields three major bands detectable with a cationic carbocyanion dye [34, 44] at 31, 20 and 16 kD. In contrast, the calcite fraction revealed six major proteins stained with this dye with apparent molecular masses of 14, 28, 35, 48 and 55 kD. The two populations of proteins also give different patterns after electrophoresis under non-denaturing conditions, although not all of the components are resolved by this procedure. Treatment of the aragonitic nacre with hypochlorite to destroy extracrystalline proteins revealed that one of the proteins released from the aragonite is intracrystalline. This protein was purified following extensive sonication of nacre in the presence of 4–6% sodium hypochlorite to disaggregate the nacre tablets and oxidatively destroy all of the extracrystalline proteins [27]. This treatment released the single tablets of aragonite, free of the surrounding elements of the protein matrix; purity of the single tablets was verified by polarizing optical microscopy and transmission electron microscopy. Demineralization of these isolated aragonite tablets then released a single polyanionic protein; gel electrophoresis under denaturing conditions indicated an apparent molecular mass of 16 kD. The amino acid compositions of two of the purified aragonitic proteins and of the entire aragonitic and calcitic fractions (Table 15.2), in conjunction with sequencing data,

Table 15.2 Amino acid composition of soluble abalone shell proteins

Amino acid	Total aragonite	Total calcite	Aragonite 16 kD	Aragonite 20 kD
Asx	30.6	34.0	15.6	21.1
Glx	5.0	2.5	3.1	3.6
Ser	6.9	1.9	1.9	3.5
Gly	29.8	44.5	57.5	44.1
His	0.4	0.3	0.5	0.9
Arg	4.0	3.4	1.9	3.7
Thr	1.4	0.7	0.6	2.5
Ala	8.8	1.0	2.7	2.9
Pro	2.7	ND	2.5	1.8
Met	1.0	3.9	ND	1.4
Tyr	1.5	2.6	0.5	0.6
Val	1.8	0.5	ND	0.7
Cys	NT	NT	NT	NT
Ile	0.8	0.2	0.6	1.5
Leu	2.2	0.8	2.7	0.6
Phe	1.8	0.2	0.5	1.6
Lys	1.5	3.5	9.5	9.6

NT = not tested; ND = none detected.

indicated that these proteins are polyanionic and glycine-rich. Staining with the cationic carbocyanin dye and the electrophoretic mobilities observed under non-denaturing conditions confirmed the polyanionic nature of these proteins.

Studies of crystal growth *in vitro* revealed the different roles of the nucleating protein sheet and the mineral-specific polyanionic soluble proteins. Crystals of calcium carbonate were grown in the presence and absence of the nucleating protein sheet and the two populations of polyanionic proteins. Growth solutions contained either 12 mM $CaCl_2$ and 12 mM NH_4HCO_3 or 20 mM $CaCl_2$ and 20 mM NH_4HCO_3 at pH 8.0. Polyanionic proteins solubilized from either the calcite or aragonite composites of native shell were added to a final concentration of 2 mg/ml; bovine serum albumen at the same concentration served as a protein control. In experiments where crystals were grown in the presence of the insoluble nucleating protein sheet, a parallel experiment was performed without this protein layer, using a glass slide as the nucleation surface instead. In each case, the nucleation density was higher, the crystals were more uniformly spaced, and the crystals were more highly organized on the protein sheet than on the glass. This suggests that the protein sheet plays an important role in the nucleation and organization of crystal growth sites.

Morphologies of the crystals grown under different conditions were visualized with crossed polarized light microscopy. Crystal habits were characterized by Raman spectroscopy and X-ray diffraction. Crystals grown in the absence of soluble protein exhibited the characteristic rhombohedral morphology of calcite; this identification was verified both by X-ray diffraction, revealing the calcite (104) reflection, and Raman microprobe analysis. These crystals were identical to those grown in the presence of bovine serum albumin under otherwise identical conditions. In contrast, crystals grown from a solution containing the polyanionic proteins that had been extracted from the calcitic composite of the shell exhibited the characteristic spherulitic morphology of calcite. Confirmation of this morphology as calcite was obtained by X-ray diffraction and Raman microprobe analysis [17].

In marked contrast to the results described above, crystals grown in the presence of the polyanionic proteins solubilized from the aragonitic composite of the shell displayed a very different morphology, forming needles of aragonite in the plane of the nucleation layer. They were always smaller in size (~ 10 mm) and showed a lower apparent density of nucleation sites compared to the crystals grown in the presence of proteins extracted from the calcitic composite. Phase and orientation of these crystals were verified by Raman microprobe, showing a red shift in the v_1 vibrational mode. The v_4 vibrational mode was not detectable because of the small crystal size, but crystal phase, morphology and orientation were verified by electron diffraction.

Crystals were also grown in the presence of a mixture of proteins from the aragonitic and calcitic composites. The predominant crystal type obtained under these conditions was composed of flat polycrystalline plates that undergo uniform extinction under cross-polarized light. The individual plates were shown by electron diffraction to be aragonite with a zone axis of (001), as found in the nacre of the shell and flat pearl. Transmission electron microscopy and electron diffraction further revealed that the plates grew in stacks, in a fashion reminiscent of the stacked

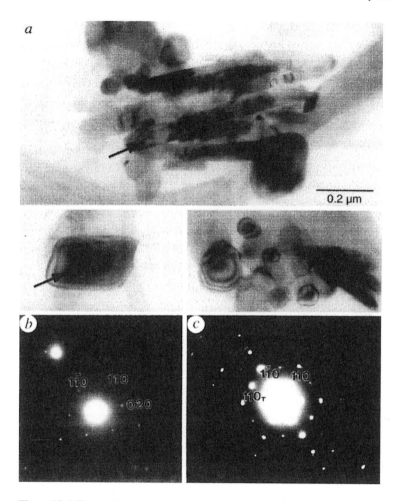

Figure 15.6 Transmission electron micrograph and electron diffraction of crystals grown *in vitro* in the presence of aragonite and calcite proteins. (a) shows the stacking plate-like morphology of the aragonite tablets. Bar = 0.2 μm. (b, c) are selected area diffraction studies showing that the plates are single crystal and c-axis oriented. Electron microscopy was performed with a Joel 2000 fx at 200 kV. Diffraction patterns were from selected areas.

plate growth of biogenic nacre (Fig. 15.6), despite the fact that the insoluble protein matrix surrounding the tablets of biogenic nacre of *Haliotis rufescens* had not been added.

Additional *in vitro* experiments were carried out to investigate the role of the soluble polyanionic proteins in the abrupt calcite-to-aragonite transition that is characteristic of abalone shell and pearl biomineralization [27, 28]. Rhombohedral calcite crystals were first grown from a saturated solution of calcium carbonate in the absence of soluble proteins. This solution was then removed and replaced with a fresh growth solution containing the same inorganics together with the soluble poly-

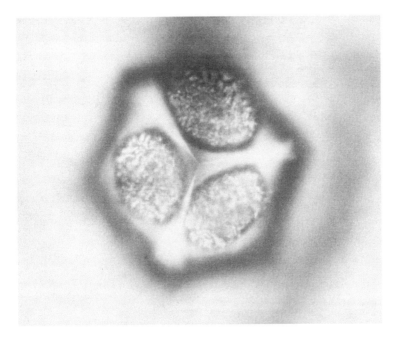

Figure 15.7 Light micrograph using cross-polarized light of aragonite needles grown on calcite in the presence of proteins isolated from nacre.

anionic proteins extracted either from the calcitic or aragonitic composites of the shell. Optical polarized microscopy and SEM revealed that the resulting new crystal growth was specific to the faces of the calcite rhombohedron and that the two families of proteins produced different crystal morphologies. X-ray diffraction studies and Raman microprobe analyses of these crystals revealed that the inorganically grown rhombohedrons were calcite, whereas the needles grown off this calcite were aragonite. These data showed that the abrupt transition from (104) calcite to aragonite seen in the biofabrication of the shell and flat pearl can de duplicated simply by addition of the aragonitic proteins to the crystal growth solution. Figure 15.7 is a calcite rhombohedron looking down the c-axis, showing aragonite needles growing off the (104) faces. The addition of a mixture of calcite and aragonite proteins to calcite rhombohedrons produces both column-like crystals off the (104) faces or flat plate-like structures.

A sequential transition of calcite to aragonite and back once again to calcite occurs when the soluble aragonitic proteins are depleted, permitting formation of calcite again as the stable phase. The addition of aragonitic proteins again to a growth solution will result in new aragonite needles on calcite rhombohedrons. This is an example of specific control over the mineral phase at the micron scale, determined by the presence or absence of the aragonite-specific polyanionic proteins.

The addition of mineral-specific polyanionic proteins solubilized from the molluscan shell is sufficient to cause a mineral phase shift in crystals growing from a solution that would otherwise yield calcite, inducing the growth of aragonite without the addition of heat, pressure or magnesium. This suggests that the abrupt calcite–aragonite transitions observed in abalone shell and pearl formation may be governed by differential secretion of two populations of soluble polyanionic proteins that control crystal orientation, morphology and mineral phase. The insoluble protein sheet that serves to nucleate the growth of highly oriented calcite in the formation of shell and pearl *in vivo* [28] also organizes nucleation *in vitro*. These observations, and the amino acid compositions, suggest that this proteinaceous layer functions analogously to the periostracum in the first events of nucleation during shell biomineralization. Moreover, duplication of the basic stacked plate structure of nacre by the addition of only these soluble proteins is possible. This suggests that this higher order crystal plate morphology is controlled, in part, by these proteins, although the precise height or thickness of the nacre plates formed during biomineralization may be governed by the fixed interlamellar spacing [45] of the insoluble proteins of the matrix.

15.4 Lustrin – a Nacre-Insoluble Matrix Protein Family

Nacre's greatly enhanced strength and toughness (compared with its geological counterpart) arises from a combination of the uniform stacking of c-axis oriented aragonite plates and their interdigitation with neighboring aragonite stacks forming a "brick wall-like" structure. Each individual plate is surrounded by an organic matrix that consists of an EDTA-soluble polyanionic and an EDTA-insoluble fraction. The effect is that the nacre is a highly organized organic–inorganic nano-composite which alternates between tablets that are 0.35–0.5 µm thick and an organic matrix that is ∼ 30 nm thick. The staggered arrangement of adjacent stacks of tablets, together with a damping effect of the intercalated organic matrix, give nacre an increased toughness and make it difficult to propagate a crack over a long distance [7, 8].

Using stained thin sections and transmission electron microscopy, Nakahara [46] showed that this matrix (in Monodonta labio) formed a box structure around mature aragonite plates. His data also suggested that this organic matrix was secreted as a pre-formed collapsed box that the mineral would grow into and would eventually be confined by the parameters of the pre-formed matrix. Weiner and Traub isolated the total matrix (from N. repertus) and with X-ray diffraction studies suggested that this matrix, which was composed of a protein and carbohydrate (in the form of chitin), formed an anti-parallel beta-pleated structure. In their model this structure served as scaffolding and a nucleation surface for nacre formation. In previous models this organic matrix, which includes the soluble and insoluble components, was thought to be secreted and assembled prior to new aragonite tablet formation. This pre-organized matrix was hypothesized to then function as a template for the next aragonite tablet. Thus each aragonite tablet nucleated independently, and was not in chemical contact with the crystal below it in the stack. Although this is true in some bivalves, recent work has shown that the growth mechanism in the gastro-pod Haliotis rufescens is quite different.

A series of experiments has shown that the tablets in a stack are crystallo-graphically vertically coherent. The first experimental evidence by Manne [18] used *in situ* imaging with an atomic force microscope (AFM). By imaging shell samples using AFM in liquids, it was possible to dissolve away the nacre layer by layer to reveal both the structure of a single tablet and its relationship to vertically adjacent tablets [18]. From etching with dilute HCl while imaging it was observed that nacre from Haliotis dissolved by widening of tablet boundaries until the tablet lifted off as a whole, revealing the underly tablet. In the underlying tablet, it was observed that the a and b axes were well aligned vertically through individual stacks of the nacre, but not laterally between stacks. This suggested that each stack represents an indi-vidual nucleation center that induces preferential orientation of the crystallographic c-axis.

The AFM data suggested that the tablets were connected by mineralized con-nection that passed between the organic matrix to form the next c-axis oriented tablet, thus no new nucleation or templating event was required between each indi-vidual tablet. The AFM data only probed the tablet orientation over a few tablets. Electron diffraction and transmission electron microscopy of transverse thin sec-tions (~ 50 nm) made through mature nacre verified the vertical coherence along single aragonite stacks [47]. The electron diffraction data confirmed that the tablets were aligned at least over a distance of 10 tablets. Over longer distances the crys-tallographic coherence was lost. This could be an artifact due to processing of the material, but could also mean that this coherence exists over this distance and is re-nucleated.

TEM images of two tablets also revealed possible mineral bridge connections between two tablets in the same stack [48]. The distance between this mineral con-nection, 50 nm, is in agreement with the distance predicted by the model developed of staggered off-sets between tablets by Schäffer [48]. This explains a possible growth mechanism but does not explain uniformness of tablet thickness. *In vitro* crystal growth data [17] elucidated a few important points concerning the nacre. One is that a phase transition can occur by the addition of mineral-specific poly-anionic proteins without any change in the type of ion or ion ratio (for example no magnesium is required). Secondly, stacked plates of c-axis oriented aragonite crys-tals can be grown *in vitro* with the addition of soluble polyanionic proteins isolated from the shell (Fig. 15.6), without the addition of the insoluble matrix sheets. This implies that the insoluble matrix sheets are not required for polymorph selection or the truncated c-axis elongated a–b plane or the stacking structure of the tablets. However, the tablet thickness, although uniform in size, is smaller than the tablet thickness of nacre. Another point is if the nacre matrix is secreted prior to new tablet formation and the aragonite is templated off from this matrix, it appears that adjacent tablets should be aligned in the a–b plane because they would be nucleated from the same organic matrix. Manne *et al.* [18] also showed using AFM that the adjacent tablets are not aligned in the a–b plane.

One hypothesis is that the nacre matrix is involved in setting the uniform tablet thickness along the c-axis. This is supported by Nakahara's TEM photographs, which show the regular spacing of this matrix at the growth edge of nacre [46]. This could also explain why the stacked tablets grown *in vitro* do not have the correct thickness [17]. Schaffer and coworkers also showed using AFM that this matrix has pore-like structures that matched the degree of lateral offset within a stack of tablets on a flat pearl. This supports the hypothesis that the tablets are connected by mineral bridges that grow through these pores. It is the distance between the pore that determines the offset of the tablets and leads to the staggered structure required for the intercalating structure between stacks, giving rise to the brick wall structure.

The EDTA-insoluble nacre matrix could serve three functions. One, it could induce tablet staggering leading to the interdigitation between the tablets. Two, it could set the tablet thickness through either the physical barriers of the insoluble sheets or by effecting ion diffusion through pores. Three, it could have an elastic function on increase toughness.

15.4.1 Lustrin Protein Matrix Isolation

Sheets of the total nacre matrix can be isolated by demineralizing pieces of nacre in acetic acid. The associated soluble proteins can be removed by extensive washing with deionized water. The remaining matrix can be stained with Coomassie blue, revealing a protein network that once surrounded the tablets, but now leaves a hole after demineralization. Nacre matrix sheets demineralized in the presence of a protein crosslinker and stained with Coomassie blue reveal blue-stained side walls where the tablets were prior to demineralization, leaving a polygonal outline. This side wall structure appears to be doubly bound, where one side would be in contact with one tablet and the other side to an adjacent tablet. These side walls also appear to be connected across the surface of the sheet, and where one wall connects to another the stain is darker. Also interesting to note is a light spot, probably a hole in the matrix, that is located inside each polygon. These holes could represent a central pore where the mineral bridge connects two tablets in the same stack.

The insoluble nacre matrix was hydrolyzed in 6 N HCl and subjected to amino acid analysis. The results are given in Table 15.3. The major mole fractions are represented by asx 17.4%, serine 11.9%, glycine 30%, alanine 15.4%. The composition of the total matrix is similar to that reported by Weiner and Traub [49] in that the serine, glycine and alanine content is greater than 50% of the total. This is similar to the composition of silk, which is known to form an anti-parallel beta-pleated structure [49, 50]. Drying of this protein matrix and analytical weighing prior to hydrolysis revealed that the matrix was 784 µg of protein per 1000 µg of matrix sheet. A major component of the remaining 22% was assumed to be carbohydrate. IR spectroscopy did confirm a large percentage of carbohydrate. Analysis of the neutral sugars by hydrolysis of the matrix followed by gas chromatography showed the presence of neutral sugars that corresponded with the standard retention factors of arabinose (1.2%), xylose (77.2%), galactose (0.67%), glucose (6.9%) and N-acetylglucoseamine (0.8%).

Table 15.3 Amino acid composition of total nacre matrix

Amino acid	Mole fraction
Asx	17.4
Glx	3.4
Ser	11.9
Gly	30.0
His	0.4
Arg	5.2
Thr	1.6
Ala	15.4
Pro	3.3
Tyr	1.3
Val	12.1
Met	1.4
Cys	NA
Ile	0.6
Leu	2.2
Phe	2.6
Lys	1.2

15.4.2 Solubilization of a Protein Component

Soluble fractions of the nacre-insoluble matrix were isolated by homogenizing the matrix in 4 M urea, 0.5 N acetic acid and 100 mM DTT. These proteins had approximate apparent molecular masses of 30, 40, 65 and 120 kD (Fig. 15.8). Amino acid composition of the resulting soluble proteins is shown in Table 15.4. There are several interesting components to these proteins. Comparing the four proteins (called nacre matrix proteins or lustrins) shows that trends in amino acid composition are the same. All are rich in the amino acids (65 kD mole fraction listed) proline ~ 16%, glycine ~ 11%, cysteine ~ 8.5%, serine 7% and arginine 8.2%. For the 40 kD proline ~ 18%, glycine ~ 16%, cysteine ~ 7%, serine 9.8%, arginine 13.6%. This can be compared to average occurrences in proteins: proline ~ 4.6%, glycine ~ 7.5%, cysteine ~ 2.8%, serine 7.1%, arginine 4.7%. The Asx composition was ~ 12.7% for the 40 kD and ~ 10.9% for the 65 kD.

Comparing the 65 kD amino acid composition with the total amino acid composition of the nacre-insoluble matrix (Tables 15.3 and 15.4), the total matrix was much higher in alanine (~ 15.4% as compared with ~ 6.6%). The total matrix was also higher in glycine (30% compared with ~ 11%), and serine (~ 12% compared with ~ 7.2%). The Asx is also greater in the whole matrix (~ 17.4% compared with ~ 11%). The higher alanine, serine and glycine ratio in the matrix compared with the ratio solubilized suggest that a silk-like component may still be in the insoluble fraction of the matrix. However, the soluble nacre matrix proteins are much higher in proline, ~ 18.3%, ~ 15.6% (40 kD, 65 kD) compared with the total matrix 3.3%. This high proline and glycine composition suggested a homology to

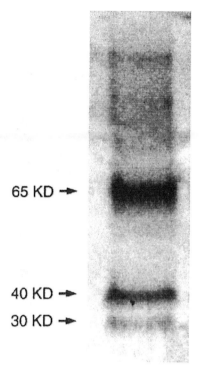

65 KD →

40 KD →

30 KD →

Figure 15.8 Protein blot from an SDS-PAGE gel used for direct sequencing of proteins isolated from the nacre insoluble of abalone shell. Proteins were solubilized using urea, acetic acid and DTT and stained with Coomassie blue.

Table 15.4 Amino acid composition of lustrin matrix proteins

Amino acid	30 kD	40 kD	65 kD	120 kD
Asx	11.5	12.7	10.9	11.7
Glx	7.2	7.2	7.0	6.0
Ser	8.2	9.8	7.2	8.4
Gly	12.1	16.0	11.2	10.7
His	0.9	0.4	0.7	1.5
Arg	13.1	13.6	18.2	0
Thr	4.5	6.0	5.6	6.7
Ala	6.8	4.9	6.0	2.7
Pro	12.1	18.3	15.6	11.1
Hypro	0.2	0.2	0.2	0.4
Tyr	ND	ND	ND	ND
Val	4.8	5.1	6.3	6.0
Cys	6.3	7.0	8.5	9.6
Ile	2.4	3.2	2.6	3.0
Leu	5.3	4.6	4.8	6.0
Phe	2.2	2.2	2.3	1.1
Lys	2.5	2.3	2.3	4.9

ND = none detected.

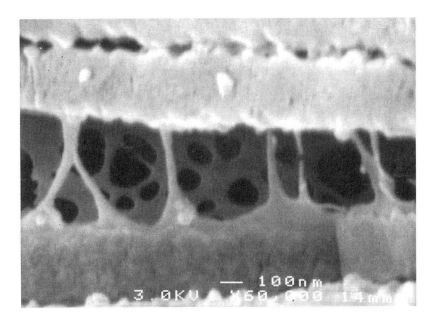

Figure 15.9 Scanning electron micrograph of abalone nacre showing antibody tagging of Lustrin A. The secondary antibody is gold and silver enhanced. This figure shows the Lustrin molecules connecting two vertically coherent nacre tablets. The c-axis of the tablets is in the plane of the paper. The fracture of the nacre is such that the tablets are pulled apart to three times their normal distance and the Lustrin A is left intact and also extends to three times its normal length, illustrating the extension and toughness of this protein.

a collagen-like protein. However the proteins were analyzed for hydroxyproline and found to have a total of $\sim 0.2\%$ hydroxyproline. The soluble nacre matrix proteins were higher in arginine, $\sim 13.6\%$, $\sim 8.2\%$ (40 kD, 65 kD) compared with the total matrix $\sim 5.2\%$ (Table 15.4) (for amino acid sequence of Lustrin A see [51, 52], for localization of Lustrin A, see Fig. 15.9).

15.5 Natural Materials to Synthetic Electronic Materials

The natural biomineralization processes described above are capable of creating structures with a range of properties not yet duplicated by known synthetic methods. It is our goal as materials scientists to attempt to harness the power of such proteins to control the nucleation and growth of specific inorganic crystals. It is both the control of crystal growth at such a microscopic level and the enhanced mechanical properties of hybrid materials that are of interest.

Shell proteins have been shown to overcome the natural thermodynamic barriers of nucleating less stable polymorphs and specific crystal orientations. In addition, these proteins have exhibited control over crystal growth on the nanometer length scale, significantly smaller than current synthetic methods of controlling small crystal growth. Direction of crystal growth at larger length scales is also possible with matrices of such proteins and other organic molecules. In shells, the organic sheets confer an unusual toughness to the inorganic layers, while in other ceramic materials organic molecules such as polymers serve the same purpose. All of these properties are quite desirable in terms of materials science. Therefore, biological routes to materials engineering are becoming important approaches to future materials synthesis. Areas where these methods are currently being applied include the synthesis of electronic devices, as well as hybrid ceramic materials.

Because traditional methods of characterizing the amino acid sequences and subsequently the structure and function of these biomineralization proteins have been proven to be extremely difficult, other methods for determining how these proteins interact with the crystals they nucleate are being investigated. One method of approach is that of working from bottom to top, using a combinatorial technique to determine the basic unit of interaction. These methods provide us with a powerful means of characterizing interactions at the molecular or even atomic level, the same level at which we wish to be able to manipulate such interactions.

The goal is to find families of amino acid sequences that can differentially bind to specific crystal faces and polymorphs of the substrates of interest. Therefore, a combinatorial approach and increasingly stringent selection techniques have been used to evolve a population of peptides with a high affinity for the investigated substrate. The vehicle for such an approach is a library of bacteriophage (phage), genetically engineered to carry random inserts of peptide sequences. These sequences are sufficiently accessible to the substrate by virtue of their location at the end of the PIII coat protein (Fig. 15.10). Each phage carries five copies of this protein, all clustered at one end of the filamentous phage. The commercially available library (New England Biolabs) contains $\sim 3 \times 10^9$ distinct random sequences, with each sequence amplified to yield roughly 55 copies per volume used in the selection steps.

The selection process begins with exposure of the substrate to the amount of phage described above (Fig. 15.10). The non-binding phage are washed away, and the bound phage are subsequently eluted from the surface by lowering the pH. After amplifying the binding phage, generation two begins with exposing the surface to the same number of phage, which are now enriched in their affinity for the surface. In order to sufficiently narrow the population and find consensus in the amino acids sequences selected, five rounds of selection are performed. The amino acid sequence of the selected phage clones is readily determined by translating the sequenced phage DNA. Once strongly binding peptide sequences are determined in this manner, engineered proteins can be developed that have the ability to pull the substrate building blocks out of an unsaturated solution and form a specific morphology of an inorganic material at environmental conditions.

The substrates studied thus far include electronic materials such as gallium arsenide, indium phosphide and zinc sulfide [53], and crystals of biomaterials such as calcite and aragonite, two common polymorphs of calcium carbonate found in many biomineralization organisms. These extensively characterized crystals provide

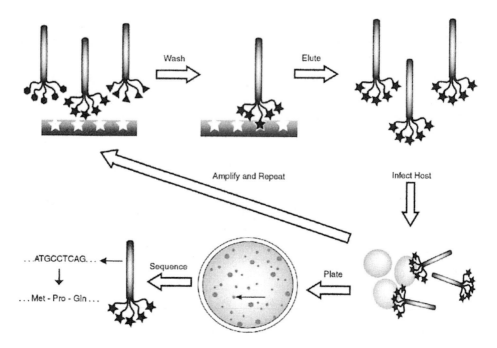

Figure 15.10 Selection process for finding proteins with an affinity for specific inorganic substrates using a random library of genetically engineered phage. Phage, which have five copies of a random protein segment at one end of the virus coat, are allowed to interact with the surface. After the non-binding phage are washed away, bound phage are eluted from the surface by disrupting the protein–surface interaction. This screening process is repeated several times in order to evolve a population of tightly binding phage. The genetically engineered random protein insert of the binding phage can be compared by translating the phage DNA isolated from a single colony of bacterial host infected by a single phage.

a good basis for understanding the mechanisms of interaction and nucleation. Another goal is to attempt to bind and nucleate crystals that are stoichiometrically similar to calcium carbonate. It has been shown that proteins isolated from abalone shells can nucleate stoichiometrically similar materials such as europium cobalt carbonate $(EuCo(CO_3)_2)$ and iron borate $(FeBO_3)$. Understanding the surface interactions and structural motifs of the proteins will facilitate the otherwise difficult engineering of proteins to be specific for different substrates. The materials being investigated are electronic, optical and magnetic materials such as iron borate $(FeBO_3)$, lead titanate $(PbTiO_3)$, barium titanate $(BaTiO_3)$ and lithium niobate $(LiNbO_3)$, and biomedical materials such as natural and synthetic bone minerals. Other work done using these same methods on non-biological inorganic crystals such as gallium arsenide and zinc sulfide have shown that it is indeed possible to screen for proteins that show face specificity and elemental discrimination.

The selected peptides that bind calcium carbonate show interesting trends in the percentage and the spacing of charged amino acids, which has been proposed to be the mechanism of face-specific nucleation for such ionic materials in biological systems. The sequences found by screening against geological calcite do not show much consensus. This lack of consensus may be due to factors such as the limitations of using a short 12 amino acid random sequence or the fact that the screenings thus far have been done at mildly basis conditions (pH 7.5 and 8.5). Further investigations are needed in order to find consensus.

Screening against geological aragonite, however, does provide trends in the amino acid sequences selected. As the selection stringency is increased, the sequences become more basic. Of course, this effect is more pronounced at pH 8.5 than at pH 7.5, because the higher pH not only affects the overall surface charge of aragonite but also the amino acid charges on the random sequences. It is not yet evident what kind of spacing of these positively-charged amino acids is necessary to interact most efficiently with aragonite, but the sequences show alternating positively-charged amino acids as well as blocks and alternating blocks of such residues. It is also interesting to note several examples of alternating positive and negative amino acids, which would be likely to interact most efficiently with such an ionic solid (Table 15.5).

Further investigations of these selected peptides include nucleation studies where the peptides are used to nucleate nanocrystals of calcium carbonate from solution. The protein-nucleated nanocrystals are currently being characterized using AFM, TEM and electron diffraction, while the nucleation thermodynamics are being

Table 15.5 Examples of *in vitro* selection of 12-mers that bind to aragonite (positively-charged residues are bold and negatively-charged residues are underlined)

Alternating positive and negative residues											
His	Ala	**His**	**Arg**	<u>Glu</u>	**His**	<u>Glu</u>	**Lys**	Pro	Met	**Lys**	Phe
Ala	Tyr	Ser	Ile	**His**	Leu	Gln	Val	**His**	<u>Glu</u>	**Arg**	Ala
Alternating positive residues											
Ile	**His**	Ile	**Lys**	Phe	**Lys**	Gln	**His**	Gln	Asn	**His**	Asn
Pro	Gln	Pro	Leu	Ala	**Lys**	Phe	**Lys**	Ala	**His**	Pro	Ile
Alternating blocks of positive residues											
Lys	**Arg**	Ser	**Lys**	Phe	Pro	**His**	**Lys**	**His**	<u>Asp</u>	Val	Ile
Gln	Leu	Phe	**Arg**	Pro	**His**	**His**	Ala	Ser	**Lys**	**His**	Ser
Blocks of positive residues											
Ala	Val	Gly	Ser	Thr	**Lys**	**His**	**Lys**	Trp	Pro	Pro	Leu
Ser	Leu	Ser	Leu	Val	Thr	**His**	**His**	**His**	**Arg**	Pro	

studied by microcalorimetry. These studies will help determine nucleation efficiency as a function of overall charge and the spacings of such charges.

The majority of proteins isolated from shells have been determined to be highly acidic. This leads to the conclusion that the protein exerts its control over the inorganic crystal face nucleated by arranging the positively-charged calcium ions in the correct two-dimensional spacing, allowing the carbonate ions to form the second layer in the desired position. However, the results of the *in vitro* interaction studies described above imply that the interaction occurs between the carbonate ions and the basic protein residues. The overall basic nature of these peptides is also consistent with several more recently characterized shell proteins. Future investigations of both *in vitro* selected peptides and natural proteins are needed to determine if there are differences in the nucleation efficiency between acidic and basic proteins.

References

[1] S. Mann, *Nature* 1988, *332*, 119.
[2] A. H. Heuer, D. J. Fink, V. J. Laraia, J. L. Arias, P. D. Calvert, K. Kendall, G. L. Messing, J. Blackwell, P. C. Rieke, *Science* 1992, *255*, 1098–1105.
[3] S. Mann, *Nature* 1993, *365*, 499–505.
[4] S. Mann, in *Inorganic Materials* (Eds D. W. Bruce and D. O' Hare), Wiley, New York, 1992.
[5] S. Mann, D. D. Archibald, J. M. Didymus, T. Douglas, B. R. Heywood, F. C. Meldrum, N. J. Reeves, *Science* 1993, *261*, 1286.
[6] A. M. Belcher, P. K. Hansma, G. D. Stucky, D. E. Morse, *Acta Mater.* 1997, *46*, 733–736.
[7] J. D. Currey, *Proc. R. Soc. Lond.* 1977, *196*, 443.
[8] A. P. Jackson, J. F. V. Vincent, R. M. Turner, *Proc. R. Soc. Lond. B* 1988, *234*, 415.
[9] A. Guthjahr, H. Dabringhaus, R. Lacmann, *J. Cryst. Growth* 1996, *158*, 296.
[10] S. Weiner, *Crit. Rev. Biochem.* 1986, *20*, 365.
[11] S. Mann, *Struct. Bond. Berlin* 1983, *54*, 125.
[12] R. J. P. Williams, *Phil. Trans. R. Soc. B* 1984, *304*, 411.
[13] S. Mann, in *Biomineralization in Lower Plants and Animals*, Vol. 30 (Ed. B. S. Leadbetter), Oxford University Press, Oxford, 1986.
[14] S. Mann, *Inorg. Biochem.* 1986, *28*, 363.
[15] A. P. Wheeler, C. S. Sykes, *Am. Zool.* 1984, *24*, 933.
[16] G. Falini, S. Albeck, S. Weiner, L. Addadi, *Science* 1996, *271*, 67–69.
[17] A. M. Belcher, X. H. Wu, R. J. Christensen, P. K. Hansma, G. D. Stucky, D. E. Morse, *Nature* 1996, *381*, 56–58.
[18] S. Manne, C. M. Zaremba, R. Giles, L. Huggins, D. A. Walters, A. Belcher, D. E. Morse, G. D. Stucky, J. M. Didymus, S. Mann, P. K. Hansma, *Proc. R. Soc. Lond. B* 1994, *356*, 17.
[19] S. W. Wise, *Ecologae Geol. Helv.* 1970, *63*, 775.
[20] K. M. Wilber, A. S. M. Saleuddin, in *The Mollusca*, Vol. 4 (Ed. K. M. Wilber), Academic Press, New York, 1983, p. 235.
[21] M. Sarikaya, K. E. Gunnison, M. Yasrebi, I. A. Askay, *Mater. Res. Soc. Symp. Proc.* 1990, *174*, 109.
[22] R. C. Bathurst, in *Developments in Sedmintology*, Vol. 12, 2nd edn, Elsevier, New York, 1975.
[23] H. Nakahara, in *Biomineralization and Biological Metal Accumulation* (Eds P. Westbroek and E. W. de Jong), Reidel, Dordrecht, 1983.
[24] H. Nakahara, in *Mechanisms and Phylogeny of Mineralization in Biological Systems* (Eds S. Sugan and H. Nakahara), Springer-Verlag, Berlin, 1989.
[25] N. Watabe, *Prog. Cryst. Growth Character.* 1981, *4*, 99.

[26] S. Weiner, Y. Talmon, W. Traub, *Int. J. Bio. Macromol.* 1983, *5*, 325.
[27] M. Fritz, A. M. Belcher, M. Radmacher, D. A. Walters, P. K. Hansma, G. D. Stucky, D. E. Morse, *Nature* 1994, *371*, 49–51.
[28] C. M. Zaremba, A. M. Belcher, M. Fritz, Y. Li, S. Mann, P. K. Hansma, D. E. Morse, J. S. Speck, G. D. Stucky, *Chem. Mater.* 1996, *8*, 679–690.
[29] N. Watabe, in *The Mollusca*, Vol. 4 (Ed. K. M. Wilber), Academic Press, New York, 1983.
[30] S. J. Suzuki, *Geol. Soc. Jpn* 1983, *89*, 433.
[31] H. Mutvei, in Origin, Evolution, and Modern Aspects of Biomineralization in Plants and Animals (Ed. R. E. Crick), Plenum Press, New York, 1991, p. 137.
[32] H. M. Rietveld, *J. Appl. Crystallogr.* 1969, *2*, 65.
[33] M. Sarikaya, I. A. Aksay, *Mater. Res. Soc. Symp. Proc.* 1992, *255*, 293–307.
[34] M. A. Cariolou, D. E. Morse, *J. Comp. Physiol. B.* 1988, *157*, 717.
[35] A. P. Wheeler, C. S. Sykes, *Am. Zool.* 1984, *24*, 933.
[36] K. Simkiss, K. M. Wilber, in *Biomineralization* (Eds K. Simkiss and K. M. Wilber), Academic Press, San Diego, 1989.
[37] L. Addadi, S. Weiner, in *Biomineralization: Chemical and Biochemical Perspectives* (Eds S. Mann, J. Webb and R. J. P. Williams), VCH, Weinheim, 1989.
[38] J. Aizenberg, S. Albeck, S. Weiner, L. Addadi, *J. Crystal Growth* 1994, *142*, 156–164.
[39] M. H. Klapper, *Biochem. Biophys. Res. Commun.* 1977, *78*, 1018–1024.
[40] H. Waite, *J. Biol. Chem.* 1982, *258*, 2911–2915.
[41] W. S. York, *Meth. Enzymol.* 1985, *118*, 3.
[42] A. Berman, D. J. Ahn, A. Lio, M. Salmeron, A. Reichert, D. Charych, *Science* 1995, *269*, 515–518.
[43] F. Lippmann, in *Sedimentary Carbonate Minerals* (Eds W. von Engelhardt and R. Roy), Spinger-Verlag, New York, 1973.
[44] D. E. Morse, M. A. Cariolou, G. D. Stucky, C. M. Zaremba, P. K. Hansma, *Mater. Res. Soc. Symp. Proc.* 1993, *292*, 59–67.
[45] H. Nakahara, G. Bevlander, M. Kakei, *Venus* 1982, *41*, 33–46.
[46] H. Nakahara, *Venus* 1979, *38*, 205.
[47] A. M. Belcher, PhD thesis, University of California at Santa Barbara, 1997.
[48] T. E. Schäffer, C. Ionescu-Zanetti, R. Proksch, M. Fritz, D. A. Walters, N. Almqvist, C. M. Zaremba, A. M. Belcher, B. L. Smith, G. D. Stucky, D. E. Morse, P. K. Hansma, *Chem. Mater.* 1996, *9*, 1731–1740.
[49] S. Weiner, W. Traub, *FEBS Lett.* 1980, *111*, 311–316.
[50] S. Hotta, *Chikyu Kagaku* 1969, *23*, 133.
[51] X. Shen, A. M. Belcher, P. K. Hansma, G. D. Stucky, D. E. Morse, *J. Biol. Chem.* 1997, *272*, 32472–32481.
[52] B. L. Smith, T. E. Schäffer, M. Viani, J. B. Thompson, N. A. Frederick, J. Kindt, A. M. Belcher, G. D. Stucky, D. E. Morse, P. K. Hansma, *Nature* 1999, *399*, 761–763.
[53] S. R. Whaley, C. Flynn, D. S. English, E. L. Hu, P. F. Barbara, A. M. Belcher, *Nature* 2000, in press.

16 Polyanions in the CaCO₃ Mineralization of Coccolithopores

M. E. Marsh

16.1 Introduction

Coccolithophores are unicellular, predominately marine algae with an outer covering of mineralized scales known as coccoliths (Figs. 16.1, 16.2). Coccolith mineralization has been examined most extensively in *Pleurochrysis carterae* and *Emiliania huxleyi*, where mineralization occurs intracellularly in the presence of acidic polysaccharides. The acidic polysaccharides are high-capacity calcium-binding macromolecules generally referred to as polyanions, in order to illustrate their relationship to phosphoprotein polyanions associated with mineralization in some metazoan tissues [1, 2]. The polyanions were first discovered in vertebrate tooth dentin [3], and since then, have generally been regarded as intermediates in the formation of calcium-based mineral phases. However, progress in discovering precisely how polyanions function has been made chiefly with coccolithophores, which are susceptible to mutations that generate defects in both the mineral phase and polyanion expression [4].

There are three fundamental processes in coccolith biomineralization: (i) ion accumulation, (ii) calcite nucleation, and (iii) crystal growth. Ion accumulation refers to processes that generate and maintain a medium supersaturated with respect to CaCO₃ at mineralizing foci throughout calcite nucleation and crystal growth, i.e. temporally and spatially controlled mechanisms for amassing calcium and bicarbonate ions and removing hydrogen ions ($Ca^{2+} + HCO_3^- = CaCO_3 + H^+$). Nucleation includes all factors other than ion accumulation which determine the rate and site of calcite nucleation and the crystallographic orientation of the initial mineral phase. Growth includes all processes subsequent to ion accumulation and nucleation which regulate crystal size and shape.

The remainder of this chapter describes first the cellular, ultrastructural and crystallographic aspects of coccolith formation in *Pleurochrysis* and *Emiliania*. Then the structure, composition, immunolocalization and physical properties of the mineral-associated polyanions in both species is presented. Finally the relationship of the polyanions to mineralization processes is described, principally as understood from studies of *Pleurochrysis* mutants with mineral and polyanion defects.

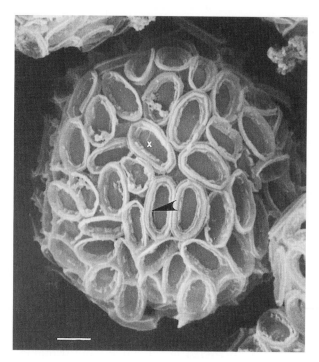

Figure 16.1 Scanning electron micrograph of a *Pleurochrysis* cell showing the mineralized scales of the coccosphere. These scales, called coccoliths, consist of organic oval-shaped bases (x) with rims of CaCO₃ crystals (arrow head). Bar = 1.0 μm. (Reproduced from *Protoplasma* [18] with permission.)

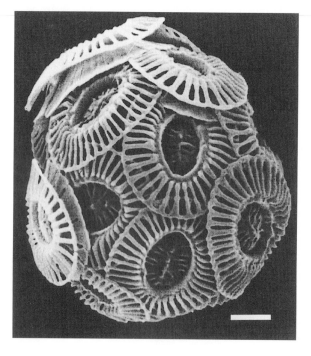

Figure 16.2 Scanning electron micrograph of *Emiliania* coccosphere. Bar = 1.0 μm. (Image courtesy of J. R. Young.)

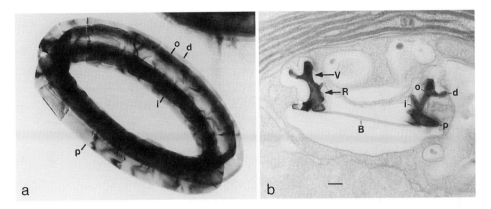

Figure 16.3 (a) Transmission electron micrograph of isolated mature *Pleurochrysis* coccolith with the coccolith plane inclined about 30° to the plane of the page. (b) Thin section showing a cross-sectional view of a mature *Pleurochrysis* coccolith in its vesicle prior to secretion into the cocco-sphere. The V and R crystal units are located on the rim of the base plate (B). The distal shield (d), proximal shield (p), inner tube (i) and outer tube (o) elements of the mineral ring are indicated. Bar = 0.1 μm. (Reproduced from *Protoplasma* [8] with permission.)

16.2 Coccoliths

16.2.1 Structure

Coccoliths are composed of calcite ($CaCO_3$) crystals attached to an underlying oval-shaped organic base plate (Fig. 16.3). Although base plates are not observed in mature coccoliths of all species such as *Emiliania*, they may be universally present during the initial phases of coccolith development. Coccoliths are classified as het-erococcoliths or holococcoliths, based on calcite crystal morphology. The crystals of holococcoliths have the simple rhombohedral or prismatic habits characteristic of inorganic sources of calcite and are generally formed extracellularly [5]. In con-trast the heterococcolith crystals have intricate complex shapes that are species-specific and not observed in inorganic sources of the mineral (Figs. 16.2–16.5). Heterococcoliths are mineralized intracellularly in a membrane-limited coccolith vesicle. The *Pleurochrysis* and *Emiliania* heterococcoliths are known as placoliths because their crystals form parallel double discs that radiate from the coccolith rim [6].

The mineral rim of *Pleurochrysis* and *Emiliania* coccoliths is composed of a ring of single interlocking crystals with alternating radial (R) and vertical (V) orienta-tion, a structure that may be common to heterococcoliths in general [7]. The V and R orientations refer to the alignment of the crystallographic c-axes of the crystal units with respect to the coccolith plane. The parallel discs of the mineral ring are

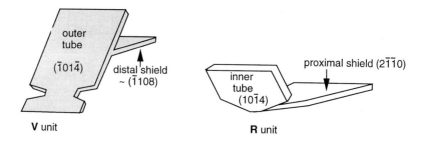

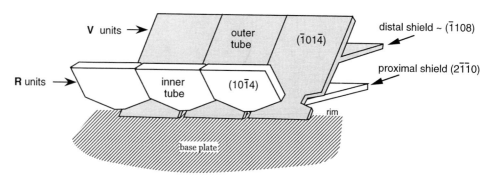

Figure 16.4 Schematic of the V and R units and their interlocking structure on the rim of the *Pleurochrysis* coccolith as viewed from the interior of the coccolith. For simplicity the crystal elements are depicted as thin plates. The crystal faces corresponding to the distal shield, proximal shield, inner tube and outer tube elements are identified. (Reproduced from *Protoplasma* [8] with permission.)

known as the distal and proximal shield elements, and the vertical or subvertical structures linking the shields are known as tube elements [6] (Figs. 16.2–16.5).

The mineral ring of *Pleurochrysis* coccoliths has four plate-like elements (Figs. 16.3, 16.4): the distal shield and outer tube elements that form the V unit, and the proximal shield and inner tube elements that form the R units [8]. The platy surfaces of both tube elements correspond to the common $(10\bar{1}4)$ rhombohedral faces of calcite, and the plates of the proximal shield element are prismatic $(2\bar{1}\bar{1}0)$ faces. The plates of the distal shield element are rather curved (Fig. 16.3) and their orientation does not correspond to a favorable calcite face, however for convenience they are described as approximately $(\bar{1}108)$ faces, faces which rarely – if ever – develop in inorganic sources of calcite.

In *Emiliania* coccoliths (Fig. 16.5), the proximal and distal shield elements and the inner and outer tube elements are all derived from the R units [9–11]. In this species the vestigial V units are only observed in very immature coccoliths as they are rapidly overgrown by the massive R units [7]. As observed in *Pleurochrysis*, the platy surfaces of the proximal shield elements correspond to $(2\bar{1}\bar{1}0)$ faces [10].

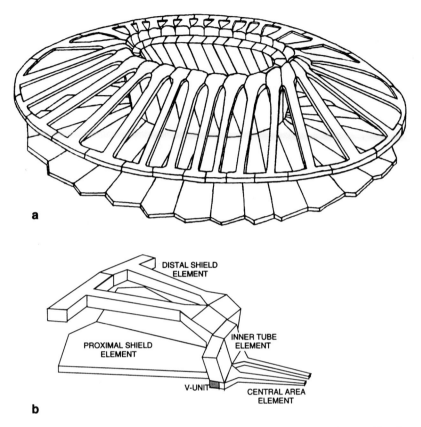

Figure 16.5 (a) An entire *Emiliania* coccolith and (b) a segment showing two R units and a vestigial V unit. The platy surfaces of the proximal shield elements correspond to $(2\bar{1}\bar{1}0)$ calcite faces. (Reprinted by permission from *Nature* [7] copyright 1992 Macmillan Magazines Ltd.)

16.2.2 Formation

16.2.2.1 Ion accumulation

Heterococcolith mineralization occurs on the rim of a preformed organic base plate in a specialized Golgi-derived structure known as the coccolith vesicle (Figs. 16.6, 16.7). After completion of the base plate in *Emiliania*, a labyrinthine membrane system known as a reticular body develops at the distal surface of the coccolith vesicle [12] and is present throughout the mineral deposition process (Fig. 16.7). The reticular body has a large membranous surface area surrounding a small internal volume. Although its precise functions are unclear, its design is beautifully suited to the rapid transport of large quantities of mineral ions into a small confined volume – the large membrane surface could accommodate vast numbers of calcium and perhaps bicarbonate ion pumps – hence providing and maintaining a super-

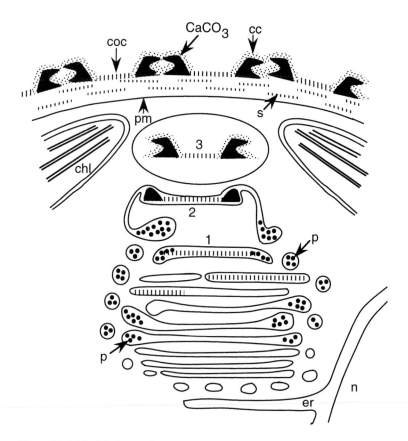

Figure 16.6 Model of coccolith formation in the *Pleurochrysis* Golgi system. The coccolith vesicle is shown before (1), during (2) and after (3) mineral deposition. Ca–PS1/PS2 complexes are synthesized in medial Golgi cisternae. They are organized in discrete particles (p) before and during mineral deposition (1 and 2) and in amorphous crystal coats (cc) after mineralization ceases (3 and coccosphere). Coccolith bases (hatch marks), coccoliths (coc), unmineralized scales (s), chloroplast (chl), endoplasmic reticulum (er), nucleus (n), plasma membrane (pm). (Reproduced from *Protoplasma* [18] with permission.)

saturated solution of mineral ions during calcite nucleation and growth. Transport enzymes associated with coccolith vesicle membranes are discussed in Chapter 17.

The *Pleurochrysis* coccolith vesicle does not develop a reticular body. Instead Golgi-derived vesicles containing large numbers of 20 nm particles (coccolithosomes) fuse with the rim of the coccolith vesicle after completion of the base plate [13, 14]. The coccolithosomes are complexes of calcium [15] and polyanions (see below) and are present throughout calcite nucleation and growth (Fig. 16.6).

In both *Pleurochrysis* and *Emiliania*, termination of mineralization coincides with a pronounced swelling of the coccolith vesicle (Figs. 16.6, 16.7). The huge influx of fluid probably quenches mineral deposition by rapidly lowering the calcium carbo-

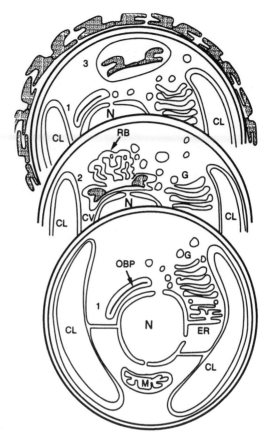

Figure 16.7 Schematic representation of *Emiliania* cells showing three stages (1–3) of coccolith formation. Upper cell also shows extracellular coccoliths. Chloroplast (CL), nucleus (N), mitochondrion (M), endoplasmic reticulum (ER), Golgi apparatus (G), organic base plate (OBP), calcifying vesicle (CV), reticular body (RB). (Reproduced with permission from the Mineralogical Society of America [26].)

nate ion product. In *Emiliania* the end of mineralization also coincides with the dissociation of the reticular body from the coccolith vesicle, while in *Pleurochrysis* the end of calcite growth is accompanied by the dissociation of the coccolithosome particles and the formation of a polyanion coat on the mineral surface (see below). The *Emiliania* crystals also acquire a polysaccharide coat but of different composition. The crystal coats in both species may inhibit further crystal growth.

16.2.2.2 Nucleation

Calcite nucleation occurs exclusively on or near the rim of the coccolith base plate in both *Pleurochrysis* and *Emiliania*. Just prior to mineralization in *Pleurochrysis*, clusters of coccolithosomes and a narrow band of organic material known as the coccolith ribbon appear on the distal rim of the base plate (Fig. 16.8, left side) [8]. Subsequently, small crystals appear within the particle clusters and in contact with the coccolith ribbon (Fig. 16.8, right side). The coccolith ribbon tethers the crystals to the base plate and is the probable site of calcite nucleation. A complete closed ring of small crystallites develops about the rim with the crystallites having alter-

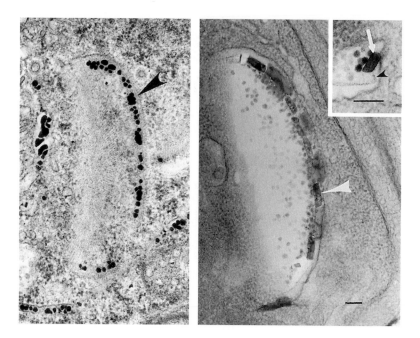

Figure 16.8 Thin sections showing successive early stages of mineralization in *Pleurochrysis* with a grazing view of the coccolith rim. (left) Base plate with dense polyanion-rich particles (coccolithosomes) associated with the rim (arrow head). (right) Base plate with a ring of small rectangular crystals among the polyanion particles on the rim (arrow head). Unstained section. Inset: cross-section showing a small crystal (arrow) above the base plate apparently attached to the coccolith ribbon (arrow head). Bars = 0.1 μm. (Reproduced from *Protoplasma* [8] with permission.)

nately radial (R) and vertical (V) orientation with respect to their crystallographic c axes (Fig. 16.9a). The initial closed ring of crystallites is known as the proto-coccolith ring [6].

Although a coccolith ribbon has not been directly observed in *Emiliania*, Young *et al.* [7] proposed a general model for alternate V and R unit nucleation based on a folded ribbon-like structure applicable to placoliths in general. Multiple folds normal to the coccolith rim [7], or a single fold parallel to the rim [8], would produce sites with similar structures but different orientations on either side of the fold. This arrangement permits the nucleation of crystals in two different orientations from the same crystallographic face, i.e. the face most compatible with the nucleation site. Although the *Pleurochrysis* crystals are bound to the coccolith ribbon presumably on (10$\bar{1}$4) faces, it is not clear whether the crystallites were actually nucleated on the ribbon or were nucleated elsewhere and subsequently bound to the ribbon.

16.2.2.3 Growth

The development of mature rim elements from the simple crystallites of the proto-coccolith ring have recently been described for *Pleurochrysis* coccoliths [8]. The

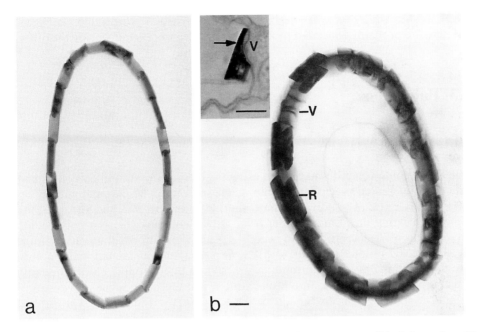

Figure 16.9 (a) Electron micrograph of an isolated *Pleurochrysis* protococcolith. It has a ring of 24 small crystals in the shape of rectangular parallelepipeds. The crystals have alternating V and R unit orientations. (b) An isolated immature coccolith at a later stage of development. The R units have double parallelogram structures as observed in projection, which will develop into the inner tube and proximal shield elements. Inset: cross-sectioned coccolith of similar stage showing an immature V unit on the rim. The ($10\bar{1}4$) plate (outer tube element) perpendicular to the section (arrow) has emerged from the initially rectangular-shaped crystal. The distal shield element of the V unit has not yet appeared. Bars = 0.1 μm. (Figure 16.9b is reproduced from *Protolasma* [8] with permission.)

earliest habits observed for both V and R units correspond to rectangular parallelepipeds (Figs. 16.8, 16.9a). Outgrowth from the initial V unit begins by expansion of a ($10\bar{1}4$) face, which forms the platy surface of the outer tube element (Fig. 16.9b, inset). Outgrowth from the initial R habit produces a double parallelogram structure when viewed in projection perpendicular to the coccolith plane (Fig. 16.9b). The outer parallelogram develops into the proximal shield element, and the inner tube element develops from the inner parallelogram. Finally the distal shield element (approximately ($\bar{1}108$) plate) emerges from the V element, completing the parallel double disc structure characteristic of the mature coccolith (Fig. 16.3).

The development of {$10\bar{1}4$} faces is not uncommon in inorganic sources of calcite, but the development of platy structures – such as the tube elements of *Pleurochrysis* crystals – as opposed to rhombohedral forms is indicative of directional growth in response to outside pressure. The spatial constraints imposed by the framework of the surrounding coccolith vesicle must limit crystal growth to specific directions. Cellular factors (e.g. cytoskeletal elements) that may influence the shape of the coccolith vesicle have not yet been identified. Crystal growth and vesicle deformation are probably concerted processes, with each process influencing the other.

16.3 Polyanions

16.3.1 Pleurochrysis

Three acidic polysaccharides have been extracted from *Pleurochrysis* coccoliths and purified by differential precipitation with magnesium ions and chromatography on DEAE-cellulose [16]. The polyanions are referred to as PS1, PS2 and PS3, with the numbers indicating their elution order from DEAE columns. PS1 is a polyuronide with a glucuronic/galacturonic acid ratio of 1:3 and contains small amounts of uncharged glycosyl residues (Table 16.1). PS3 is a galacturonomannan and the only polyanion containing significant amounts of sulfate ester groups. PS2 is the most abundant polyanion and has an unusual structure, a repeating sequence consisting of D-glucuronic, *meso*-tartaric and glyoxylic acid residues (Fig. 16.10). The tartrate and glyoxylate residues are probably introduced in a post-polymerization process by oxidative cleavage of C2–C3 bonds in alternate residues of a nascent polyuronide. With a net ionic charge of –4 per repeating unit, PS2 is the most acidic mineral-associated polyanion yet described.

PS1 and PS3 molecules are expressed in a narrow range of molecular weights. By contrast PS2 is very heterogeneous and its degree of polymerization ranges from less than 10 to over 100 repeating units per molecule (lane 1, Fig. 16.12a). PS1, PS2 and PS3 represent about 22, 76 and 2%, respectively, of the coccolith polyanions. Coccoliths contain about 6.2 calcium ions per polyanion carboxyl group [17]. Because the polyanions can bind up to 0.5 calcium ions per carboxyl group, based on electrostatic stoichiometry, about 8% of the coccolith calcium is bound to the polyanions. *In vitro* the polyanions are aggregated by calcium ions.

Table 16.1 Composition of *Pleurochrysis* polyanions expressed as mol ratios

Residue	PS1	PS2	PS3
Glucuronic acid	0.30	1.04	0.16
Galacturonic acid	1.00	0.10	1.00
Galactose	0.050		
Rhamnose	0.035		
Arabinose	0.10		
Tartaric acid		1.00	
Glyoxylic acid		1.01	
Mannose			0.53
Xylose			0.09
SO₄	trace		0.27

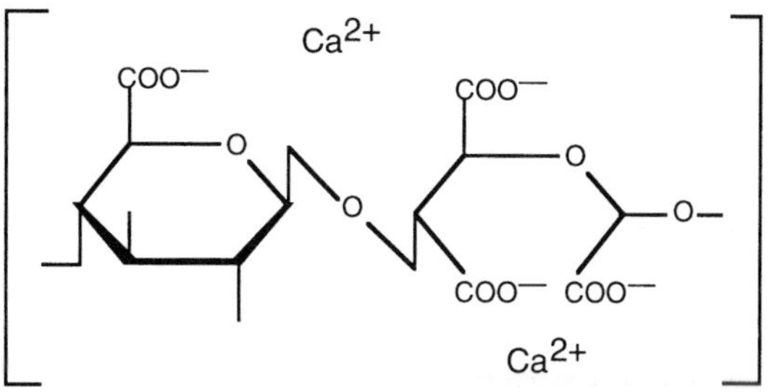

Figure 16.10 Structure of the calcium salt of PS2.

Well characterized antibodies have been used to immunolocalize PS1 and PS2 at the ultrastructural level during the nucleation of calcite and subsequent growth of the mineral phase [18]. Both polyanions are synthesized in medial Golgi cisternae where they coaggregate with calcium ions to form 20 nm particles – the coccolithosomes. The particles are released in small vesicles that fuse with the coccolith vesicle containing the base plate (Fig. 16.6). After fusion, the polyanion particles are localized on the base plate where calcite nucleation will occur (Figs. 16.6, 16.8). Nucleation and crystal growth occur in the presence of the PS1- and PS2-containing particles. During the final phase of coccolith formation, the polyanion particles are reorganized into an amorphous organic coat that surrounds the mature crystals and remains with the mineral phase after the coccoliths are extruded from the cell into the coccosphere.

Well characterized antibodies are not yet available for immunolocalization of PS3, but kinetic studies have shown that it is synthesized and secreted by a pathway different from PS1 and PS2 [17].

16.3.2 Emiliania

Coccolith polyanions have been isolated from several strains of *Emiliania huxleyi* [19–21]. The polyanions are acidic polysaccharides that migrate in a single band or two discrete bands during polyacrylamide gel electrophoresis; the number of bands depends upon both the *Emiliania* strain and culture conditions [21]. The polyanions from strain A92 migrate in a single band and those from strain 92D migrate in two bands. The polyanions of strain L migrate as a single band when the cells are cultured under normal growth conditions, but they migrate in two distinct bands when cultured at an elevated growth rate. When two polyanion bands are present, both fractions have the same glycosyl composition. There is, however, some difference in the glycosyl composition from strain to strain (Table 16.2).

The *Emiliania* polyanions are galacturonomannans with high levels of rhamnose and xylose residues and sulfate ester groups [20, 21]. A large fraction of the mannose, rhamnose and xylose residues are methylated in strains L and A92, but not in strain 92D. The galactosyl residue content in strain 92D is higher than observed in the other strains. The primary structure of the A92 polysaccharide has been partially determined [22]. Its backbone consists of $(1 \rightarrow 3)$-linked D-mannosyl residues

Table 16.2 Composition of *Emiliania* polyanions from different strains expressed as mol ratios

Residue	92[a]	A92[a]	92D[b]	L[b]
L-Rhamnose	1.8	1.4	2.7	1.5
D-Ribose	0.6	0.5	–	0.5
D-Xylose	1.2	1.4	1.0	1.3
Mannose	3.0	3.0	3.0	3.0
Galactose	0.5	0.5	2.3	–
D-Galacturonic acid	2.5	2.1	2.8	2.3
L-Arabinose	0.4	0.4	0.3	0.2
6-O-Methylmannose	0.6	0.2	trace	0.9
2,3-Di-O-methyl-L-rhamnose	1.1	0.3	–	0.8
3-O-Methyl-D-xylose	1.2	0.5	0.2	0.7

[a] From Fichtinger-Schepman *et al.* [22].
[b] From Borman *et al.* [21].

with ester sulfate groups and many side chains. The galacturonic acid residues occur in short homopolymeric side chains of 2–3 residues and in complex heteropolymeric side chains.

Compositionally the *Emiliania* polyanion most closely resembles the *Pleurochrysis* polyanion PS3. However, the ratio of galacturonic acid to mannose is considerably higher in the latter. In fact all of the *Pleurochrysis* polyanions are much more acidic than the *Emiliania* polyanion, probably explaining why the *Pleurochrysis* polyanions, but not the Emiliania polyanion, aggregate in the presence of calcium ions. The *Emiliania* polyanion inhibits the rate of CaCO$_3$ precipitation *in vitro* from supersaturated solutions [23].

The *Emiliania* polyanion has been localized in the Golgi apparatus, the coccolith vesicle, and in a coat surrounding the mature calcite crystals [24, 25]. The distribution is similar to that observed for PS1 and PS2 in *Pleurochrysis* cells. However the *Emiliania* polyanions do not form calcium-laden particles (coccolithosomes) similar to PS1 and PS2, which is consistent with the observation that the *Emiliania* polyanions do not aggregate in the presence of calcium ions *in vitro* [26].

Recently an acidic calcium-binding protein GPA has been isolated from *Emiliania* [27]. At the light microscope level GPA has been immunolocalized in a layer surrounding the cell, but its intracellular distribution has not yet been determined.

16.4 Mutants

Coccolithophores have a much higher density than non-mineralizing phytoflagellates due to the presence of calcite crystals (density $= 2.7$ g ml^{-1}) in the former. Hence mutants with reduced mineralization (Fig. 16.11) have been separated

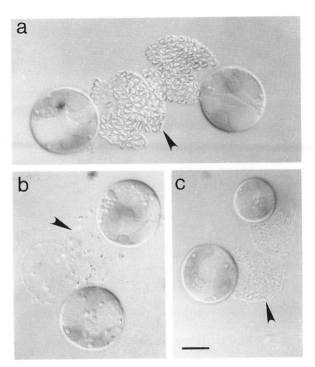

Figure 16.11 Nomarski photomicrographs of *Pleurochrysis* cells and their coccospheres (arrow heads) which were shed during microscopy. (a) The coccosphere of the wild type cell is composed of oval refractile structures representing the calcite rims of the coccoliths. (b) The coccospheres of the *ps2*-cells contain little refractile material. (c) The coccoliths of the *ps3*-cells are barely discernible as their calcite rims are quite narrow in comparison to the wild type rim. Bar = 5 µm. ((a) and (b) reproduced from *Protoplasma* [4] with permission.)

from wild type cells by density gradient centrifugation and subsequently isolated on agar plates [4]. Over 100 spontaneous and chemically-induced *Pleurochrysis* mutants have been isolated. Each mutant has been assayed for polyanion expression by pulse labeling the cells with ^{14}C-bicarbonate or ^{35}S-sulfate, and then isolating the polyanions by extraction with trichloroacetic acid followed by precipitation with calcium ions from alkaline solutions. The polyanions have been resolved on polyacrylamide gels and detected by autoradiography (Fig. 16.12).

16.4.1 PS1

One chemically-induced mutant has been isolated that expresses little, if any, PS1 (unpublished data). Its radiolabeled polyanion fraction displays either a faint band (^{14}C-labeled) or no band (^{35}S-labeled) with the mobility of PS1 on polyacrylamide gels (lane 2, Fig. 16.12a, b). The coccoliths of the *ps1*-mutant appear similar to the wild type coccolith in all respects; hence little if any PS1 is required for the formation of coccoliths with a wild type morphology. The number of coccoliths formed per cell is variable in the *ps1*-mutant; some individuals have a complete coccosphere like the wild type cell while others have fewer coccoliths. However, this phenomenon is probably unrelated to the expression of PS1.

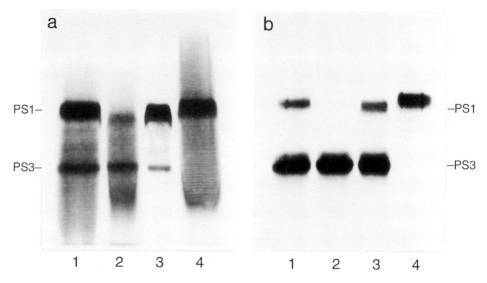

Figure 16.12 Polyacrylamide gel electrophoresis/autoradiography of the polyanion fractions from *Pleurochrysis* cells pulse labeled with ^{14}C-bicarbonate (a) or ^{35}S-sulfate (b). The PS1 and PS3 bands are indicated. The ladders in (a) are the PS2 bands; they are not observed in (b) as PS2 is not sulfated. The apparent difference in molecular weight distribution of PS2 in lanes a1, a2 and a4 is not real; the PS2 mobility is very sensitive to the ionic composition of the sample. Lanes 1a, b: wild type cells. Lanes 2a, b: mutant expressing little if any PS1. Lanes 3a, b: mutant which does not express PS2. Lanes 4a, b: mutant which does not express PS3.

16.4.2 PS2

One chemically-induced mutant and three independent spontaneous mutants that do not express PS2 have been isolated (lane 3, Fig. 16.12a) [4]. Phenotypically the spontaneous and chemically-induced *ps2*-mutants are identical. The calcium content of the mutants is less than 5% of the wild type level, and their coccospheres contain little crystalline material (Fig. 16.11b). Most *ps2*-coccoliths are unmineralized, but when crystals are present, they occur on the coccolith rim as in wild type cells (Fig. 16.13a). When the *ps2*-rim elements are large enough, they display V and R unit morphology characteristic of wild type crystals.

The obvious defect in *ps2*-cells is the low level of calcium carbonate deposition. However, the underlying cause of reduced mineralization is not as clear. If the supply of mineral ions to the coccolith-forming vesicles was impaired, reduced calcium carbonate deposition would occur. The polyanion particles are calcium carriers, but are not the sole source of calcium for coccolith mineralization. The amount of calcium transported to the coccolith vesicle sequestered in polyanion particles is less then 10% of the total calcium deposited in the coccolith crystals [17]. Because the calcium dissociation constant of PS2 is about 1.0 mM, its major func-

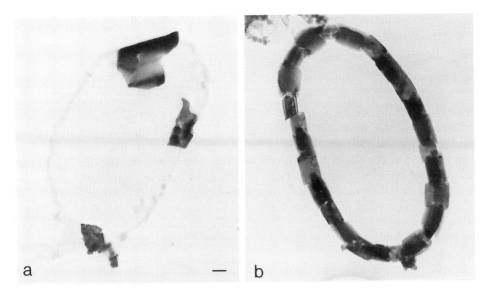

Figure 16.13 Transmission electron micrographs of isolated coccoliths from *Pleurochrysis* mutants. Compare with wild type coccolith (Fig. 16.3a). (a) *ps2*-mutant. A few rim elements of variable size are observed. (Reproduced from Marsh and Dickinson [4] with permission.) (b) *ps3*-mutant. Complete rims with immature parallelepiped-shaped crystals are observed. Compare with protococcolith rim in Fig. 16.9a. Bar = 0.1 µm.

tion is probably to buffer the free calcium ion concentration in the coccolith vesicle [18].

Because PS2 is present in massive quantities during calcite nucleation and growth, the mineralizing foci have a polyanion-bound calcium concentration of up to 6 M [15]. Hence PS2 can buffer the calcium ions at the mineralization front, keeping the calcium carbonate ion product high enough to support sustained $CaCO_3$ deposition. The presence of an efficient calcium buffer may compensate for less efficient calcium uptake into the coccolith vesicle. A buffering function for PS2 in *Pleurochrysis* cells may also account for the absence of a PS2-like polyanion in *Emiliania*. *Emiliania* coccoliths are mineralized in a chamber connected to a labyrinthine membrane system that may take up calcium ions so efficiently that calcium buffering is not required.

The small amount of mineral deposited in *ps2*– cells occurs on the coccolith rim as in wild type cells, demonstrating that PS2 is not absolutely required for calcium carbonate nucleation. However, PS2 may have a significant influence on the rate of calcite nucleation. Poorly mineralized coccoliths would be produced if the rate of nucleation in *ps2*– cells were slow compared to other processes, such as the residence time of the coccolith base within the mineralizing vesicle. Hence PS2 may affect the level of *Pleurochrysis* mineralization by (a) buffering the calcium ion concentration in the coccolith-forming vesicle, (b) accelerating the rate of calcite nucleation or (c) both. However, PS2 does not influence crystal morphology, because the more mature crystals in *ps2–* cells exhibit the V or R unit morphology of wild type crystals.

16.4.3 PS3

One chemically-induced mutant has been isolated that does not express PS3 (unpublished data, see lane 4 in Fig. 16.12a, b). This is perhaps the most interesting mutant as it is the only one found so far in which growth of the mineral rim has been arrested at the stage of the protococcolith ring (Figs. 16.11c, 16.13b). In the *ps3–* mutant the morphology of the V and R units does not vary much from the rectangular parallelepiped structure observed in the protococcolith ring of wild type cells. After deposition of the initial ring in the mutant, the V and R units simply increase in size with little change in shape. Hence the *ps3–* coccoliths do not develop the parallel double discs characteristic of the wild type coccolith (compare Figs. 16.3 and 16.13b).

In wild type cells, PS3 is associated with the coccolith crystals in the coccosphere [16], and its presence in the coccolith-forming vesicle can be inferred because uptake of PS3 from the extracellular milieu after secretion of the coccolith is unlikely. Although not proven from the mutagenesis experiments alone, PS3 is probably an essential part of the machinery controlling the morphology of growing crystals in *Pleurochrysis*. Although there is no data yet to indicate how PS3 might function, one possible scenario is that PS3 binds to the protococcolith crystals, and perhaps to an integral membrane protein of the coccolith forming vesicle; once activated by PS3, the protein may recruit elements from the cytoplasm (e.g. clatherin-like proteins or cytoskeletal elements) capable of deforming the vesicle and thereby influencing the direction of crystal growth. PS3 is the only one of the three *Pleurochrysis* polyanions that shows any relationship to the *Emiliania* polyanion; both molecules are galacturonomannans. Hence the galacturonomannans may be part of the apparatus for determining crystal shape in both species.

16.5 Summary

Three polyanions are present during coccolith mineralization in *Pleurochrysis* while only one has been identified in the coccolith-forming vesicle of *Emiliania*. The *Pleurochrysis* polyanion PS2 is a calcium ion buffer and its major function is probably maintaining the calcium carbonate ion product at a high enough level to support crystal nucleation and sustained crystal growth within the coccolith-forming vesicle. *Pleurochrysis* mutants that do not express PS2 have a calcite content less than 5% of the wild type level. *Emiliania* does not have PS2-like molecules; apparently its coccolith-forming vesicle has an efficient calcium uptake system requiring little if any calcium buffering. The *Pleurochrysis* polyanion PS3 is apparently an essential element of the machinery regulating development of the calcite rim from its initial protococcolith form to its final double disc morphology. Calcite morphology in the *ps3*-mutant does not evolve beyond the initial parallelepiped form observed in the protococcolith ring. The *Emiliania* polyanion is a sulfated galacturonomannan similar to PS3 and may have a similar function. The *Pleurochrysis* polyanion PS1 appears to have little influence on the nucleation or growth of the coccolith mineral. Mutants expressing little if any PS1 produce coccoliths with wild type morphology.

Acknowledgements

The author's contributions to the studies summarized is this review were supported by grant N00014-96-1-0565 from the Office of Naval Research and grant AR 36239 from the National Institutes of Health.

References

[1] A. George, L. Bannon, B. Sabsay, D. W. Dillon, J. Malone, N. A. Jenkins, D. J. Gilbert, N. G. Copeland, A. Veis, *J. Biol. Chem.* 1996, *271*, 32869–32873.
[2] M. E. Marsh, R. L. Sass, *J. Exp. Zool.* 1983, *226*, 193–203.
[3] A. Veis, A. Perry, *Biochemistry* 1967, 6, 2409–2416.
[4] M. E. Marsh, D. P. Dickinson, *Protoplasma* 1997, *199*, 9–17.
[5] J. D. Rowson, B. S. C. Leadbeater, J. C. Green, *Br. Phycol. J.* 1986, *21*, 359–370.
[6] J. R. Young, J. A. Bergen, P. R. Bown, J. A. Burnett, A. Fiorentino, R. W. Jordan, A. Kleijne, B. E. Van Niel, A. J. Ton Romein, K. Von Salis, *Palaeontology* 1997, *40*, 875–912.
[7] J. R. Young, J. M. Didymus, P. R. Bown, B. Prins, S. Mann, *Nature* 1992, *356*, 516–518.
[8] M. E. Marsh, *Protoplasma* 1999, *207*, 54–66.
[9] N. Watabe, *Calcif. Tissue Res.* 1967, *1*, 114–121.
[10] S. Mann, N. H. C. Sparks, *Proc. R. Soc. Lond. B* 1988, *234*, 441–453.
[11] S. A. Davis, J. R. Young, S. Mann, *Botanica Marina* 1995, *38*, 493–497.
[12] D. Klaveness, *Protistologica* 1972, 8, 346–355.
[13] D. E. Outka, D. C. Williams, *J. Protozool.* 1971, *18*, 285–297.
[14] P. van der Wal, E. W. de Jong, P. Westbroek, W. C. de Bruijn, A. A. Mulder-Stapel, *J. Ultrastruct. Res.* 1983, *85*, 139–158.
[15] P. van der Wal, E. W. de Jong, P. Westbroek, W. C. de Bruijn, in *Environmental Biogeochemistry*, Vol. 35 (Ed. R. Hallberg), Swedish Research Council, Stockholm, 1983, pp. 251–258.
[16] M. E. Marsh, D. K. Chang, G. C. King, *J. Biol. Chem.* 1992, *267*, 20507–20512.
[17] M. E. Marsh, *Protoplasma* 1996, *190*, 181–188.
[18] M. E. Marsh, *Protoplasma* 1994, *177*, 108–122.
[19] E. W. de Jong, L. Bosch, P. Westbroek, *Eur. J. Biochem.* 1976, *70*, 611–621.
[20] A. M. J. Fichtinger-Shepman, J. P. Kamerling, J. F. G. Vliegenthart, *Carbohyd. Res.* 1979, *69*, 181–189.
[21] A. H. Borman, E. W. de Jong, R. Thierry, P. Westbroek, L. Bosch, *J. Phycol.* 1987, *23*, 118–123.
[22] A. M. J. Fichtinger-Schepman, J. P. Kamerling, C. Versluis, J. F. G. Vliegenthart, *Carbohyd. Res.* 1981, *93*, 105–123.

[23] A. H. Borman, E. W. de Jong, M. Huizinga, D. J. Kok, P. Westbroek, L. Bosch, *Eur. J. Biochem.* 1982, *129*, 179–183.

[24] P. R. van Emberg, E. W. deVrind-de Jong, W. T. Daems, *J. Ultrastruct. Molec. Struct. Res.* 1986, *94*, 246–259.

[25] P. van der Wal, E. W. de Jong, P. Westbroek, W. C. de Bruijn, A. A. Mulder-Stapel, *Protoplasma* 1983, *118*, 157–168.

[26] E. W. de Vrind-de Jong, J. P. M. de Vrind, in *Geomicrobiology: Interactions Between Microbes and Minerals* (Eds J. F. Banfield and K. H. Nealson), The Mineralogical Society of America, Washington DC, 1997, pp. 267–307.

[27] P. L. A. M. Corstjens, A. van der Kooij, C. Linschooten, G. J. Brouwers, P. Westbroek, E. W. de Vrind-de Jong, *J. Phycol.* 1988, *43*, 622–630.

17 The Calcifying Vesicle Membrane of the Coccolithophore, *Pleurochrysis* sp.

Elma L. González

17.1 Introduction

The coccolithophorids are ubiquitous unicellular components of the open ocean flora. Their presence is most notable when blooms in the northern and southern oceans cover hundreds of thousands of square kilometers, with cell densities during a bloom reaching levels of 10^7 cells L^{-1} [1]. The calcitic coccoliths and coccospheres that these organisms elaborate are the most important source of carbonate for deep ocean sediments and are thought to be important components of global carbon cycles [2]. Coccolithophorid mineralization occurs subcellularly in a regulated series of steps that include (i) ion accumulation in the coccolith vesicle, (ii) calcite nucleation, (iii) crystal growth and (iv) exocytosis (steps ii and iii are discussed in Chapter 16).

The ion substrates of coccolithophorid mineralization, inorganic carbon and calcium, are present in the water column at ~ 2 mM and 10 mM concentration, respectively. The organism controls the flux of these ions into the coccolith vesicle in response to environmental cues such as light intensity, pH, alkalinity, temperature and nutrient availability [3]. The coccolithophore's ability to exercise precise control over mineralization activity is presumed to have conferred selective advantage for life in the oceans from the Triassic period to the present (pH 8.2–8.3, low nutrient levels and ion concentrations as noted above). It is thought that the two important benefits of calcification are the production of protons for metabolic processes and the capture of liths to form an unique "cell wall" (coccosphere) featuring organic and inorganic components [4, 5].

This chapter examines the current evidence for molecular mechanisms of ion transport at the coccolith vesicle membrane of *Pleurochrysis* sp., CCMP299. This is the only species whose coccolith vesicle has been isolated.

17.2 The Coccolith Vesicle Membrane

17.2.1 Isolation

Isolation of organelles from *Pleurochrysis* in general, and the coccolith vesicle in particular, is complicated by several features unique to the coccolithophores. The first challenge for biochemical studies is the absence of known, unique enzymatic markers for the coccolith vesicle. Although V-Type ATPase has now been detected on the coccolith vesicle, its activity is also resident on Golgi membranes [6], thus reducing the value of this enzyme as a diagnostic marker for the coccolith vesicle. Another obstacle to organelle isolation is due to the cytoskeletal framework that interconnects the coccolithophorid organelles. Fortunately, the microtubules can be disassembled by introducing 0.3% colchicine into the grinding medium cocktail. Isolation of the coccolith vesicle by sedimentation under rate zonal conditions is facilitated by the relatively higher density of the coccolith within the vesicle. Usually short spins at low speed in a swinging bucket rotor are sufficient for separation, but conditions for centrifugation should be established empirically. Although discontinuous gradients could be useful, linear sucrose gradients seem to minimize aggregation artifacts at sucrose layer interfaces [7].

Presumptive coccolith vesicle fractions should meet the following criteria: an internal lith that conforms to coccolith size (~ 1 µm in diameter), presence of a baseplate that stains with calcofluor (detects complex polysaccharides), and the presence of a membrane (labeled with NBD-ceramide) to distinguish the structure from the outer coccoliths. The isolated vesicle containing a protococcolith should have a three-dimensional appearance under Nomarsky optics (Corstjens *et al.*, submitted).

Cells are best ruptured by cavitation in either a French Pressure Cell Press or by passage through a narrow gauge needle attached to a 50 mL hypodermic syringe [7]. Low yields are balanced against a high degree of organelle integrity. Although free coccoliths (released from the coccosphere) do not interfere with analyses of the coccolith vesicle membrane, the outer coccosphere can be easily dissolved within 1–2 min of adjusting the cell suspension to pH 5. Cells broken after acid treatment will yield undamaged membrane-bound coccoliths and the coccolith vesicle membrane from this preparation can be run on gels. Brief acid treatment, as described above, does not damage the cell's ability to remineralize if, after dissolution of the coccosphere carbonate, the pH of the cell suspension is restored to original values [8].

17.2.2 Characterization of Membrane Polypeptides

Carefully isolated vesicle membranes show a distribution and complexity of polypeptides consistent with the presence of the V-ATPase complex. Although not all of the polypeptides have been identified, the polypeptides marked (Fig. 17.1) have apparent molecular weights that closely match the principal subunits of both the V_1 (A–E) and V_o (a, c) domains of the ATPase (Table 17.1). The entire vacuolar ATPase complex has been extensively studied in many species [9]. Subunit composition varies between 12 and 13 subunits for nearly all the organisms examined [9, 10]. The number of subunits in the coccolithophorid V-ATPase complex has not been determined.

In all, the purified coccolith vesicle membrane contains approximately 20 relatively abundant polypeptides as detected in silver-stained SDS-PAGE (Fig. 17.1). We have confirmed the identity of only two of these. The polypeptide with an ap-

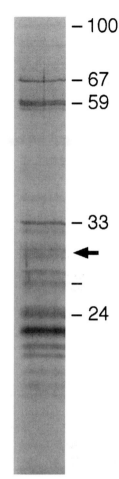

Figure 17.1 SDS-polyacrylamide gel electrophoresis of the isolated coccolith vesicle membrane polypeptides. Polypeptides at 59 kDa and 24 kDa have been identified on Western blots as subunits of the V-ATPase complex. Other polypeptides, particularly at 100, 67 and 33 kDa may also be components of the V-ATPase complex. The arrow indicates the position of a polysaccharide that is strongly reactive to ruthenium red. (Modified from [7].)

parent M_r of 59 kDa corresponds to the 60 kDa subunit B because it specifically reacts with the 2E7 monoclonal antibody raised against subunit B, the ATP-binding subunit, of oat roots [11] (Fig. 17.2). Immunofluoresence microscopy also shows that this antibody reacts with the membrane of aldehyde-fixed, isolated coccolith vesicles (Fig. 17.3).

The "16 kDa proteolipid" has also been identified in this preparation. This polypeptide, also known as subunit c of the V_o domain [9], has been referred to as the "16 kDa proteolipid" [12] despite the fact that covalent modification of the polypeptide in eukaryotes has never been verified. However, subunit c in *Pleurochrysis* presumably undergoes post-translational modification because it has an apparent M_r of 24 kDa (Corstjens *et al.*, in review). This is a higher molecular mass than would be expected from the coding capacity of the corresponding cDNA (16.2 kDa) from *Pleurochrysis* [13]. In *Pleurochrysis*, the gene encoding subunit c is present as a single genomic copy (Corstjens *et al.*, in review).

Table 17.1 Subunit composition of the vacuolar ATPase holoenzyme

	Oat [10]	Yeast [9]	Pleurochrysis [7]
V_1			
A	70	69	(67)
B	60	60	59
H	44	54	(48)
C	42	42	(42)
D	32	32	(33)
E	29	27	(28)
F	13	14	
G	12	13	
V_0			
a	100	100	(100)
d	36	36	
c	16	17	
c'		17	
c''	24	23	24

Subunit M_r are shown $\times\ 10^{-3}$. Parentheses indicate polypeptides (Fig. 17.1) whose identity has not been confirmed.

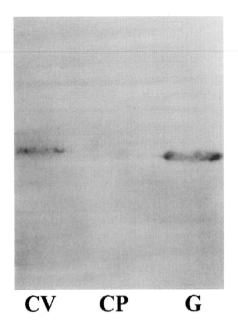

CV   CP   G

Figure 17.2 Immunoblot of coccolith vesicle membrane polypeptides. One polypeptide, 59 kDa, cross-reacted with the 2E7 monoclonal antibody against the 60 kDa subunit from oat roots [11] in preparations from Golgi (G) and coccolith vesicle (CV) membranes, but not chloroplast (CP) membranes. (Modified from [7].)

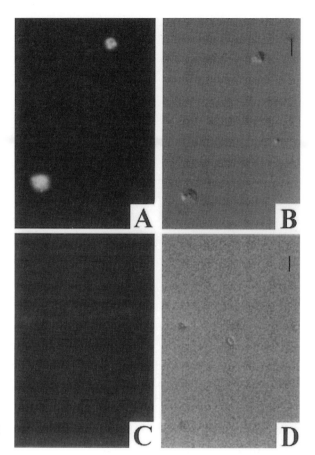

Figure 17.3 Immunofluoresence microscopy of coccolith vesicles. Isolated aldehyde-fixed vesicles were exposed to the 2E7 monoclonal (see Fig. 17.2) and secondary antibody labeled with FITC fluorochrome. (A, B) vesicles exposed to the 2E7 monoclonal viewed with epifluoresence (A) and Nomarsky optics (B). (C, D) Vesicles exposed to non-immune ascites fluid; epifluoresence (C) and Nomarsky (D). Bar = 1 μM.

Another interesting polypeptide of the coccolithophorid calcifying vesicle membrane is a band that migrates between 26 and 28 kDa. As shown in Fig. 17.4, this polypeptide is the only one found to have strong affinity for ruthenium red [7]. This reagent is thought to react with calcium-binding oligosaccharides or glycopeptides.

Remarkably, neither the 26–28 kDa nor any other polypeptide profiled on SDS-PAGE from isolated coccolith vesicle membranes were seen to bind to ^{45}Ca, despite the strong binding observed for the positive control (calmodulin, a calcium-binding protein) that was run on parallel tracks on the same gel [7]. Nevertheless, some material at ~ 150 kDa present in both the coccolith vesicle membrane and extracts of free outer coccolith material does show ^{45}Ca binding. This material does not stain with protein stains (such as amido black) suggesting that it is not proteinaceous (or below detection). It may well be residual calcium-binding polysaccharides that are primarily components of the completed coccoliths but may also have some association to the coccolith vesicle membrane [14].

Because the coccolith vesicle derives from the trans-Golgi, not surprisingly both subunits B and c have been detected in highly enriched Golgi preparations, in-

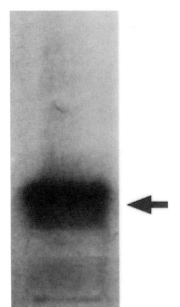

Figure 17.4 SDS-polyacrylamide gel of the coccolith vesicle membrane polypeptides stained with ruthenium red. The arrow at approximately 28 kDa indicates the position of a strongly staining band.

dicating the presence of the vacuolar ATPase on the Golgi [6]. Interestingly, when Golgi preparations were compared to the highly purified coccolith vesicle membranes, the latter preparations revealed additional polypeptides at 55–56 kDa and 28 kDa [6]. Thus, at least these two polypeptides appear to be exclusive to the coccolith vesicle membrane and probably not part of the V-ATPase complex.

17.3 Proton Transport

Wholly intracellular calcification is rare among organisms, and is particularly interesting because it permits an organism to exercise precise control over mineralization, chemistry and lithomorphogenesis [14]. However, although calcification is a source of protons, the coccolithophorid cell must maintain a defined, alkaline microenvironment sufficient to permit crystal nucleation, growth and the integrity of fully mineralized structures until their eventual exocytosis. It is evident that these conditions are met because the organism exhibits a high rate of calcite production and excretion (20–40 pg per cell per day) [1] and it may therefore be inferred that the resulting protons are neutralized or secreted from the calcifying (coccolith) vesicle (cv). Various mechanisms have been proposed but the situation is still not fully resolved.

One possible mechanism was suggested by reports of Ca-stimulated ATPase activity in coccolithophore extracts [15]. The coccolithophorid cell has abundant

Table 17.2 Diagnostic responses of the Ca^{2+}-stimulated ATPases of the plasma membrane (PM), Golgi, and coccolith vesicle (CV) to various effectors. Organelles of *Pleurochrysis* were isolated from sucrose gradients. Values represent the activity remaining after treatment as % of the control (sample without inhibitor) activity. Reproduced from [6] by permission of the Journal of Phycology.

	PM	Golgi	CV
P-type Inhibitor			
0.1 mM Na_3VO_4	27.4 ± 3.6	94.3 ± 0.2	104.4 ± 4.6
0.1 mM NEM	37.0 ± 9.8	89.0 ± 6.4	46.9 ± 2.04
ER P-type Inhibitor			
10 µM Thapsigargin	123.0 ± 0.1	89.5 ± 0.1	90.9 ± 9.7
20 µM CPA	86.7 ± 4.8	85.8 ± 2.0	72.5 ± 2.3
V-type Inhibitor			
10 mM $NaNO_3$	93.3 ± 14.5	73.4 ± 0.0	44.9 ± 4.9
10 mM KSCN	83.9 ± 0.5	55.7 ± 5.7	62.9 ± 10.0
10 mM NaN_3	65.7 ± 0.6	5.0 ± 1.7	6.8 ± 1.8
1 µM NEM	96.6 ± 3.3	99.7 ± 2.4	106.9 ± 11.2
1 µM NBD-Cl	44.8 ± 13.4	96.2 ± 6.1	70.7 ± 7.8
F-type Inhibitor			
(Mitochondrial)			
0.1 mM NaN_3	121.7 ± 9.3	94.5 ± 1.9	97.6 ± 2.6
P-type Stimulator			
0.5 µM Calmodulin	121.0 ± 0.2	101.3 ± 0.0	102.0 ± 0.0

ATPase activity in all or nearly all membranous organelles [6, 8]. We have examined the activities of the principal organelle fractions in the presence of inhibitors and other pharmacological agents to assess organelle-specific diversity of the ATPase activities including V-, P- and F-type enzymes (Table 17.2). Energy-dependent transport of protons and calcium, an ubiquitous process in all cells, is achieved by genetically and structurally distinct ATP-dependent enzymes [9, 16].

The *Pleurochrysis* plasma membrane (PM) demonstrates a P-type, calcium-stimulated ATPase that has very different inhibitor sensitivity from that of the coccolith vesicle V-type enzyme [6]. The isolated PM is capable of ATP-dependent ^{45}Ca transport that is sensitive to well-characterized inhibitors of P-type ATPase (Fig. 17.5). In contrast to the PM, the coccolith vesicle membrane shows a V (vacuolar)-type ATPase [6]. The V-ATPase is also present on the Golgi, which is not unexpected because the coccolith vesicle is derived from the trans-Golgi [17]. We have examined isolated coccolith vesicles for proton-pumping and calcium-pumping activities.

Purified coccolith vesicles are capable of ATP-dependent proton transport (Fig. 17.6). This activity was inhibited by nitrate but not vanadate, which is consistent with a V-type activity as compared to the P-type activity present on the PM (Table 17.2). In contrast to the PM, $^{45}Ca^{2+}$ transport across the coccolith vesicle mem-

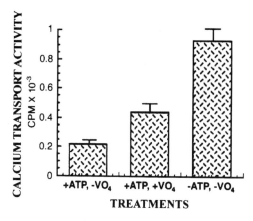

Figure 17.5 $^{45}Ca^{2+}$ efflux across plasma membrane. The plasma membrane was isolated in sucrose gradients. Radioactivity remaining in pelleted microsomes was measured after incubation in the assay mixture. The graph shows that $^{45}Ca^{2+}$ efflux was dependent on ATP and sensitive to vanadate. (Reproduced from [6] by permission of the Journal of Phycology.)

brane was not observed, although the coccolith vesicle binds Ca^{2+} with high affinity and such a high background may mask low level transport activity [6].

In addition to the enzymatic localization studies described above, the localization of the coccolithophorid V-ATPase has been confirmed by means of an antibody raised against a peptide synthesized from the amino acid sequence for the V_o subunit c encoded by a cDNA clone from *Pleurochrysis* (Corstjens et al., in review). (The cDNA library represents genes expressed during calcification [13]). The antibody to *Pleurochrysis* subunit c was reactive against aldehyde-fixed isolated coccolith vesicles and against a 24 kDa polypeptide in SDS-polyacrylamide gels of highly enriched coccolith vesicle membranes (Corstjens et al., in review).

Despite the presence of the V-ATPase on the coccolith vesicle membrane, its contribution to proton secretion from the coccolith vesicle has not been fully explained. The orientation of the holoenzyme in the membrane appears to be typical of V-ATPases of the endomembrane system [9]. In other words, its orientation indicates that protons are pumped into the vesicle. This finding implicates yet another membrane polypeptide, or polypeptide complex, that could act as a potential ion exchanger across the coccolith vesicle membrane. Antiporters have been described

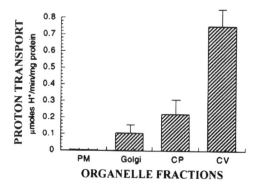

Figure 17.6 ATP-dependent proton-pumping specific activity in organelle fractions. Organelles were isolated in sucrose gradients. These activities were sensitive to nitrate and insensitive to vanadate; Table 17.2. PM, plasma membrane; CP, chloroplast; CV, coccolith vesicle. (Reproduced from [6] by permission of the Journal of Phycology.)

in organisms where luminal spaces sequester ions by exchanging protons. The midgut of *Manduca* is an example where highly conserved V-ATPase acts in concert with an ion antiporter to maintain a highly alkaline pH [18]. The vacuolar membrane of *M. crystallinum*, a halophyte with inducible crassulacean acid metabolism, exhibits a proton pumping ATPase, and a separately inducible ion/proton antiporter that work together to raise Na$^+$ ion concentrations to high levels within the tonoplast (vacuole) [19].

17.4 Inorganic Carbon (C$_i$) Transport

In the coccolithophorids, calcification takes place in a membrane-delimited vesicle completely isolated within the cellular cytosol. Therefore, protons generated by calcification have an immediate impact on metabolic processes in the cytosol and organelles. Protons from calcification are thought to be an essential requirement of a carbon-concentrating mechanism (CCM) in the chloroplast, as depicted by the model in Fig. 17.7 [20, 21]. An active CCM permits the chloroplast to attain rates of photosynthesis higher than those predicted at the low levels of dissolved inorganic carbon (DIC) in the ocean [22]. In another role, protons of calcification are postulated to be the necessary counterion in an exchange of nutrients at the plasma membrane (Fig. 17.8) [4]. Whether one or the other fate is achieved would depend on the regulated expression of enzymes and ion carriers resident on the plasma membrane and the coccolith vesicle membrane.

Valuable insights have been gained from experiments with *Emiliania* cells in batch culture [23, 24], but very little is known about the actual molecular mechanism of DIC transport across the calcifying vesicle membrane. Although there are only two options, CO$_2$ and HCO$_3$$^-$, there are several interesting speculations about what might happen in the cytosol and at the plasma membrane should one or the other carbon species be the one taken up at the calcifying vesicle. Furthermore,

Figure 17.7 The coccolithophorid carbon concentrating mechanism (CCM) and calcification. This model features interdependent, separately sequestered pathways. CCM is dependent on protons from calcification. This model takes into account carbonic anhydrase activity, CA, in the chloroplast, CP, and predicts at least one DIC transporter (dark oval) at the plasma membrane and possibly others at the organelle membranes.

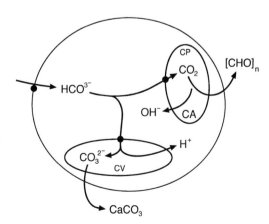

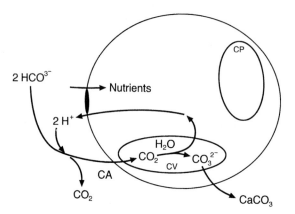

Figure 17.8 Nutrient uptake and calcification. In this hypothetical scheme, low nutrient levels may induce cell membrane transporters (dark oval) that exchange protons for nutrients. Consequent acidification of the pericellular space would result in increased levels of external CO_2. According to this model, carbonic anhydrase, CA, is periplasmic and CO_2 diffusing in would also be available to the chloroplast. Calcification would not necessarily be linked to photosynthesis and the calcification:photosynthesis ratio could be greater than 1.

because calcium carbonate is formed regardless of the substrate species, one (if HCO_3^-) or two protons (if CO_2) per carbonate would be released into the calcifying vesicle. A discussion of these potential scenarios is useful, if for no other reason than for predicting the possible biochemistry of the coccolith vesicle membrane. That is, HCO_3^- and CO_2 are likely to enter the coccolith vesicle via different paths, whether the entry of CO_2 is facilitated, or is diffused across the lipid bilayer.

The case has been made for an active CCM in the coccolithophorids [20]. This is supported by the presence of carbonic anhydrase in the chloroplast stroma [25]. The possible consequences of a stromal CCM where $HCO_3^- \rightarrow CO_2 + OH^-$ may be a pronounced proton deficit beyond normal levels reached within the stroma during photosynthesis. The coccolithophore solution to this problem may have been to maintain neutrality by facilitating a flow of protons from calcification across the coccolith vesicle membrane (Fig. 17.7). This may be why there is an observed $1:1$ stoichiometric ratio between calcification and photosynthesis [24]. Several advantages to the cell are envisaged: maintenance of "normal" levels of ATP synthesis in the chloroplast, and lower energy expenditure to remove alkaline equivalents against a charge gradient across the plasma membrane.

It is logical to assume that after a period of rapid cell proliferation in the ocean, local nutrient concentration may be reduced. A decrease in nutrients may signal a transition from high CCM activity expressed in cells during an early exponential stage of cell proliferation, i.e. early in the bloom, to calcification-driven proton/nutrient exchange as a response to low nutrient levels at the very end of the bloom. This idea is consistent with results of work on *Emiliania* [26]. Under the latter conditions, a flow of protons toward the pericellular compartment may be envisaged, promoting nutrient uptake (Fig. 17.8) [4]. Acidification of the pericellular space should then increase the level of CO_2 available to the cell and the coccolith vesicle. Assuming an active proton efflux mechanism at the coccolith vesicle membrane, the resulting alkaline vesicle lumen would trap CO_2 influx to precipitate $CaCO_3$ and generate two protons. This could result in a net increase in external CO_2 ($2HCO_3^{2-} + 2H^+ \rightarrow 2CO_2$) and a tendency toward a more acidic boundary layer (Fig. 17.8). This may explain reports of small, localized increases in CO_2 concentration observed at the end of *E. huxleyi* blooms [27]. However, even casual observation of laboratory cultures of *Pleurochrysis* suggests that calcification is not favored in mildly acidic medium when the culture has adequate nutrients and is growing rapidly. In fact, only non-calcifying cells do well in acidic medium [22]. At pH 8.2 and high nutrient levels, calcification (measured as calcium solubilized from calcite bound in internal, coccosphere, and free coccoliths) is diminished but not abolished (Fig. 17.9).

The evidence for inorganic carbon carriers at the plasma membrane or coccolith vesicle is inferential. An allele for a bicarbonate-requiring phenotype has been described in *Scenedesmus*, a marine alga, and is believed to encode an anion transporter [28]. Similarly, work by Rosenberg *et al.* has identified cDNA clones that encode polypeptides believed to function as anion transporters in several higher plants [29]. Reports by an Israeli group described cross-reactivity of the plasma membrane of the marine green alga, *Ulva*, to a monoclonal antibody raised against the anion transporter (band 3) of the human red blood cell membrane [30]. Clearly, the biochemistry of DIC transport at the coccolith vesicle should be a priority for future research.

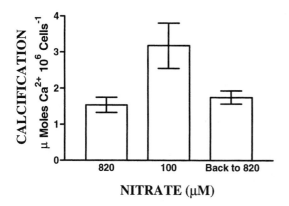

Figure 17.9 Nutrient levels affect calcification. Cells were grown for 7 days at high nitrate (820 μM) and replicates assayed for total calcium. The remainder of the cells were then transferred to low nitrate (100 μM) for 7 days and replicates taken for assay. Remaining cells were finally returned to high nitrate (820 μM) for 7 more days and re-assayed. (Calcite in internal, external and free coccoliths was solubilized and calcium determined with the BAPTA assay [22].)

17.5 Calcium Transport

The mechanism by which calcium is taken into the coccolith vesicle has not been directly examined in the coccolithophorids. There are basically only two potential pathways by which the coccolith vesicle could take up calcium. From the biochemical perspective, the simplest pathway would consist of membrane calcium transporters at the plasma and coccolith vesicle membranes [31]. This option would entail active calcium uptake at the cell membrane, diffusion of calcium through the cytosol, and calcium ion uptake at the coccolith vesicle membrane. The second scenario that can be imagined would feature membranous sequestration of calcium during its passage through the cytosol. A brief discussion of these two hypothetical scenarios is warranted.

The presumed role for hypothetical membrane calcium transporters is a logical hypothesis stemming from the almost ubiquitous presence of Ca^{2+}-stimulated ATPases in the coccolithophorid cell [6, 8]. Ca^{2+}-stimulated ATPase activity in the various membranes and organelles of *Pleurochrysis* has been characterized by sensitivity to inhibitors and other pharmacological agents (Table 17.2). The two Ca^{2+}-stimulated ATPase activities found on the plasma membrane and coccolith vesicle are of the P (plasma membrane)-type and V (vacuolar)-type families of ATPases, respectively [6]. The P-type Ca^{2+} ATPase of the plasma membrane is a highly conserved enzyme found in all the eukaryotes [16]. This enzyme typically functions to remove calcium from the cell and maintains the necessary low levels ($\sim 10^{-7}$–10^{-6} M) of calcium ions in the cytosol [32]. In *Pleurochrysis*, this enzyme transports $^{45}Ca^{2+}$ across the membrane, but does not require Mg^{2+}, nor does it demonstrate ATP-dependent proton transport [6]. It is not, apparently, a Ca^{2+}/H^+ antiporter [33]. Thus, there is no evidence that the entry of calcium into the coccolithophorid cell is mediated by an energy-dependent enzyme carrier.

The Ca^{2+}-stimulated ATPase of the coccolith vesicle membrane is a proton pump. It is a V-type enzyme whose subunits are highly conserved. For example, its 16 kDa subunit c proteolipid is encoded by a highly conserved gene (*vap* [13]) encoding 164 amino acids (75% similar in composition to the higher plant *Beta vulgaris*). Coccolith vesicle membranes also react with a monoclonal antibody raised against the 60 kDa subunit (subunit B) from oat roots, also indicating a high degree of similarity to the V-ATPase proton pump of higher plants [10, 11]. Notwithstanding the positive effect of calcium on the proton pumping activity, transport of $^{45}Ca^{2+}$ across the coccolith vesicle membrane could not be demonstrated [6]. Furthermore, the K_m of this enzyme for calcium is $\sim 10^{-3}$ M [6], considerably higher than the 10^{-7}–10^{-6} M that is the typical concentration of Ca^{2+} in the cytosol [32]. This finding suggests that in *Pleurochrysis*, calcium may have an allosteric effect on the proton pump, but it does not appear to be a substrate. Thus, the available evidence suggests that the ATPase of the coccolith vesicle membrane is a typical proton pump without a function in calcium transport.

Finally, it may be worth speculating about the possibility of a "sequestered" pathway for calcium. This hypothesis is suggested by the findings of Marsh, which clearly show that calcium precipitation is associated with complex polysaccharides [14]. The polyanions originate in the medial Golgi, which suggests that these polysaccharides are packaged within membrane-bound vesicles. Coccolith particles (coccolithosomes), containing polyanions and calcium, eventually fuse with the coccolith vesicle [34]. These coccolithosomes are well known for *Pleurochrysis* but not for *Emiliania*. This absence may hint at differences in the cell biology of coccolithogenesis among the coccolithophores.

Where do the membrane-bound polyanion-containing structures take up calcium? Do they pick it up from the cytosol immediately after being formed? Is this a passive scavenging mechanism that maintains the low concentrations in the cytosol, consistent with its second messenger function? If the implied scenario were accurate, one would predict an increase in the numbers of coccolith particles during periods of darkness when calcification rates are low. There are no reports of this.

An alternative model introduces a modification to cytosolic uptake suggested in the discussion above. We have observed a ruthenium red staining polysaccharide associated with the coccolith vesicle membrane (Fig. 17.4). This finding suggests that the association of polysaccharide and membrane may persist, even if the newly formed polyanion-containing Golgi vesicles were targeted and fused to the plasma membrane. Calcium loading might trigger recruitment of targeting proteins, recovery of the loaded vesicle into the cytosol and subsequent delivery to the coccolith vesicle. The advantage of this model is that it is consistent with known facts, i.e. (i) calcification is correlated with production of polyanions, and (ii) the eukaryotic cytosol is intolerant of large-scale fluctuations in calcium ion concentrations. Only future investigation at the cell and molecular level will provide answers to the many questions that remain.

17.6 Summary

The rarity of subcellular calcification among eukaryotes suggests that the coccolithophorids have uniquely evolved from a pre-existing vacuolar mechanism to create the conditions that permit subcellular calcification in the coccolith vesicle. The coccolith vesicle has been isolated from actively calcifying cells of *Pleurochrysis*. The calcifying vesicle membrane polypeptides have been characterized on SDS-PAGE. These membrane preparations are reactive to antibodies raised against the 60 kDa subunit of a higher plant proton pump, and against the 16 kDa proteolipid subunit c of the *Pleurochrysis* V-ATPase. A V-type ATPase with proton pumping activity has been detected on isolated coccolith vesicle membranes. Both the ATPase and proton pumping activities of the enzyme are sensitive to inhibitors known to inhibit V-ATPases. The enzyme did not catalyze Ca^{2+} transport across the coccolith membrane. There is no direct evidence bearing on mechanisms of either C_i or calcium transport across coccolith vesicle membranes. DIC transport across the coccolith vesicle may be either in the form of CO_2 or HCO_3^-, with interesting consequences for the expression of enzymes on the coccolith vesicle and plasma membranes. Inference from existing data suggests that calcium transport through the cytosol may be sequestered to minimize potential disruption in the cytosol. It is obvious that much more work is needed to permit a clear understanding of the ion transport mechanisms of the coccolith vesicle.

Acknowledgments

Work in the author's laboratory was supported by the Office of Naval Research, Department of Defense, USA.

References

[1] W. M. Balch, P. M. Holligan, K. A. Kilpatrick, *Continental Shelf Res.* 1992, *12*, 1353–1374.
[2] T. A. McConnaughey, in *Past and Present Biomineralization Processes. Consideration about the Carbonate Cycle*, Vol. 13 (Eds E. Soumenge, D. Allemand and A. Toulemont), Bulletin de l'Institute Oceanographique, Monaco, 1994, pp. 137–156.
[3] (a) E. Paasche, *Annu. Rev. Microbiol.* 1968, *22*, 71–86.
    (b) C. Linschooten, J. D. L. van Bleijswijk, P. R. van Emburg, J. P. M. de Vrind, E. S. Kempers, P. Westbroek, *J. Phycol.* 1991, *27*, 82–86.
    (c) J. Beardall, A. Johnston, J. Raven, *Can. J. Bot.* 1998 *76*, 1010–1017.
[4] T. A. McConnaughey, J. F. Whelan, *Earth Sci. Rev.* 1997, *42*, 95–117.
[5] M. R. Brown, *Portugaliae Acta Biol.* 1974, *95*, 369–384.
[6] Y. Araki, E. L. González, *J. Phycol.* 1998, *34*, 79–88.
[7] Y. Araki, PhD thesis, University of California, Los Angeles, 1997.
[8] D.-K. Kwon, E. L. González, *J. Phycol.* 1994, *30*, 689–695.
[9] M. E. Finbow, M. A. Harrison, *Biochem. J.* 1997, *324*, 697–712.
[10] H. Sze, J. M. Ward, S. Lai, *J. Bioenerg. Biomembr.* 1992, *24*, 371–381.
[11] J. M. Ward, A. Reinders, H.-T. Hsu, H. Sze, *Plant Physiol.* 1992, *99*, 161–169.
[12] M. Mandel, Y. Moriyama, J. D. Hulmes, Y.-C. E. Pan, H. Nelson, N. Nelson, *Proc. Natl Acad. Sci USA* 1988, *85*, 5521–5524.
[13] P. L. A. M. Corstjens, Y. Araki, P. Westbroek, E. L. González, *Plant Physiol.* 1996 *111*, 652.
[14] M. E. Marsh, *Protoplasma* 1994, *177*, 108–122; *Protoplasma* 1996, *190*, 181–188.
[15] M. Okazaki, M. Fujii, Y. Usuda, K. Furuya, *Botanica Marina* 1984, *27*, 285–297.
[16] P. L. Pedersen, E. Carafoli, *Trends Biochem. Sci.* 1987, *12*, 146–150.
[17] M. R. Brown, D. K. Romanovicz, *Appl. Polym. Symp.* 1976, *28*, 537–585.
[18] J. A. T. Dow, *J. Exp. Biol.* 1992, *172*, 355–375.
[19] (a) M. S. Tsiantis, D. M. Bartholomew, J. A. C. Smith, *Plant J.* 1996, *9*, 729–736.
    (b) B. J. Barkla, L. Zingarelli, E. Blumwald, J. A. C. Smith, *Plant Physiol.* 1995, *109*, 549–556.
[20] T. A. McConnaughey, *Can. J. Bot.* 1998, *76*, 1119–1126.
[21] (a) N. A. Nimer, M. D. Iglesias-Rodriguez, M. J. Merrett, *J. Phycol.* 1997, *33*, 625–631.
    (b) K. Sekino, Y. Shiraiwa, *Plant Cell Physiol.* 1994, *35*, 353–361.
[22] A. A. Israel, E. L. González, *Mar. Ecol. Prog. Ser.* 1996, *137*, 243–250.
[23] (a) N. A. Nimer, M. J. Merrett, *New Phytol.* 1992, *121*, 173–177.
    (b) N. A. Nimer, C. Brownlee, M. J. Merrett, *Mar. Ecol. Prog. Ser.* 1994, *109*, 257–262.
[24] C. S. Sikes, R. D. Roer, K. M. Wilbur, *Limnol. Oceanogr.* 1980, *25*, 248–261.

[25] O. Quiroga, E. L. González, *J. Phycol.* 1993, *29*, 321–324.
[26] N. A. Nimer, M. J. Merrett, C. Brownlee, *J. Phycol.* 1996, *32*, 813–818.
[27] D. W. Crawford, D. A. Purdie, *Limnol. Oceanogr.* 1997, *42*, 365–372.
[28] J. Theilmann, N. E. Tolbert, A. Goyal, H. Senger, *Plant Physiol.* 1990, *92*, 622–629.
[29] L. A. Rosenberg, P. E. Padgett, S. M. Assman, L. L. Walling, R. T. Leonard, *J. Exp. Bot.* 1987, *48*, 857–868.
[30] R. Sharkia, S. Beer, Z. I. Cabantchik, *Planta* 1994, *194*, 247–249.
[31] D. P. Briskin, *Plant Physiol.* 1990, *94*, 397–400.
[32] (a) N. S. Dhalla, D. Zhao, *Prog. Biophys. Molec. Biol.* 1988, *52*, 1–37.
      (b) B. W. Poovaiah, A. S. N. Reddy, *CRC Crit. Rev. Plant Sci.* 1987, *6*, 47–103.
      (c) G. Wagner, U. Russ, H. Quader, in *The Cytoskeleton of the Algae* (Ed. D. Menzel), CRC Press, New York, 1992.
[33] (a) V. Niggli, J. T. Penniston, E. Carafoli, *J. Biol. Chem.* 1979, *254*, 9955–9958.
      (b) I. M. Andreev, V. Korenkov, Y. G. Molotkovsky, *Plant Physiol.* 1990, *136*, 3–7.
[34] D. E. Outka, D. C. Williams, *J. Protozool.* 1971, *18*, 285–297.

Index

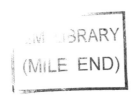